The Essential Criteria of Graph Databases

The Essential Criteria of Graph Databases

Ricky Sun
Ultipa

ELSEVIER

Elsevier
Radarweg 29, PO Box 211, 1000 AE Amsterdam, Netherlands
50 Hampshire Street, 5th Floor, Cambridge, MA 02139, United States

Notices
Knowledge and best practice in this field are constantly changing. As new research and experience broaden our understanding, changes in research methods, professional practices, or medical treatment may become necessary.

Practitioners and researchers must always rely on their own experience and knowledge in evaluating and using any information, methods, compounds, or experiments described herein. In using such information or methods they should be mindful of their own safety and the safety of others, including parties for whom they have a professional responsibility.

To the fullest extent of the law, neither the Publisher nor the authors, contributors, or editors, assume any liability for any injury and/or damage to persons or property as a matter of products liability, negligence or otherwise, or from any use or operation of any methods, products, instructions, or ideas contained in the material herein.

ISBN: 978-0-443-14162-1

For information on all Elsevier publications
visit our website at https://www.elsevier.com/books-and-journals

Publisher: Matthew Deans
Acquisitions Editor: GlynJones
Editorial Project Manager: Naomi Robertson
Production Project Manager: Gomathi Sugumar
Cover Designer: Vicky Pearson Esser

Typeset by STRAIVE, India

Working together
to grow libraries in
developing countries

www.elsevier.com • www.bookaid.org

Contents

1. **History of graph computing and graph databases**

  1.1 **What exactly is a graph?**   1
      1.1.1 The forgotten art of graph thinking, part I   1
      1.1.2 The forgotten art of graph thinking II   4
      1.1.3 A brief history of graph technology development   13
  1.2 **The evolution of big data and database technologies**   19
      1.2.1 From data to big data to deep data   19
      1.2.2 Relational database vs graph database   22
  1.3 **Graph computing in the Internet of things (IoT) era**   23
      1.3.1 Unprecedented capabilities   24
      1.3.2 Differences between graph computing and graph
              database   29

2. **Graph database basics and principles**

  2.1 **Graph computing**   33
      2.1.1 Basic concepts of graph computing   34
      2.1.2 Applicable scenarios of graph computing   56
  2.2 **Graph storage**   60
      2.2.1 Basic concepts of graph storage   60
      2.2.2 Graph storage data structure and modeling   68
  2.3 **Evolution of graph query language**   81
      2.3.1 Basic concepts of database query language   81
      2.3.2 Graph query language   87

3. **Graph database architecture design**

  3.1 **High-performance graph storage architecture**   101
      3.1.1 Key features of high-performance storage systems   102
      3.1.2 High-performance storage architecture design
              ideas   105
  3.2 **High-performance graph computing architecture**   114
      3.2.1 Real-time graph computing system architecture   115
      3.2.2 Graph database schema and data model   118
      3.2.3 How the core engine handles different data types   122
      3.2.4 Data structure in graph computing engine   124
      3.2.5 How to partition (shard) a large graph   125
      3.2.6 High availability and scalability   129
      3.2.7 Failure and recovery   132

3.3   **Graph query and analysis framework design**   134
      3.3.1   Graph database query language design ideas   135
      3.3.2   Graph visualization   150

4.   Graph algorithms

4.1   **Degree**   162
4.2   **Centrality**   166
      4.2.1   Graph centrality algorithm   167
      4.2.2   Closeness centrality   168
      4.2.3   Betweenness centrality   170
4.3   **Similarity**   172
      4.3.1   Jaccard similarity   173
      4.3.2   Cosine similarity   176
4.4   **Connectivity calculation**   178
      4.4.1   Connected component   179
      4.4.2   Minimum spanning tree   181
4.5   **Ranking**   184
      4.5.1   PageRank   184
      4.5.2   SybilRank   187
4.6   **Propagation computing**   189
      4.6.1   Label propagation algorithm   189
      4.6.2   HANP algorithm   192
4.7   **Community computing**   194
      4.7.1   Triangle counting   194
      4.7.2   Louvain community recognition   197
4.8   **Graph embedding computing**   206
      4.8.1   Complete random walk algorithm   207
      4.8.2   Struc2Vec algorithm   210
4.9   **Graph algorithms and interpretability**   213

5.   Scalable graphs

5.1   **Scalable graph database design**   223
      5.1.1   Vertical scalability   224
      5.1.2   Horizontal scalability   233
5.2   **Highly available distributed architecture design**   244
      5.2.1   Master-slave HA   244
      5.2.2   Distributed consensus system   248
      5.2.3   Horizontally distributed system   256

6.   A world empowered by graphs

6.1   **Real-time business decision-making and intelligence**   272
      6.1.1   Credit loan application   272
      6.1.2   Online fraud detection   276
6.2   **Ultimate beneficiary owner**   278
6.3   **Fraud detection**   282

6.4 Knowledge graph and AML   284
    6.4.1 Antimoney laundering scenario   287
    6.4.2 Intelligent recommendation   290
6.5 Asset and liability management and liquidity
    management   297
6.6 Interconnected risk identification and measurement   309
    6.6.1 Innovation of graph computing in interconnected
           financial risk management   309
    6.6.2 The broad prospects of graph computing in the financial
           field   312

7. Planning, benchmarking and optimization of graph systems

7.1 Planning your graph system   317
    7.1.1 Data and modeling   319
    7.1.2 Capacity planning   324
7.2 Benchmarking graph systems   330
    7.2.1 Benchmark environment   332
    7.2.2 Benchmark content   338
    7.2.3 Graph query/algorithm results validation   357
    7.2.4 Graph system optimization   366

Index   381

Chapter 1

History of graph computing and graph databases

1.1 What exactly is a graph?

What is a "graph?" Where does the "graph" come from? Why has graph technology represented by graph computing or graph databases been widely used in many fields such as finance, medical care, supply chain, oil, media, etc.? Why is it prospering at an explosive speed? The great thing about the research and thinking on graph databases is that it is not a new thing, but a great revival of graph thinking in the pursuit of scientific and technological advancement.

1.1.1 The forgotten art of graph thinking, part I

In the preface, we mentioned a bold idea and prediction for the future: Graph database technology is the only way and important tool for artificial intelligence (AI) to move toward artificial general intelligence (AGI). Because the graph database (including knowledge graph) restores (simulates) human thinking and way of thinking to the greatest extent.

So, how do human beings think?

This is a question without a standard answer (or it is difficult to form a consensus). Some would say that most of us think linearly; some say we think nonlinearly; some say focused thinking; some say divergent thinking. If we refine this problem into a mathematical problem and describe it in the language of mathematics, human beings essentially think in terms of graphs. The world we live in is high-dimensional, connected, and constantly expanding. From the moment we come to this world to the moment we leave, we have been interacting with this world—all the entities (a person, a thing, a piece of news or old news, a knowledge point, a book, or even a strand of emotion) that we come into contact with every moment are stored in our brain (memory). The human brain is very much like a well-designed computer. When we need to extract any piece of information or a knowledge point from it, we can quickly locate and obtain it. When we diverge our thinking, we start from one knowledge point or multiple knowledge points, follow the path(s) between knowledge points, traverse the

The Essential Criteria of Graph Databases. https://doi.org/10.1016/B978-0-443-14162-1.00004-0

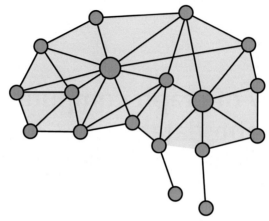

FIG. 1 The 1940s network model of human brains.

knowledge network, and reach the destiny points as we need to. It is a network of intertwined information. God has endowed human beings with the ability to think boundlessly (far and wide). What is "boundless?" That is, no matter how far the place is in our mind, our thoughts can reach there. This is an ultra-deep graph search and traversal capabilities. As early as the 1940s, before the concept of social network was invented, researchers had tried to use the graph network model to describe and explain the working mechanism of the human brain, as shown in Fig. 1.

When we need to describe any knowledge point in detail, we can give it many attributes, and the relationship between knowledge points can also have attributes. Through these attributes, we have a deeper understanding. For example, we have filled out many family relationship forms since childhood such as father, mother, brothers and sisters, ancestral home, age, gender, contact information, education level, etc. When we fill in these forms, we call out a "family relationship subgraph" to put information in the right places. When the forms are submitted electronically, the service providers (schools, healthcare providers, government agencies, etc.) can automatically extract entities and relationships out of the form and construct a small graph for us and our family, plus all their customers' family information to form a gigantic graph. Do you now see how insurance companies can do fraud detection or product recommendations based on the graph data? (We'll show you some graph-powered real-life use cases in Chapter 6).

The network composed of these entities and relationships is called a graph. When the nodes and edges in this kind of network graph have some attributes, that can help us conveniently use the attributes to perform information filtering, aggregation, or conductive computing, this kind of graph is called a property graph (PG), a concept we'll revisit later in this chapter.

Property graphs can be used to express everything in the world, whether they are related or discrete. When things are related, they form a network; when they are separated, they are just a list of things (nodes), like rows of data in a table of a relational database. The main point to be expressed here is graphs are high-dimensional. High dimensional model can be downward compatible and express the content of low-dimensional space, but the opposite can be very difficult if not impossible—that is the case with low-dimensional relational databases—often half the result with twice the effort. In the following chapters, we will specifically analyze why relational databases run into serious efficiency problems when dealing with complex scenarios.

This way of expressing graphs has great similarities with the storage mechanism of neuron networks in the human brain and its cognition of things. We are always connecting, diverging, reconnecting, and re-diverging. When we need to locate and search for someone or something, finding it does not mean the end of the search, but often the beginning of a series of searches. For example, when we make inferences about divergent thinking, we are to perform some kind of real-time filtering or dynamic traversal search on graphs. When we proclaim that a person is a polymath, when we cite extensively from humongous sources, we seem to let our thoughts jump from one graph to another. Our brains store a lot of graphs, these graphs are either interlinked or interactively accessible and are ready to provide services at any time according to our needs. If the same operational mode of the human brain can be realized on a graph database, why wouldn't we believe that a graph database is the ultimate database? Of course, the premise is that we must agree on this point: The human brain is the ultimate database. We can even say that on the path to the realization of artificial general intelligence, making graph databases the ultimate database is the way to go.

Let's use a real-world scenario to illustrate how human brains work in a "graphic" way. One of your favorite dishes comes to mind—Dongpo Pork, how did you come up with it? According to modern web search engine technology, if you first enter the word "Dongpo," the engine will suggest "pork" on top of the recommended list of words—may be human brains don't work in this inverted index search way, but it is not important, because locating "dongpo pork" is only a starting point for us. In the graph thinking mode, how to extend to the subsequent nodes is the key. Starting with dongpo pork, you may think of the famous Song Dynasty poet Su Dongpo, and of course, Song Dynasty Poem, and one of the few female contemporary poets Li Qingzhao, who suffered from the Humiliation of Jingkang (1127), and the paragon of loyalty General Yue Fei, and loyal politician Wen Tianxiang, who looked up to Yue Fei and led the resistance to Kublai Khan's invasion, particularly led and failed the Song Dynasty's last major battle—Naval Battle of Yashan, and the Yuan Dynasty was finally able to conquer the entirety of China after such decisive win, which marks the greatest conquest of human history by Mongols started by Genghis Khan some 70 years earlier—the Mongol's Invasion of Central Asia and Europe, over the course, many great nations and cities were doomed, one of the most famous one was the Siege of Baghdad in 1258, led by Kublai Khan's younger brother.

What we just brainstormed is a long path of 11 steps (12 major knowledge points) concerning ancient East Asia history and gourmet (illustrated in Fig. 2). What's illustrated in Fig. 2 is but only four paths (2×2) out of an infinite number of possible paths between the starting point and the ending point. The most appropriate tool to handle the query and traversal of such a vast knowledge network, in real-time, is a graph database.

If you feel the above example is overwhelming, let's use a much simpler one, exponentially simpler. When I first joined the Silicon Valley IT workforce in the early 2000s, I heard hosts on NPR (National Public Radio) talking about the "Butterfly Effect," and specifically they wondered if there are any relationships between Newton and Genghis Khan.

With the help of graph database (storage and computing power) and knowledge graph (visualization and explainability), we can articulate their butterfly-ish relationships—what's illustrated in Fig. 3 is almost a "causality relationship" spanning across east and west, time and space for 400 years and thousands of miles.

Every piece of knowledge we have learned is not isolated. These ever-increasing knowledge points are weaved into a huge knowledge network, from which we can extract, summarize, organize, expand, deduce, and relate at any time. All the wise men, writers, geniuses, peddlers, passers-by, or anyone in the history of mankind, every time he (she) had a shocking flash of inspiration or merely followed the rules of the ordinary, he (she) was practicing graph thinking. Shocking the world or having a flash of inspiration is just because he (she) traverses deeper, wider, and faster with graph thinking; being predictable, mundane, naïve or lack of innovation is just because he (she) has been too shallow, too slow in applying graph thinking.

In essence, every network is a graph. Every person who knows the past and the present is full of graphs in his mind, and good at putting graphs to work—to think, diverge, summarize, integrate, and extend endlessly. If one graph doesn't do the trick, add another!

1.1.2 The forgotten art of graph thinking II

In real life, do most people have a way of thinking in graphs? Let's take a look at the following problem together: Construct a graph (a graph composed of vertices and edges), assuming that there are 5 triangles in the graph, and only add 1 edge to multiply the number of triangles by 10. Please draw a picture of this graph.

Prerequisites: In this graph, assuming the vertices are accounts, and the edges are transactions between accounts. There can be multiple transactions between any two accounts. The so-called triangle is an atomic-level triangle formed by transfer transactions between three accounts. An atomic triangle

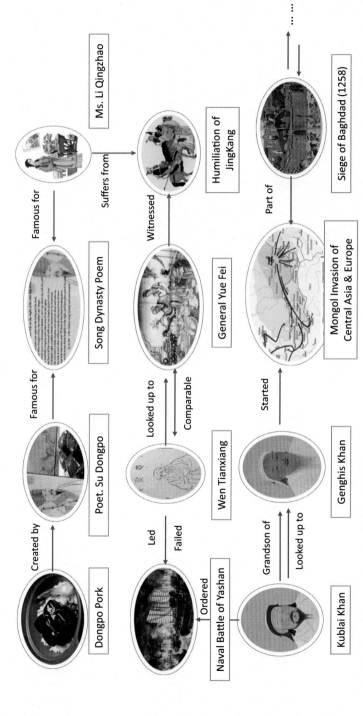

FIG. 2 From Dongpo pork to the siege of Baghdad.

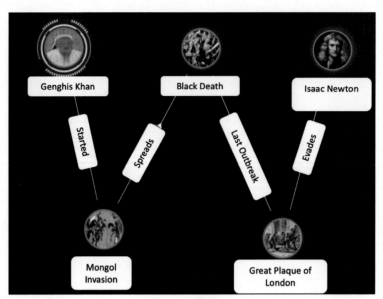

FIG. 3 Relationships from Genghis Khan to Newton.

cannot be nested. Each edge of the triangle is atomic and indivisible (transaction is atomic).

Let's analyze the key point of this problem: Only one edge is added, but the number of triangles needs to be increased by 5. There were 5 ∠ previously, if they share an edge to be added, they will form 5 new triangles. It doesn't matter what happened to the previous 5 triangles. The shared new edge means that the 5 ∠ will also share two vertices. Given the analysis up until now, is the answer ready to come out?

This is an interview question that the author comprehended and refined from real business scenarios. For test takers who are good at Googling around, they probably can't find a neat solution for this. Ninety percent of candidates for technical positions fell silent after seeing this question in the online interview stage.

The remaining 9% of candidates will come up with a graph as shown in Fig. 4. The vertical line in this answer introduces multiple edges and multiple vertices. It does not conform to the restriction of adding only one edge, and it also destroys the existing (atomic, inseparable) triangles and edges. These candidates didn't read the question carefully. Let us attribute this problem to this impetuous era.

The best thing about this question is that it can have many solutions, just like there are many roads leading to Rome (yes, this is the beauty of graph thinking too). The solution in Fig. 5 is correct, well-intentioned, and perfectly compliant with the question requirements, although a bit complicated—the newly added edge AB will create 5 new triangles, on top of 5 existing triangles, making the total a perfect 10.

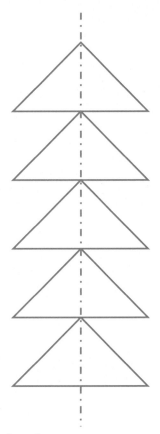

FIG. 4 A Typical no-brainer (wrong) answer.

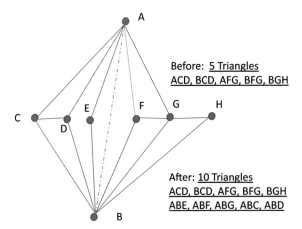

FIG. 5 A correct-yet-complicated solution.

The solution in Fig. 5 involves 8 vertices and $14+1$ edges, and the topology is not easy (intuitive) to construct in one's mind. It essentially is a simple graph (a graph theory concept that will be introduced later in this chapter). Can we push a bit forward to come up with a more intuitive and concise solution with fewer vertices and edges?

Fig. 6 is a solution involving 7 vertices and $11+1$ edges, about 20% less than the total number of vertices and edges from the previous solution. As we are representing the high-dimensional space solution in a two-dimensional plane, we use dashed lines to represent the covered edges. The newly added edge (") between vertex F and vertex G magically forms 5 new triangles. Many people find this difficult to understand, but if you read the prerequisites in the question carefully if A and B are two accounts, there can be multiple transactions between the accounts, which is already hinting to the reader that adding an edge can accelerate solving the problem.

Now, if we can push a little bit forward, can we come up with solutions that use even fewer number of vertices and edges? (Or put it another way, can fewer accounts and transactions generate more triangles?).

Fig. 7 was a solution proposed by an undergraduate intern (He was hired right away as soon as the interviewer saw the solution he scribbled on the white-board—the topology he constructed was quite skillful.). This solution requires only 4 vertices and $7+1$ edges, almost a 40% drop from the previous solution (Fig. 6). The trick here is about how we count triangles—each unique triangle is composed of a unique composition of vertices and edges—for the three vertices A, B, and C, there are eight possible compositions after the new (dashed) edge between B and C is added.

Fig. 8 shows the most streamlined topology known so far, with only 3 vertices and $7+1$ edges. If you, the reader, can think of more succinct and extreme solutions, please contact the author.

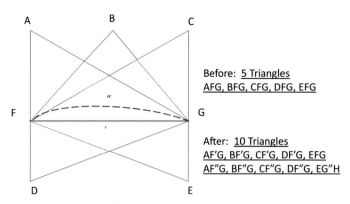

FIG. 6 The extra edge creating 5 new triangles.

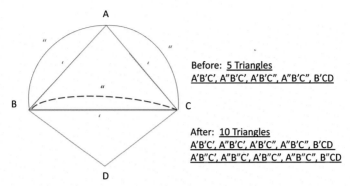

Before: <u>5 Triangles</u>
A'B'C', A"B'C', A'B'C", A"B'C", B'CD

After: <u>10 Triangles</u>
A'B'C', A"B'C', A'B'C", A"B'C", B'CD
A'B"C', A"B"C', A'B"C", A"B"C", B"CD

FIG. 7   The solution with 4 vertices and 7 + 1 edges.

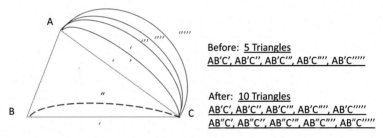

Before: <u>5 Triangles</u>
AB'C', AB'C", AB'C"', AB'C"", AB'C""'

After: <u>10 Triangles</u>
AB'C', AB'C", AB'C"', AB'C"", AB'C""'
AB"C', AB"C", AB"C"', AB"C"", AB"C""'

FIG. 8   One of the candidates' answers to the questions.

Let us now explore the meaning of the graph behind this interview question. In the banking industry, taking retail wire transfers as an example, there are hundreds of millions of debit card accounts in large banks, and these accounts have hundreds of millions of transactions per month (usually the number of transactions will be several times greater than the number of accounts), if we model card account as vertices and transactions between accounts as edges, a large graph with hundreds of millions of vertices and edges is constructed. How to measure the degree of closeness (connectivity) of the graph? Or how to judge the topological structure of this graph? Similarly, in an SNS social network graph, the vertices are users, and the edges are the associations between users. How to measure the topology or closeness of the social graph?

A triangle is the most basic structure to express a multiparty association relationship. In a large graph (network), the number of triangles can reflect how closely (or loosely) the entities in the graph relate to each other.

Fig. 9 captures a portion of a social network, a total of 32 triangles (2 * (2 * 2 * 2 + 2 * 2 * 2)) are formed, concerning only two groups of four closely connected vertices.

The number of triangles formed among accounts can be astronomical. In 2020, we counted the total number of triangles from a major bank's 3-month

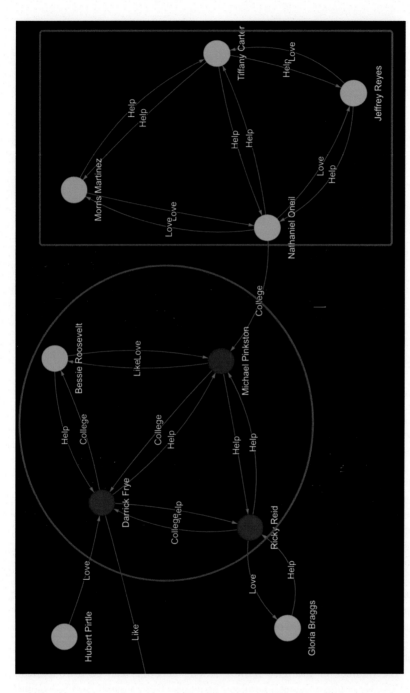

FIG. 9 Social network graph (partial).

worth of transactions by $\sim$200 million accounts, and the result is a staggering 2 trillion triangles! The total number of transactions is less than 1 billion. In other words, on average, each vertex participates in thousands of transfer triangles! This number is unbelievable, but if we analyze it carefully, we will find that when there are multiple transfer relationships between a small number of vertices, for example, there are 100 edges between any two of the three vertices A, B, and C, then a total of 1 million triangle relations are constituted. Similarly, there is a relationship between the two vertices A and B, and they both have a transfer relationship with another 1000 accounts. At this time, as long as one more edge is added between A and B, 1000 more triangles will be added to the graph. Isn't it a little magical?

In our language inheritance, a triangle is widely used to express a close and equal relationship (seemingly unbreakable), which is the case in all Western and Eastern cultures. For example, the seven-year political alliance between Caesar and Pompey and Crassus in the Roman Republic era—The former Triumvirate (First Triumvirate, 60–53 BCE), and the last Triumvirate (Second Triumvirate, 43–33 BCE) formed by Octavian, Mark Antony, and Lepidus 10 years later in the early Roman Empire era. The stories in the classic Chinese novel "Romance of the Three Kingdoms" are unforgettable, probably because of the rises and falls of the heroes and villains from the three kingdoms of Wei, Shu, and Wu.

Of course, whether it is financial transactions, social networks, or e-commerce data, the network (graph) formed by the data is defined by mathematical language as a topological space. In topology, what we care about is the correlation, connectivity, continuity, convergence, similarity (which nodes have more similar behaviors and characteristics), centrality, influence, and propagation strength (depth, breadth), etc. The graph constructed by the method of data association in the topological space can restore 100% and reflect how we record and perceive the real world. If one topological space (a graph) cannot satisfy the description of heterogeneous data, construct another graph to express it. Each graph can be regarded as the clustering and association of a data set from one (or more) dimensions or domains (such as knowledge domains, subject knowledge bases, etc.). Different from the two-dimensional tables of relational databases, each graph can be an organic integration of multiple relational tables.

The table join operation in the traditional relational database is no longer needed on the graph, path finding or neighborhood traversal is used on the graph instead. We know that the biggest problem with table joins is the inevitable challenge of Cartesian Product, especially on large tables. The calculation cost of the Cartesian Product is extremely high. The more tables involved in table joins, the larger the product (for example, three tables X, Y, Z, the amount of calculation of their full cross-join product is $\{X\} * \{Y\} * \{Z\}$, if X, Y, Z each have 1 million, 100,000, 10,000 rows, the amount of calculation is astronomical $1,000,000,000,000,000 = 1000$ trillion), in many cases, one of the main reasons for the slow batch processing of relational databases is the need to deal with various multitable join (and table scans) problems. This inefficiency is rooted in the inability of relational databases (and its supporting query language SQL) to reflect the real world 100% at the data structure level.

The storage and computing of the human brain never waste time on Cartesian Products and endless table scans, which are extremely inefficient—but this is exactly the case when we are working on relational databases and Excels. Graph computing, however, may need to traverse layer after layer of neighbors in a brute-force manner, but its complexity is much lower than the Cartesian Product. In a graph with 10 million nodes and edge points, the maximum complexity of traversing it is 10 million—any computing complexity higher than its maximum number of vertices and edges is a reflection of inefficient and badly designed data structure, algorithm, or system architecture.

In Section 1.1.1, we mentioned a key concept and a vision: The graph database is the ultimate database, and the ultimate destiny of all artificial intelligence is AGI, that is, it has human-comparable intelligence. What we can be sure of is that the way of thinking of humans is the way of graph thinking—a graph database or a graph computing method that can compare and restore the human way of thinking is called a native graph. Through native graph computing and analysis, we can enable machines to have the same efficient association, divergence, derivation, and iteration capabilities as humans.

The so-called native graph is relative to a nonnative graph and essentially refers to how graph data are stored and computed more efficiently. Nonnative graphs may use relational databases, columnar databases, document databases, or key-value databases to store graph data; native graphs need to use more efficient storage (and computing) methods to serve graph computing and queries.

The construction of a native graph bears the brunt of data structure, and a new concept is introduced here—Index-Free Adjacency (IFA). This concept is as much about storage as it is about computing. In the following chapters, we will introduce this logical data structure in detail. Here, in short, the biggest advantage of the IFA data structure over other data structures is that the minimal time complexity required to access any data in the graph is $O(1)$. For example, starting from any data point, the time complexity of visiting its adjacent neighbors is $O(1)$. This kind of data access with the lowest time complexity is exactly the way humans use it when they search for any knowledge points in their brains and associate them with each other. This data structure is different from the tree index-based data structure commonly found in traditional databases, and it is $O(1)$ instead of $O(\log N)$ in terms of time complexity. In more complex queries or algorithm implementations, this difference will be magnified to $O(K)$ vs $O(N * \log N)$ or greater differences (assuming K is the number of operations, where $K \geq 1$, usually less than 10 or 20, but much smaller than N, assuming N is the number of vertices or entities in the graph). This means that there will be an exponential difference in the turnaround time of complex operations. If a graph needs 1 s to complete a task, traditional databases may take 1 h, 1 day, or longer (or cannot be completed at all)—this means that the batch-processing operations of $T + 1$ in traditional databases or data warehouses can be completed in $T + 0$ or even pure real-time!

Of course, the data structure is only one aspect of solving the problem. We also need to make the (native graph) graph database take off from multiple dimensions such as architecture (such as concurrency, high-density computing), algorithm concurrency optimization, and code engineering optimization. With the support of native graph storage, high-concurrency and high-density computing on the underlying computing power, the traversal, query, calculation, and analysis of the graph can take a further leap forward! If the last generation of graph database claims to be 5–1000 times faster than the relational database, then the newer generation graph database after the leap is one million times faster!

Welcome to the world of graph databases.

1.1.3 A brief history of graph technology development

The foundation of graph database technology is graph computing and storage technology (for ease of discussion, throughout this chapter, we use graph computing interchangeably with graph database, the logic behind this is that computing is substantially more important than storage in graph database comparing with other types of databases), while the theoretical basis of graph computing (graph analytics) is graph theory. In this section, we review the development history of graph theory-related disciplines and technologies to better understand graph technology.

The origin of graph computing can be traced back to the famous mathematician Leonhard Euler from almost 300 years ago. Euler is considered to be the greatest mathematician in human history. He is the founder of graph theory and topology. He was born in Basel, Switzerland (yes, Basel-III in the financial world is born in Basel and named after it—this small knowledge points association is a typical 2-hop path linkage), and made his fame in Russia.

Euler started the discipline of graph theory by describing the problem of the Seven Bridges of Königsberg. In a park in Königsberg (Kaliningrad, Russia today, on the verge of the Baltic Sea, before World War II, it was the largest city in eastern Germany, occupied by the Soviet Red Army in 1945, and changed to its present name in 1946), there were seven bridges, two islands joining the river banks (landmass) along a river call Pregel (again, the famous edge-centric distributed graph computing frameworks is named after the river). The locals were puzzled on whether it was possible to start from any one of the four landmasses, cross each bridge exactly once, and return to the starting point. Euler studied and proved this problem in 1736. He attributed the problem to be a "one stroke" problem, proving that it is impossible to complete in one stroke (Fig. 16).

Euler's invention of graph theory (and geometric topology) lies with a simple topological graph (topological network) formed by abstracting land and

a
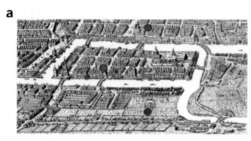

b
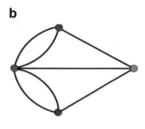

FIG. 10   Seven bridges of Königsberg and graph theory.

bridges into nodes and edges, respectively. In Fig. 10, the one-stroke problem is abstracted and simplified to find whether there is an "Euler Path," that is, a path that can traverse each edge in the graph only once (Euler Circuit or Euler Ring, is a special case of Euler Path, which is defined as the existence of a path that can traverse all edges in the graph from a starting point and return to that very point).

How did Euler falsify this one-stroke problem? According to the definition of the Euler Path, only the start point and the end point may have odd-numbered edges, and all other path nodes may only have even-numbered edges. The abstracted topo-structure in Fig. 10 does not conform to this feature, so the one-stroke solution cannot be found. If it is extended to the problem of the Euler Ring, then the condition will be harsher that all vertices must have an even number of edges. According to the sufficient and necessary conditions, if there is no Euler Path, there must be no Euler Ring, and the proof is completed. In this process, a basic concept of graph theory, degree, is introduced. That is, the concept of connected edges of each vertex if it is an undirected graph (ignoring the direction of the edges in the graph), then the number of connected edges of each vertex is its degree. In a directed graph, the degree of each vertex is equal to the sum of the number of in-degree edges + the number of out-degree edges.

A major scenario for early applications of graph theory was map rendering (coloring). With the advent of the Age of Discovery in the 15th to 17th century and the rise of the concept of nation-states after the French Revolution (1789–1799), countries all over the world began to draw higher-precision maps, and how to use the least number of colors in the drawing to ensure that two adjacent areas (countries, states, provinces, or regions) with different colors is a classic graph theory problem – as illustrated in Fig. 11—how to abstract the territories on the map and the border relationship between them into vertices and edges).

In the middle of the 19th century, mathematicians proved the problem of the "five-color map" by manual calculation, but it was not until 1976, a full century later, that the "four-color map" was initially proved with the help of a computer (four color map theorem). Moreover, this kind of proof has been continuously evolving with the improvement of computer computing power. In 2005, the feasibility of the four-color map was proved in a general way with the help of

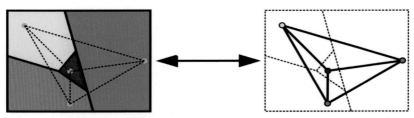

FIG. 11 Abstraction of territory and border into graph nodes and edges.

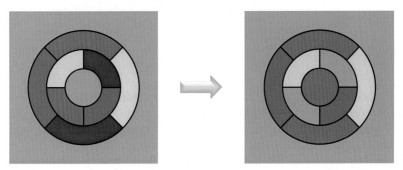

FIG. 12 NP-complete problem of map coloring.

complex theorem-proving software (Fig. 12). This is also the first mainstream theory to be assisted by "exhaustive calculations" of computer programs.

The map coloring problem is a typical NP-complete (or NPC, full name as Nondeterministic Polynomial-time Complete) problem in mathematics, that is, the most difficult decision problem in NP.

After 110 years, Euler solved the seven-bridge problem, another German mathematician, Johann B. Listing (1808–1882), first proposed the concept of topology in 1847. Topology research areas include dimensionality, connectivity, and compactness. For example, the study of the Möbius Strip was initiated by Listing in 1858 and has been widely used in daily life—the Google Driven logo is a Möbius Strip. The belt of power machinery is made into a "Möbius Strip" shape so that the belt will not only wear on one side. The strip also has many magical properties—if you cut a Möbius strip from the middle, you don't get two narrow strips, but a loop where the ends of the strip are twisted twice and rejoined. If you divide the width of the strip into thirds and cut it along the dividing line, you will get two loops, one is a narrower Möbius strip, and the other is a loop that has been rotated twice and rejoined (Fig. 13).

In 1878, the British mathematician Sylvester (J.J Sylvester) first proposed the concept of the graph in a paper published in Nature that a graph consists of some vertices (entities) and edges (relationships) connecting these vertices. Readers will never be unfamiliar with another term invented by Sylvester in 1850—Matrix.

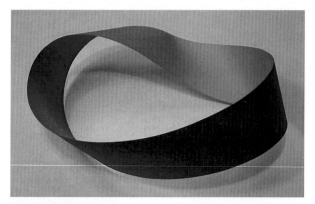

FIG. 13 Möbius strip. *((Photo taken by David Benbennick.))*

Around the 1960s, Hungarian mathematicians Erdős and Rényi established the Random Graph Theory, pioneering the systematic study of random graph theory in mathematics. In the following 40 years, random graph theory has also been the basic theory for studying complex networks.

In the early 1990s, Tim B. Lee, the father of the Internet, proposed the concept of the semantic web, which means that all pages on the entire WWW are regarded as resource nodes in a huge network, and these nodes are interrelated. This idea of graph-theorizing Internet resources led to the launch of W3C's RDF (Resource Description Framework) standard, which is regarded as the starting point of graph computing. RDF came from a strong academic atmosphere, PGs, however, is more practical and efficient in describing the entities and relationships in the graph. Although the Semantic Web has considerable academic vigor, it has not achieved great success in the industry. Instead, it has given birth to the giants of the Internet, the first Yahoo! in the mid-to-late 1990s, the second Google in the late 1990s, and the third Facebook in the early 2000s, a social platform built on top of social graphs. It is no exaggeration to say that the core concept of Facebook's social graph is the 6-degree of separation theory, which means that the distance between any two people in the world will not exceed six steps (hops or degrees)—This concept is very important in the vigorous development of social networks. It can be said that Twitter, Weibo, WeChat, LinkedIn, eBay, and Paypal are all built on this concept.

Graph computing systems or graph databases are generally considered to be a subset of NoSQL databases. NoSQL is relative to structured query language (SQL)-centric relational databases. Its exact meaning is Not Only SQL, which means that the vast world other than SQL is also covered by NoSQL databases. As we all know, since the 1980s, SQL-speaking relational database has become the mainstream, and it is still widely used in the IT environment of most companies around the world. The core concept of SQL and relational database management system (RDBMS) is to use relational tables to model complex problems. At the very core of SQL/RDBMS, you will find only two-dimensional tabular data. However, the theoretical basis of the graph database is graph theory. Its core idea is to use high-dimensional graphs to naturally express and restore the same high-dimensional world—using the simplest vertices and edges to express any complex relationship. Graph theory has many application scenarios such as navigation, map coloring, resource scheduling, and search and recommendation engines. However, many solutions corresponding to these scenarios do not use native graphs in terms of storage and computing. In other words, people are still using relational databases, columnar databases, and even document databases to solve graph theory problems, which means inefficient, low-dimensional tools are used to forcibly solve complex, high-dimensional problems, so that user experience may be poor, and return over investment (ROI) may be terrible. In recent years, 40 years after the invention of the Internet, as knowledge graphs gradually became popular, the development of graph databases and graph computing has only begun to receive attention again.

In the past half-century, many algorithms for graph computing have come out, from the well-known Dijkstra algorithm (1956, the shortest path problem), to PageRank invented by Google's co-founder Lawrence Page in 1998, and more complex various community discovery algorithms (used to detect associations between communities, customer groups, suspects, etc.). In short, many large Internet companies and financial technology companies today were born or dwell with graph computing technologies such as:

- Google: PageRank is a large-scale web page (or link) sorting algorithm. It can be said that the core technology of the early days of Google was a shallow concurrent graph computing technology.
- Facebook: The core of Facebook's technical framework is its Social Graph, that is, friends connect with friends and then connect with friends. If you've ever heard of the "six degrees of separation"—Facebook has built a strong network of social relationships, between any two people, it only takes five or six people to connect. Facebook has open-sourced many things, but the core graph computing engine and architecture have never been open-sourced.
- Twitter and Weibo: Twitter is the Weibo of the United States, and Weibo is the Chinese version of Twitter. Twitter briefly open-sourced Flock DB on GitHub in 2014, but then it was rescinded. The reason is simple. Graph computing is the core of Twitter's business and technology, and the open-source model does not increase its commercial value—in other words, if any commercial company's core technology and secrets are built on open source, its commercial value is nil.
- LinkedIn: LinkedIn is a professional workplace social network. The core social feature is to recommend experts who are 2 to 3 hops away from you. To provide this kind of recommendation service, you must use a graph computing engine (or graph database).

- Goldman Sachs: If you recall during the global financial crisis that broke out in 2007–2008, Lehman Brothers went bankrupt, and Goldman Sachs was the first to retreat. The real reason behind it was that Goldman Sachs applied a powerful graph database system—SecDB, which successfully calculated and predicted the impending financial crisis.
- Aladdin, the core IT system of Blackstone, the world's largest private equity fund management organization—Aladdin, the asset and debt management system, essentially manages more than 20 trillion US dollars of global assets (exceeding 10% of all global financial assets) by building a dependency graph among liquidity risk elements.
- Paypal, eBay, and many other financial or e-commerce companies, for these technology-driven new Internet companies, graph computing is not uncommon—the core competency of graphs can help them to reveal the internal associations of data, while traditional relational databases or big data technologies are too slow, and they were not designed to deal with deep associations between data at the beginning.

The development of any technology usually goes through a curved cycle of technology germination, development, expansion, overheating, cooling, and further development. During this process, there are usually some norms or de-facto standards established to regulate the development of technology and to enhance the cooperation and interoperability of the industry. There are two types of graph computing standards by now:

- RDF: W3C specification
- GQL: ISO/IEC International Standard

W3C's RDF specification (v1.0 in 2004 and v1.1 in 2014) was originally used to describe the metadata model, which is usually used for knowledge management. Today, academia and a considerable number of knowledge graph companies are using RDF to describe the "metadata" in the graph. The default query language of RDF is SPARQL Protocol and RDF Query Language (SPARQL). But the problem with RDF and SPARQL is that the logic is complex, verbose, and hard to maintain—developers don't like it. For example, do you prefer eXtensible Markup Language (XML) or JavaScript Object Notation (JSON)? Probably JSON, right? Because it is simpler and more convenient. After all, being lightweight and fast is the main theme of the Internet age.

The GQL (Graph Query Language) standard is due to be released by late 2023 or 2024, it is currently in DIS (Draft International Standard) mode, and it can trace its origin to PGs and LPG (Labeled Property Graphs), whereas LPG is considered a special form of PG, and popularized by Neo4j.

As the name implies, PG is a graph with properties or attributes. For the two types of basic data types (meta-data) in the graph: Both nodes and edges can have attributes such as name, type, weight, timestamp, etc. Other than Neo4j, there are a handful of PG graph players such as TitanDB (withdrawn from the market in 2016), JanusGraph (a derivative of Titan), Neptune of AWS, DGraph, TigerGraph, ArangoDB, Memgraph, Ultipa Graph, and so on.

The characteristics of these graph database products are different, such as the architectural construction methods they adopt at the bottom of their respective technology stacks, the service models, business models, programmable APIs, and SDKs are all different. The current development of graph databases is at a stage where a hundred flowers are blooming, because the market is developing extremely rapidly, and user needs are diverse. If a certain graph database solution is bespoke and built to address certain specific scenarios, it will inevitably run into issues dealing with general scenarios.

The good news is that 40 years after SQL dominated the database world, the second set of international standards GQL will finally be ushered in. Interestingly, the development of NoSQL in the field of big data in the past 20 years has not spawned any international standards. On the contrary, the development of graph databases will usher in their international standards, which just shows the converging trend toward a standardized future of the graph database industry.

If Human brains were to be the ultimate database,
Graph Database is the shortest path to be there!

1.2 The evolution of big data and database technologies

Today, big data are everywhere—every industry is affected by it. But at the same time, we also found that the real world is completely different from the table-based model behind traditional relational databases. It is very rich, high-dimensional, and interrelated. In addition, once we understand the evolution of big data and the strong demand for advanced data processing capabilities, we will truly understand why "graphs" are ubiquitous, and why they will have sustainable competitive advantages and become the ultimate next-generation mainstream database standard.

1.2.1 From data to big data to deep data

The development of big data is in the ascendant. We usually think the big data era started in 2012, but the emergence of big data-related technologies was much earlier than in 2012. For example, Hadoop was released by Yahoo! in 2006 and donated to the Apache Foundation, and the Hadoop project was inspired by Google's two famous inventions (and papers)—the GFS (Google File System) in 2003 and MapReduce in 2004. If we go back further, the reason why GFS and MapReduce can appear is because of the development of Google's Internet search engine business, and the core technology of its search engine is probably PageRank (an algorithm for sorting web pages). The PageRank algorithm, named after Google co-founder Lawrence Page (and a pun on the Web page), is a typical graph algorithm. We are circling—the development of big data technology comes from a kind of

Trend: Data to Big Data to Deep Data

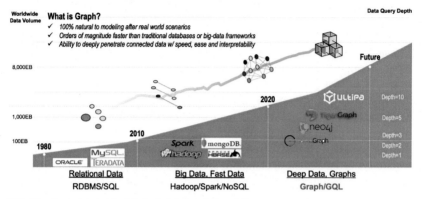

FIG. 14   From data to big data to fast data and deep data.

graph technology, and its development trend is to go bigger, faster, and eventually deeper (which is graph per se).

Looking at the development history of data processing technologies over the past half century, it can be roughly divided into three stages (Fig. 14):

1. The era of relational databases as the core (1975-present)
2. The era of nonrelational database and big-data frameworks (2003-present)
3. A New Era Beyond Relational or Nonrelational Databases—Postrelational Database Era (2020 and beyond)

These three stages have each produced a query language for data operations. The corresponding relationships are as follows:

1. Relational Database = SQL (First standard in 1983)
2. Nonrelational Database = NoSQL (Extensions to SQL, not standardized)
3. Postrelational Database = NewSQL (No standard) and GQL (Draft standard in 2023)

If you measure it according to the data characteristics and dimensions corresponding to each stage, you can interpret Fig. 20 in the following three-generation way:

1. Relational Database = Data, Prebig Data Era
2. Nonrelational Database = Big Data, Fast Data Era
3. Postrelational Database = Deep Data, or Graph Data Era

Each generation surpasses the previous generation. When we talk about big data, it contains the characteristics of the previous era, but there are also the so-called 4V special characteristics, which were coined by IBM and widely spread by the industry:

- Volume
- Variety

- Velocity (timeliness, speed)
- Veracity

In the era of graph data, on top of the 4V characteristics, $d$eep analysis for maximum value extraction is much needed, perhaps we can summarize it as 4V+D.

Why do we care so much about the deep(er) relationship among data? There are two dimensions that best explain the challenges across industries:

1. Business Dimension: Business values embedded within a network of relations.
2. Technology Dimension: Traditional databases lack the capabilities of network-based value extraction.

With the development of big data, more and more dimensional data are collected, and these multidimensional data need to be analyzed in more commercial scenarios, such as in risk control, antimoney laundering, antifraud, intelligent recommendation, smart marketing, or user behavior pattern analysis, only when the data are combined in the form of a network and the relationship between them is deeply analyzed, can we get rid of the shackles of the lack of in-time intelligence or computing power in the traditional databases—traditional database architectures cannot quickly discover deep associations between entities through tabular multitable joins.

During the period from 2004 to 2006 when the Hadoop project was incubated internally at Yahoo!, there were other mass data processing projects parallel to Hadoop. In 2004, Yahoo! still had the world's largest server clusters, with tens of thousands of Apache Web servers producing many terabytes of weblogs waiting to be analyzed and processed daily. Interestingly, from the perspective of distributed system processing capabilities (data throughput rate, operation delay, functionality, etc.), Hadoop has no advantage over other systems (one thing that needs to be clarified is that the goal of Hadoop at the beginning is to use a bunch of cheap and low-configuration PC machines to process data in a highly distributed manner, but it has never been efficient. Many so-called distributed systems cannot process data efficiently and promptly), which directly led to Yahoo!'s decision to donate the Hadoop project, which could not find a way to work out internally, to the Apache Foundation and its open-source community. This incident tells us that if a system has intrinsic vitality, high performance, and can create huge commercial value, it will unlikely be open-sourced. Of course, if we view the problem from another dimension, Hadoop solves the problem of data volume (Volume) and data diversity (Variety) storage and analysis. Especially for the cluster utilization of low-configuration machines, it is the biggest advantage of Hadoop. However, it is extremely lacking in terms of data speed (Velocity) and in-depth (Depth) processing capabilities.

In 2014, Apache Spark was born. The core development team of Spark from the University of California, Berkeley has a lot of experience in Hadoop performance problems that are widely criticized in the industry. In terms of distributed memory-accelerated data processing performance, Spark can achieve a speed-up over Hadoop for as much as 100 times. It also integrates components such as GraphX to realize some graph analysis capabilities, such as PageRank (webpage sorting algorithm), Connected Component (connected subgraph), and Triangle Counting (triangle calculation) are typical graph algorithms. Compared with the Hadoop framework, Spark has made great progress in speed, especially for shallow graph calculation and analysis. However, Spark was not designed to handle online data processing. It is not good at handling dynamic and real-time changing data sets, which limits it to be only an Online Analytical Processing (OLAP) system with offline analysis capabilities. There is still a capability gap between Spark and what we call the ultimate goal of real-time, dynamic, and deep data processing.

The so-called deep data processing capability is *to dig out the value hidden in the data by mining the linkage between multilayer and multidimensional data in the shortest time*. Especially in this era of data interconnection, the platform that can realize deep data network analysis in a general way is the protagonist of this book—graph database. In different scenarios, we may also call it a graph analytics system, graph platform, or graph computing engine. All of these refer to one thing—to construct data according to graph theory principles, and to perform network analysis on it. We believe that graph database is the crown jewel of future generations of databases. You cannot find real-time, deep data association solutions from other types of NoSQL databases, big data frameworks, or relational databases. In other words, whenever business demands change, the underlying architecture needs to be substantially rearchitected to support them, and a lot of bespoke systems were created to address such changing demands. How can a bespoke system have long-term vitality? Key-value stores, columnar databases, Hadoop, Spark, MongoDB, Elastic Search, and Snowflake are all inappropriate for dealing with connected data analysis challenges. It is these bottlenecks and challenges mentioned above that enable the graph database market to be born and flourish.

In this book, the author will invite the readers to have an overview of the past, present, and future development trends of the graph databases from the perspective of being close to the market and business needs, and close to the essence of technology implementation.

1.2.2   Relational database vs graph database

The purpose of the graph database was very simple when it was invented and created, which is to capture the huge value contained in the data association through (deep) mining of data from different sources and network analysis. The data storage of a relational database is to cluster (similar) data in tables, aggregate the same type of entities in the same table, and store different types of entities in different tables. When entity association is required, the connection between multiple tables is required. However, due to the design concept of relational database and the limitation of its underpinning data structures, when

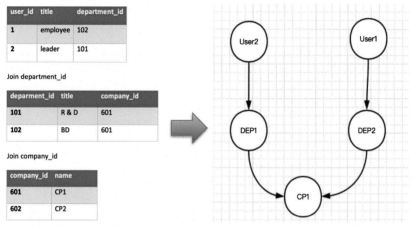

user_id	title	department_id
1	employee	102
2	leader	101

Join department_id

deparment_id	title	company_id
101	R & D	601
102	BD	601

Join company_id

company_id	name
601	CP1
602	CP2

FIG. 15 Faster data association and value extraction using graph.

performing table join operations, especially when the content of the table is large, the time consumption of the operation will increase exponentially compared with the operation of a single table. When multiple large tables are joined, the relational system either crashes or cannot return due to excessive time consumption and system resource consumption.

To give a classic example: Employees, departments, and companies, three tables are used to store these different types of entities in a relational database. If you want to find out which employees work in which company, this requires three tables to be joined to each other—a process that is not only inefficient but also highly counter-intuitive. As shown in Fig. 15, you can find out which department each employee belongs to from the employee table, then you can find out which company each department belongs to from the department table, and finally find out the company corresponding to each company ID from the company table. Perhaps, you want to say that all this information can be compressed into one table, but the price of doing so is to create a large and wide table with lots of redundant information, and this design concept is not consistent with the normalization principle in relational data modeling.

If we were to do the employee-department-company data modeling using a graph, this query is a typical and simple path query with a depth of only 2 in the graph (see Fig. 15 to the right). When there are a large number of employees, departments, and companies, the efficiency of graph traversals will be tens of thousands of times higher than that of SQL-based relational queries.

1.3   Graph computing in the Internet of things (IoT) era

Graph database (graph computing) responds to the general trend of today's macrobusiness world. It relies on the mining, analysis, and association of massive, complex, and dynamic data to gain insights. Although it cannot completely replace the other data platforms that have been fully understood and used by users in a short period, the market demand for this technology is constantly stimulating the endogenous power of graph databases. In this section, we will talk about how linked data challenges previous generations of technologies, and at the same time discuss in detail the differences between graph databases and graph computing.

1.3.1 Unprecedented capabilities

If you still doubt the limitations of relational databases (NoSQL and even most AI systems as well), then imagine using a relational database management system to solve the following problems:

- How to find the friends of friends of a friend's friends?
- How to (in real-time) find the connecting paths between an account and multiple blacklisted accounts?
- How to judge whether the transaction (or transaction networks) between two accounts in a certain period is normal?
- In a supply chain network, if a manufacturing plant/factory in North America shuts down, what kind of impact will it have on the flagship store of a department store in South Korea?
- In freight, power transmission, and communication networks, if a network node or service goes down (goes offline), what is the extent of the ripple effect?
- In healthcare, if a user submits his electronic medical records and health records, can a service provider offer real-time personalized insurance recommendations for major diseases?
- In an antimoney laundering scenario, how do you find out that an account holder transfers his funds through multiple intermediary accounts, and finally remits them to his own (or other related party) accounts?
- Today's search engines, or even ChatGPT-like LLM (Large Language Models) based AI systems, can only perform one-dimensional, keyword-based searches. For example, if you enter: "Are there any relationships between Newton and Genghis Khan," how can you return any meaningful, connecting, and even causal results?
- In a knowledge graph, how to find the network of pairwise associations formed between multiple knowledge entities? Similarly, in a large network comprised call records, how to find the correlation network formed by 100 criminal suspects within six degrees of separation between each pair?

The above nine problems are just a few of the many challenges that today's traditional data management systems (also including NoSQLs, Big Data, and AI frameworks) or search engines cannot accomplish efficiently and cost-effectively. Fig. 16 illustrates the use of graphs for network analysis, which

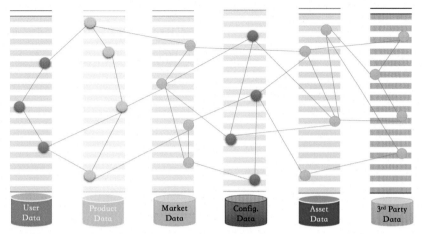

FIG. 16   Graph database for connected/networked data analysis.

differs from other types of data processing frameworks in that it can deeply penetrate (for unknown and indirect relationships), and quickly respond to changes.

With the help of real-time graph databases and graph computing engines, we can find deep correlations between different data in real-time. For example, a knowledge graph dataset can be built on top of Wikipedia articles (with URLs/articles being entities or vertices, and relationships between these articles being edges, a gigantic networked graph of knowledge can be formed automatically). The graph set can be ingested into the graph database for real-time query processing. Fig. 17 illustrates how a graph database can take care of real-time path finding between two seemingly unrelated entities. However, the same cannot be achieved using even the latest cutting-edge ChatGPT (Fig. 18).

We know the system components underpinning OpenAI/ChatGPT include graphs (some graph neural networks and some knowledge graphs), but those graph components are not designed to handle really deep and white-box causality path findings. That's why they fail to subjugate meaningful information when being queried (Fig. 18).

Different from traditional search engines, when you search in a high-dimensional data processing framework like a graph database, what you expect to return is no longer a list of ranked pages (articles or URLs), but more

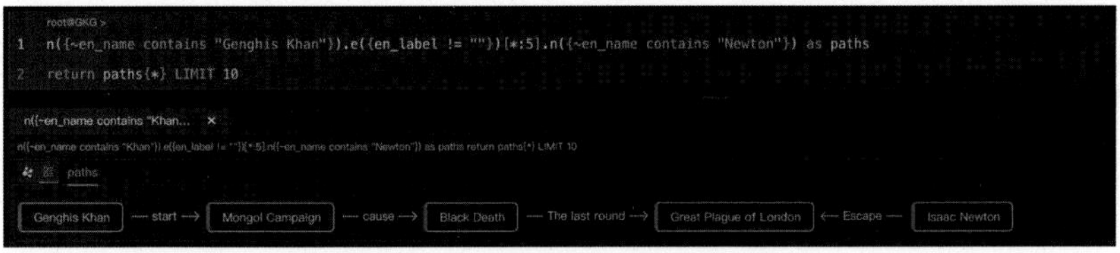

FIG. 17   Real-time shortest path between Newton to Genghis Khan.

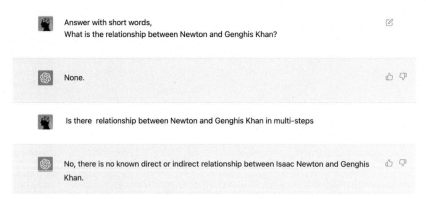

FIG. 18   OpenAI's ChatGPT on Newton and Genghis Khan.

sophisticated, causal paths, or even intelligent relationship networks. With the support of the real-time graph database, it is possible to assemble intelligent paths that are beyond the reach of the human brain! If you look closely at Fig. 17, the deep linkage between the two entities can be interpreted in the following ways:

- Genghis Khan started the quest to conquer Europe (in early 13th century).
- The Mongol invasion of Europe led to the outbreak of the Black Death.
- The last major outbreak of the Black Death (17th century): The Great Plague of London.
- Newton, who was studying at Cambridge University at the time, returned to his hometown to escape the Great Plague in London (and subsequently invented calculus, optics, and Newton's laws…).

The search result illustrated in Fig. 17 was only one shortest path within 5 hops. If we deepen (and broaden) the search to 5+ hops, there could be many more paths, some may make sense (strong causality for easier human cognition), some may not. The beauty of a graph database is that you can specify how you would like to customize your search such as filtering by types of relationships, direction of connectivity, entity or edge attributes, traversal depth and breadth of the query, etc. As the filtering conditions become more stringent, the number (breadth) of paths will reduce, but the depth may go deeper, and the search turnaround time shouldn't increase exponentially. This is a very important characteristic of a real-time graph database—the ability to dynamically prune or trim traversal paths. Graph databases lacking such capability make it impossible to handle commercial application scenarios like real-time fraud detection or online knowledge graph search.

Fig. 19 visually illustrates why there is a need and a challenge to traverse densely populated graph datasets. The most advanced AI systems today still do not have the capacity as we human beings do—we reason logically, in a rigorous step-by-step manner, connect one thing to another, and eventually

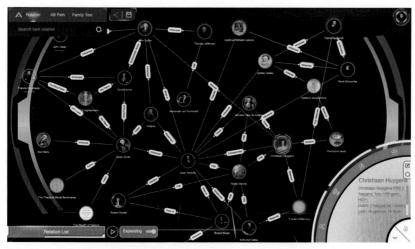

FIG. 19   A densely connected knowledge network.

forming a gigantic knowledge repository. All the artificial knowledge graphs built are to replicate the human thinking process, and the graph database is the most appropriate utensil to facilitate queries against the knowledge repository. However, it is not without challenges. As a query starts traversal from a certain node in the graph, the deeper it goes, theoretically speaking, the computing complexity grows exponentially (this is especially the case from SQL and RDBMS' table-join perspective—Cartesian Effect, but graph computing alleviates this problem that we will address in the following chapters). If on average a node has 100 adjacent neighbors, hopping for four layers from the node would be equal to traversing 100^4 ($100,000,000 = 100$ million) nodes (and associated edges, in multigraph the challenge is even more as some neighbors may have multiple edges connecting). The hotspot super node would only complicate the traversal. Fig. 20 illustrates a super node that is highly connected (100 times or more connected than average nodes, making the traversal proportionally more challenging and slower).

Another example is AML (Antimoney Laundering). Regulators, banks, and corporations around the world are working on tracking the flow of transactions such as transfers, remittances, and cash withdrawals to identify if there are any illegal activities. The parties involved in money laundering do not make direct transactions that would expose them too directly and too soon, instead, they disguise and make layers of intermediary transactions, and you will not see the whole picture unless you use a graph database. Fig. 21 shows that after digging the transaction network for 10 layers, all the money from the account of the left-most red-dot entity flows into the account of the right-most tiny red-dot entity. The significance of real-time graph databases is self-evident—when criminals evade supervision in the form of graphs, they will construct deep graph models to avoid antimoney

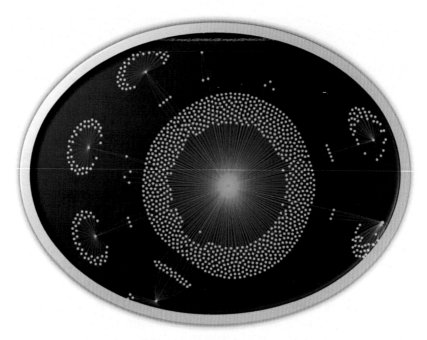

FIG. 20 Super node(s) in a graph.

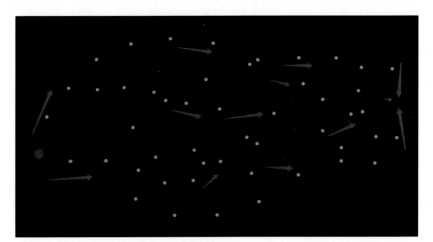

FIG. 21 Money flow pathways between entities.

laundering tracking, and regulators should adopt technologies like graph databases with deep penetration analysis capabilities to expose illegal activities.

Most people know the "butterfly effect," which is to capture the subtle relationship between one (or more) entity and another distant (or multiple) entities in the ocean of data and information. From the perspective of the data processing framework, the butterfly effect is extremely difficult to detect without the help of graph computing. Some people will say that with the exponential increase in computing power (quantum computing, for instance), we will realize it 1 day in the future. The author thinks that this day has come! A real-time graph database is the best tool for demystifying (quantifying through deep traversal) the butterfly effect!

In 2017, Gartner, a well-known data analysis company, proposed a five-layer network analytics model (Fig. 22). In the model, the future of data analysis lies in "network analysis," or entity link analysis. This general, cross-department, and cross-dataset multidimensional data association analysis requirements can only be realized with efficacy and productivity by graph databases—the reason for this is revealed in the previous sections—a graph system converts data into network topology and is built-in a way to search for associations in the network, with efficiency far exceeding that of relational database management systems.

1.3.2   Differences between graph computing and graph database

The difference between graph computing and graph database is not easy to tell for many people who are new to graph. Although in many casual circumstances, graph computing can be mixed and used interchangeably with graph databases, there are many differences between them. It is necessary to dedicate a separate section to clarify their differences and applicable scenarios that each is dedicated to.

Graph computing is the short form of Graph Processing Frameworks and Graph Computing Engines. Its main job is to calculate and analyze historical (or offline) data. Most of the graph computing frameworks can be traced back to their academic roots back in the 1960s to 1990s, which is related to the cross-disciplinary research between graph theory, computer science, and social behavior analysis.

The main focus of graph computing frameworks in the past 20 years is to perform batch data processing (bulk analysis) in OLAP scenarios.

The emergence of graph databases came much later. The earliest form of graph database didn't come to exist until the late 1990s, while real attribute graphs or native graph technologies did not appear until 2011.

The core functions of graph databases can be divided into three parts: Storage, computing, and application-oriented services. Unlike traditional databases such as RDBMS or other types of NoSQLs, the storage component of a graph database desires to store and process data in a graph-native way, which can be un-tabular. The computing component is a critical new addition, in comparison with traditional databases, as computing can hardly be performed solely by the storage component, which would be too shallow and too slow. The latest trend in graph databases takes into account both OLAP and online transactional processing (OLTP) capabilities, which form a new type of hybrid transactional and analytical processing (HTAP) graph database (we'll cover this in the following chapters). The app services component (or layer) contains many features such as graph-query engine, software development kit (SDK) and application programmable interfaces (APIs), graph DBMS, etc.

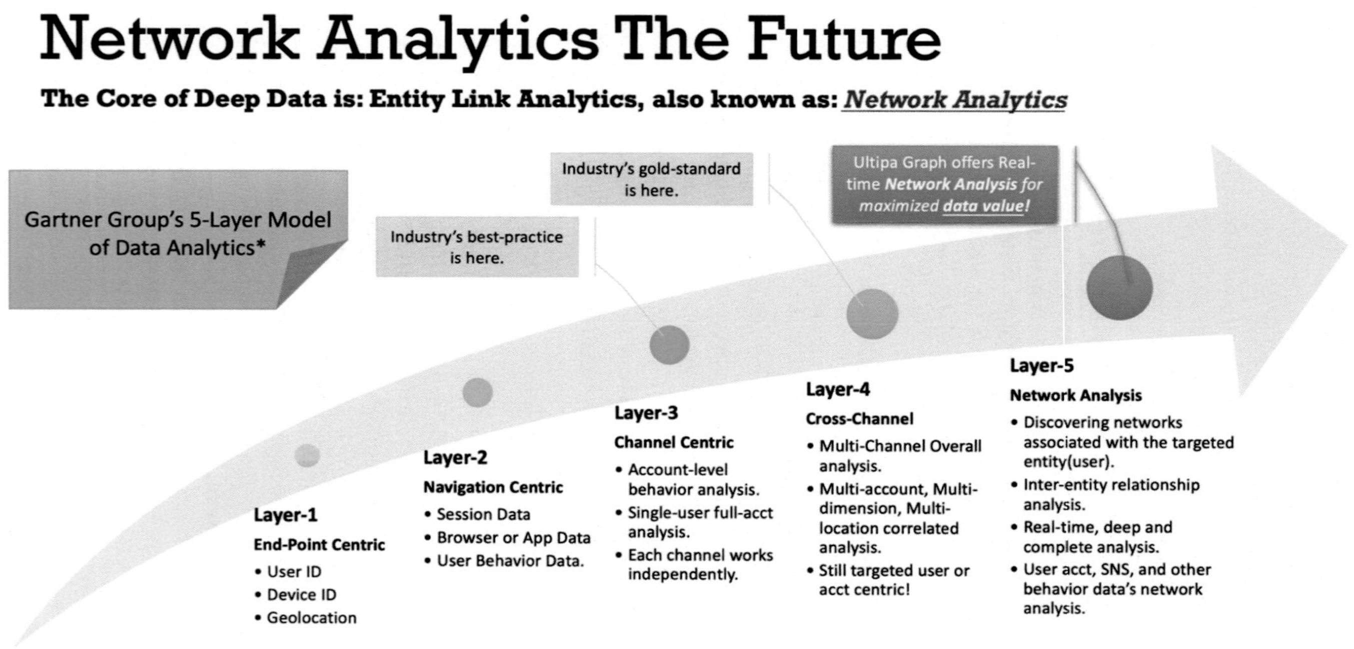

Network Analytics The Future

The Core of Deep Data is: Entity Link Analytics, also known as: _Network Analytics_

Gartner Group's 5-Layer Model of Data Analytics*

Industry's best-practice is here.

Industry's gold-standard is here.

Ultipa Graph offers Real-time _Network Analysis for maximized data value!_

Layer-1
End-Point Centric
- User ID
- Device ID
- Geolocation

Layer-2
Navigation Centric
- Session Data
- Browser or App Data
- User Behavior Data.

Layer-3
Channel Centric
- Account-level behavior analysis.
- Single-user full-acct analysis.
- Each channel works independently.

Layer-4
Cross-Channel
- Multi-Channel Overall analysis.
- Multi-account, Multi-dimension, Multi-location correlated analysis.
- Still targeted user or acct centric!

Layer-5
Network Analysis
- Discovering networks associated with the targeted entity(user).
- Inter-entity relationship analysis.
- Real-time, deep and complete analysis.
- User acct, SNS, and other behavior data's network analysis.

FIG. 22   Gartner five-layer model—network analytics the future.

Roughly speaking, from a functionality point of view, a graph database is a superset of graph computing. The computing component (or engine) of a graph database is roughly equivalent (but more feature-complete) to a graph computing framework.

However, there is an important difference between graph computing and graph databases: Graph computing usually only focuses on and processes static (offline) data, while graph databases must be able to handle dynamic (online) data. In other words, graph databases must guarantee data consistency as data changes so that business requirements can be fulfilled. The difference between the two is the difference between analytical processing (AP) and transactional processing (TP) types of operations.

The difference between static and dynamic data has its historical reasons. Most graph computing frameworks originate from academia, and their focuses and scenarios are very different from graph databases in the industrial world. At the beginning of creation, many of the former are oriented to static disk files, which are preprocessed, loaded into disk, or postprocessed in memory; the latter, especially in financial, communication, and Internet of Things scenarios, data are constantly flowing and frequently updated. A static graph computing framework cannot meet the demands of such business scenarios. This also catalyzes the continuous iteration of the graph databases, starting from the OLAP scenarios, until it develops to realize the real-time and dynamic data processing of the OLTP type. But OLTP is not the end of evolvement for graph databases, a major milestone is the ability to handle HTAP operations concurrently within a distributed graph database system, which means such a system not only ensures data consistency but also takes care of sophisticated data analytics tasks.

The problems solved by the graph computing framework and the data sets faced are usually road networks and social networks due to historical reasons. Especially for the latter, the relationship types in social networks are very simple (for example: following), and there is only one edge of the same type between any two users, this kind of graph is called a simple graph (single-edge graph) in graph theory. In a financial transaction network, the transfer relationship between two accounts can form a lot of edges (each edge represents a transaction), we call this graph a multigraph (multiedge graph). Using a simple graph to express a multi-edge graph will result in missing information, or it is necessary to add a large number of intermediate vertices and edges to achieve the same effect and it will cause high(er) storage cost and lower processing efficiency on the graph.

Furthermore, the graph computing framework generally only focuses on the topological structure of the graph itself and does not need to pay attention to the complex attributes of vertices and edges on the graph. For graph databases, this is unacceptable. For example, in the nine "impossible" scenarios we mentioned in Section 1.3.1, almost all have the need to filter (or prune) by vertices and edges (and their attached attributes).

Two other differences between graph computing and graph databases include:

1. The algorithms provided by graph computing frameworks are relatively simple and limited. In other words, the processing depth in the graph is

relatively shallow, such as single-source shortest paths (SSSP), Betweenness Centrality (BC), PageRank, LPA, Connected Components, Triangle Counting, and only a few others. Many of these algorithms were designed to run either deeply on a small amount of data over a single instance (such as SSSP and BC), or shallowly on a massive amount of data over a highly distributed cluster framework (such as the famous PageRank). The graph databases, however, must take a balanced approach in handling the algorithm complexities, query depth, and the need for a rich collection of algorithms, and the ability to handle real-world data sizes (usually somewhere in between the small academic research dataset and excessively large and synthetic dataset) plus integration with application business logics.

2. The operating interface of the graph computing framework is usually an API call, while the graph database needs to provide richer programming interfaces, such as APIs, SDKs in various languages, visual graph database management and operation interfaces, and most importantly—graph query language. Readers who are familiar with relational databases must not be unfamiliar with SQL, and the query language corresponding to graph databases is GQL, through which complex queries, calculations, algorithmic calls, and business logic can be realized.

We hereby summarize the key differences between graph computing and graph database in Table 1.

I hope that the extensive background introduction in this chapter can better prepare you for entering the world of graph databases.

TABLE 1 Differences between graph computing and graph database.

	Graph computing	Graph database
Static vs dynamic data	Mainly static	Need to support real-time changing data
OLAP vs OLTP	OLAP	There are both OLTP+OLAP scenarios
Simple-graph vs multigraph	Mainly simple-graph	Need support multigraph
Attribute filtering	Usually not	Must support
Data persistency	Not a concern	Must support
Application scenarios	Academia focused	Industry oriented
Data consistency	N/A	ACID or serialization
Graph algorithms	Common/simple graph algorithms	Richer and more complex graph algorithms and queries
Query language	API only	GQL/SDK/APIs

Chapter 2

Graph database basics and principles

This chapter aims to help readers clarify the basic concepts and principles in the graph database framework. Different from traditional relational databases, data warehouses, and even other types of NoSQL databases, a graph database has three major components—graph computing, graph storage, and graph query language, of which the graph computing component is crucial. Traditional database architectures are storage-centric, and computing (functionality or capability) is often embedded within the storage engine. In graph databases, to solve the efficiency problems of high-performance computing and graph traversal query, the importance of graph computing is self-evident, which we will focus on in the first section. In the second section, we mainly introduce graph storage concepts and related data structures and data-modeling (graph composition) logic. In the third section, we introduce the reader to the concept of GQL, the graph database query language, it is worth pointing out that in the past 40–50 years of database development, GQL has been the second and the only international database standard after SQL, which is a proof to tell the importance of graph database.

2.1 Graph computing

In general, graph computing can often be equated with graph databases. After all, the most important work that graph databases need to accomplish is graph computing. However, from the perspective of graph theory and database development, graph computing frameworks have developed independently for quite a long period, while graph databases have only been formed in the last 10–20 years. The content of the earliest graph computing is to study some algorithms based on graph theory, such as the Dijkstra algorithm published in 1959 (an algorithm for finding the shortest path in a road network). In the context of graph databases, graph computing focuses on how to efficiently complete database query, calculation, analysis, and dynamic adjustment of data. The most important point that distinguishes the computing engine of a graph database from a traditional graph computing framework is that traditional frameworks deal with static data, which is also one of the core differences between all computing frameworks and databases.

The Essential Criteria of Graph Databases. https://doi.org/10.1016/B978-0-443-14162-1.00001-5

2.1.1 Basic concepts of graph computing

Before understanding the basic concepts of graph computing, you need to understand the basic concepts of graph theory. Chapter 1 introduced that Euler created graph theory in the first half of the 18th century (1736). The object of graph theory research is graph—a spatial topology composed of vertices and edges connecting the vertices. This graph is used to describe a certain relationship (edge) between things (vertices). What we call a graph is a network, and the essence of graph computing is complex network-oriented computing, or computing for linked data and relationships.

Many of the data types we encounter in our daily lives are essentially the research objects of graph theory, such as sets, trees, lists, stacks, queues, arrays, or vectors.

Careful readers will ask: Arrays are discrete data, why are they also the research object of graph theory? Although graph theory studies complex networks, in extreme cases, when there is only a set of vertices in the network and with no edge connecting them, these discrete vertices are best represented as a set of arrays. By the same logic, a relational database based on a table structure can be regarded as a two-dimensional database, while a graph database is essentially a high-dimensional database. High-dimensionality can be downwardly compatible with low-dimensionality, but the other way around can be quite challenging.

The most concerned points in graph computing are as follows:

- Basic computing and query mode
- Common data structures
- Discussion on computational efficiency

1. Query mode

Let's first look at the basic query modes in graph computing. There are different classification methods from different angles.

(1) From the perspective of association or discreteness, it can be divided into discrete query (Metadata Query) and associative query (Connected Query).

The most typical discrete queries are metadata-oriented queries. In the context of graph computing and graph databases, metadata refers to vertices or edges, because these two are the smallest granularity of graph data, usually with a unique ID to identify them, no further subdivision is required. All operations that only operate on vertices or edges are typical discrete-type operations.

Some readers expressed confusion about whether an edge counts as metadata, thinking that an edge does not exist independently and that it looks like a composite data structure because it associates vertices. It should be pointed out that, on the one hand, the definition of metadata is the smallest granularity and cannot be further divided, and on the other hand, it has a unique matching ID to locate it. In a graph database, only vertices and edges meet these two points. However, the expression of the edge is indeed more complicated, because, besides the ID, the edge also associates the start vertex and end vertex, direction, and other attributes. In some distributed graph systems, edges are cut, and the two vertices associated with an edge may be split and stored in different partitions (or shards), this kind of edge-cutting design may cause a lot of trouble with computing efficiency (We will get to this in great details in Chapter 5).

The associative query is the most characteristic graph computing operation in graph databases compared to other databases. For example, path query, K-hop query, subgraph query, networking query, and various graph algorithms are typical associative queries. Associative queries usually start from a certain vertex (or multiple vertices), and return associated datasets by filtering edges, vertices, and their respective attributes. From the perspective of a relational database, this kind of associative query is similar to the operation of table joins. However, the flexibility and high performance brought by graph computing are unmatched by relational databases.

(2) From the perspective of the how query (traversal operation) is conducted on the graph, it can be divided into breadth-first search (BFS) and depth-first search (DFS).

The most typical breadth-first traversal is K-hop and shortest path query. Taking K-hop as an example, its original definition is to start from a certain vertex and find all unique vertices whose shortest path distance from the original vertex is K hops (steps or layers). The logic of K-hop calculation is that if you want to know all the neighbors of a vertex on the K-th layer, you need to know all its K-1-layer neighbors first, and so on. Describe it from another angle: Starting from the vertex, find all its first-layer neighbors, and proceed to the second layer until all the K-th layer neighbors are found.

The computational complexity of the K-hop query on a huge and highly connected data set may be very high, because starting from any vertex, as long as the K value is large enough, it can be connected to all nodes. In fintech (Financial Technology) scenarios, taking credit or debit card transaction data as an example, the transaction network formed between all cards and their counterparty POS machines is almost completely connected (the premise of this assumption is that each card, as long as it is active [swiped], it will be associated with at least one POS machine, and this POS machine will also be interacting with other cards, and these cards will be associated with more POS machines). Such a network is highly connected, and there are no isolated entities. (In actual operation, isolated cards or POS machines may be eliminated or omitted for network analysis.) Fig. 1 illustrates how the two planes of POSes vs Cards are interacting with each other.

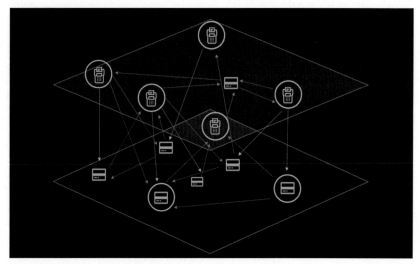

FIG. 1 Transaction network between cards and POS machines.

Fig. 2 demonstrates three different types of K-hop operations:

1. Return only the K-th hop neighbors
2. Return all neighbors from the first to the K-th hops
3. Return neighbors from the M-to-N hops for $0 \leq M < N \leq K$

Note that the difference between the above operations is whether the neighbor of K-hop contains only the neighbors on exactly the K-th hop or neighbors from other hops, say, the 1st hop to the K-th hop. For example, when $K = 3$, only the third-layer neighbors are returned, but when K is a range value of [1–3] or [2–3], the return value of this operation is greater than the former, because it also additionally contains the number of shortest-path neighbors in other hop(s).

It is possible to filter the results by vertex or edge attributes, which is a cool and practical way to narrow the graph search, and we will cover that in the following chapters.

If we add up the returned values of the 1st, 2nd, and 3rd hop neighbors, the sum should be equal to the K-hop of $K = [1\text{–}3]$—$1,001,159 + 19,875,048 + 20,220,738 = 41,096,945$. This is a typical means to verify the accuracy of a graph query or algorithm. Due to the innate complexity of graph computing, even the most basic operations like K-hop may go wrong in many ways, such as the vertex of K-1 layer may appear on the Kth layer, or the same vertex may appear multiple times across different layers, or a certain vertex never shows on in the results. Most of these mistakes are caused by either wrongful way of traversal or, in rare cases, incompletely loaded data. In Chapter 7, we will dedicate a section on how to verify the accuracy of any graph query and algorithm.

FIG. 2   Typical K-hop operations (three variations).

Talking about the wrongful way of graph traversal, we should focus on examining the two fundamental types of traversals: BFS vs DFS, as illustrated in Fig. 3.

K-hop, by definition, is ideally implemented using BFS—neighbors from the 1st hop are fully traversed before the 2nd hop neighbors can be visited, and so on. It is, theoretically speaking, possible to implement K-hop using DFS, but in a more convoluted way, which means a much more complicated

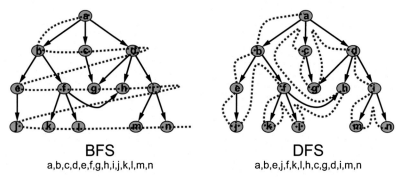

BFS

a,b,c,d,e,f,g,h,i,j,k,l,m,n

DFS

a,b,e,j,f,k,l,h,c,g,d,i,m,n

FIG. 3 BFS and DFS.

and counter-intuitive deduplication process must be done before any results can be returned correctly. We have seen some graph systems ask the user to put a limit on the number of K-hop vertices to return or to query for, this can be troublesome (and implies that the underpinning implementation method might be DFS based), because K-hop, by definition, must traversal all vertices across all qualified hops, no matter how many vertices on each hop. If you recall the 3rd hop neighbors of ID = 12 from Fig. 2, there are over 20 million neighbors, limiting the number of return values is pointless. In real-world practice, returning the counted number of K-hop neighbors helps quantify things like the scope of impact in risk management, stress testing, or scenario simulation.

The DFS algorithm is more common in finding a connected path (for instance, a loop, circle, or ring) from a certain vertex to another one or more vertices in the graph according to a specific filtering rule. For example, in the bank's transaction network, finding all transfer paths in a certain direction in descending or ascending order of time can reveal an account's behavior pattern and help analysts gain insights into its holder.

In theory, both breadth-first and depth-first algorithms can fulfill the same query requirements, the difference lies in the overall complexity and efficiency of the algorithm, as well as the consumption of computing resources. In addition, the original BFS/DFS traversal algorithms are implemented in a single-threaded fashion. In the case of concurrent traversal by multiple threads and against a vast amount of data (10,000s of times larger dataset than the days the algorithms were first invented), the complexity can be overwhelming. Though the topic will be dissected in later chapters, it is beneficial for us to start talking about data structures and computational efficiency.

2. Data structures and computational efficiency

Many kinds of data structures can be used for graph computing, some of which have been mentioned above, and we can try to sort them out more clearly here. We usually divide data structures into primitive data structures and nonprimitive user-defined data structures, as shown in Fig. 4.

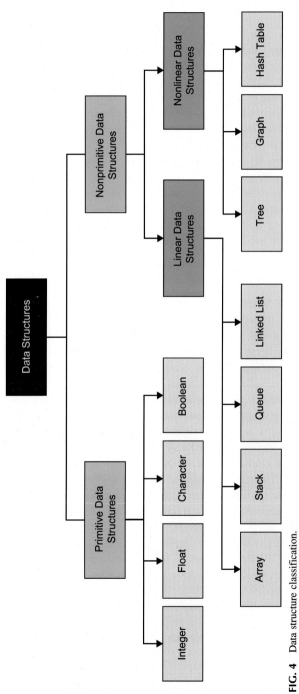

FIG. 4 Data structure classification.

The primitive data structure is the basis for constructing user-defined data structures. The definition of the primitive data structure is different in different programming languages, for example, short integer (short), integer (int, int32, int64), unsigned integer (uint), floating point number (float, double), pointer, character (char), character string (string, str), Boolean type (bool), etc. Interested readers can query related reference books for expanded discussions. This book focuses more on the linear or nonlinear data structures defined by graph databases.

Almost any primitive data structure can be used to assemble and perform any computation regardless of efficiency, but the difference in efficiency between them will be exponential. As shown in Fig. 4, the graph data structure is considered to be a composite, nonlinear, high-dimensional data structure. Many primitive or nonprimitive data structures can be used to construct graph data structures, such as arrays, stacks, queues, linked lists, vectors, matrices, hash tables, maps, HashMap's, trees, and graphs.

In a specific graph computing scenario, we need to analyze which data structures need to be used, mainly considering the following dimensions:

- Efficiency and algorithm complexity.
- Read and write requirements.

The above two dimensions are often intertwined. For example, under read-only condition, data is static, continuous memory storage can achieve more efficient data throughput and processing efficiency; if under read-n-write-mixed condition, data is dynamic, and a more complex data structure is needed to support addition, deletion, modification, and query, which also means that the computing efficiency will be reduced (compared to ready-only). The term "space-time trade-off" is used to depict the differences between the two scenarios we just described. The graph computing engine component of the graph database undoubtedly needs to support dynamic and changing data, yet most of the graph computing frameworks implemented in academia only consider static data. The applicable scenarios of these two kinds of graph computing and the work they can accomplish are very different. The content involved in this book belongs to the former—graph database. Readers who are interested in learning more about the latter can refer to GAP Benchmark[a] and other graph computing implementations.

Many real-world application scenarios are expressed by graph data structures, especially when these applications can be expressed as networked models, from traffic road networks to telephone switch networks, power grids, social networks, or financial transaction networks. Many well-known companies in the industry (such as Google, Facebook, Goldman Sachs, and Blackstone) are built or based on graph technology.

Fig. 5 shows a typical social graph network. It is dynamically generated as an instant result of a real-time path computing query against a large graph

---

a. GAP Benchmark Suite: http://gap.cs.berkeley.edu/benchmark.html.

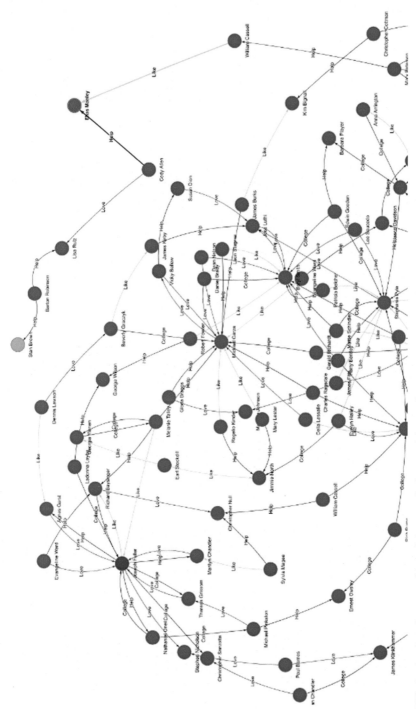

FIG. 5 A subgraph generated in real time out of a large social network.

dataset. The green node is the starting node and the purple node is the ending node, there are 15 hops in between the pair of nodes, and over 100 paths are found in between. Along each path, there are different types of edges connecting the adjacent nodes, with edges colored differently to indicate different types of social relationships (help, like, love, cooperation, competition, etc.).

Graph data structures consist of three main types of components:

1. A set of **vertices** that are also called nodes
2. A set of **edges** where each edge usually connects a pair of nodes (note there are more complicated scenarios that an edge may connect multiple nodes which are called hyperedge, it is rare and can be implemented using just edges and not worth expanded discussion here)
3. A set of paths that are a combination of nodes and edges, essentially, a path is a compound structure that can be boiled down to just edges and nodes, for ease of discussion, we will just use the first two main types: nodes and edges in the following context

Graph data representations:

1. **Vertex**: v, v, w, a, b, c …
2. **Edge**: (u, v)
3. **Path**: (u, v), v, (v, w), w, (w, a), a, (a, j)… … (Note the *vertices* are omittable if any two consecutive edges are juxtaposed next to each other)

Note that an edge in the form of (u, v) represents a directed graph, where u is the out-node, and v is the in-node. The so-called undirected graph is best represented as a bidirectional graph, that every edge needs to be stored twice (in a bidirectional way), for instance, we can use (u, v, 1) and (v, u, −1) to differentiate that u → v is the original and obvious direction, and v ← u is the inferred and not-so-obvious direction. The reason for storing bidirectionally is that: if we do not do it this way, we will not be able to go from v to u, which means missing or broken path.

Traditionally, there are three types of data structures used to express graphs: Adjacency List, Adjacency Matrix, and Incidence Matrix.

Adjacency lists use linked lists as the basic data structure to express the association relationship of graph data, as shown in Fig. 6, the directed graph on the left (note the edges with weights) is expressed by the adjacent linked list on the right, which contains the "array or vector" of the first layer, in which each element corresponds to a vertex in the graph, and the data structure of the second layer is a linked list composed of vertices directly associated with the outgoing edges of each vertex.

Note that the adjacent table on the right in Fig. 6 only expresses one-way edges in the directed graph. If you start from vertex *E*, you can only reach vertex *F*, but you have no way of knowing that vertex *D* is pointing at vertex *E* unless you scan through the entire graph, in that case, the efficiency will be quite low. Of course, another way to solve this problem is to insert reverse edges in the

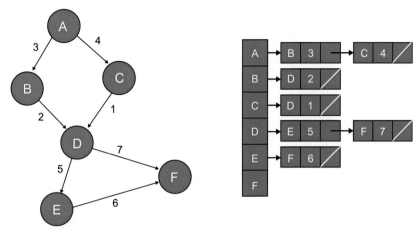

FIG. 6 Adjacency list for directed graph.

linked list, similar to how to use additional fields to express the direction of the edge mentioned above.

The adjacency matrix is a two-dimensional matrix, which we can express with a two-dimensional array data structure, where each element represents whether there is an edge between two vertices in the graph. For example, the directed graph in Table 1 is expressed by the adjacent matrix AM. Each edge needs to use an element in the matrix to correspond to a vertex in the matching row and column. The matrix is 6×6 in size, and there are only seven elements are assigned (seven edges). This is a fairly sparse matrix, with an occupancy rate of only $(7/36) < 20\%$, but the minimum storage space it requires is 36 bytes (assuming that each byte can express its corresponding edge weights). If it is a graph with 1 million vertices, the required storage space is 1000 GB ($1 M \times 1 M = 1$ trillion bytes). In real-world scenarios, many graphs have

TABLE 1 Adjacency matrix for directed graph.

Adjacency matrix (AM)	A	B	C	D	E	F
A		3	4			
B			2			
C				1		
D					5	7
E						6
F						

hundreds of millions of vertices, it is quite a challenge to use an adjacency matrix to express such kind of large graphs.

Perhaps readers may question that the estimation of the storage space of the adjacent matrix above is exaggerated: if the elements in each matrix can be expressed by 1 bit, then the storage space of the whole graph of 1 million vertices can be reduced to 125 GB. However, if the value range of weights exceeds 256, we may need 2 bytes, 4 bytes, or even 8 bytes to store correspondent information. If the edge has more than one attribute, then there will be a greater or even unimaginable demand for storage space. Modern GPUs are known for being good at handling matrix operations, but usually, the size of a two-dimensional matrix is limited to less than 32 K (32,768) vertices. This is understandable because an adjacency matrix of 32 K vertices would require a memory space of 1 GB, which is about 25% of a modern GPU memory. In other words, GPU is not suitable for operations on large graphs, unless a sophisticated divide-and-conquer mechanism is applied to cut and slice large graphs into smaller ones for further processing with GPUs, or some compression is done to squeeze out those unoccupied interactional elements. But either improvement still has to face the ultimate reality—is the adjacency matrix adequate to represent real-world content-rich graphs? Is the map-reduce-like mechanism productive? How to represent the properties associated with each vertex and edge? What if more than one edge exists between a pair of vertices?

An incidence matrix is a typical logic matrix, which can associate vertices and edges in a graph. For example, the head of each row corresponds to a vertex, and the head of each column corresponds to an edge. Taking the above-directed graph as an example, we can design a two-dimensional weighted association matrix with $6 \times 7 = 42$ elements, as shown in Table 2.

The two-dimensional matrix in Table 2 can only express undirected graphs or unidirectional graphs in directed graphs. If you want to express reverse edges or metadata attributes, this data structure is not enough.

TABLE 2 Incidence matrix for directed graph.

Incidence matrix (IM)	E1	E2	E3	E4	E5	E6	E7
A	3	4					
B			2				
C				1			
D					7	5	
E							6
F							

Graph databases in real industrial scenarios rarely use the above three data structures, the reasons for their inability to fully address the following needs:

- Express the attributes of vertices and edges
- Effective use of storage space (reduced storage volume)
- Perform high-performance (low-latency) computing
- Support dynamic data operations
- Support high concurrency for complex graph queries

Storage inefficiency is perhaps the worst enemy of a data structure such as an adjacency matrix or an incidence matrix, despite their $O(1)$ access time complexity—the time required to locate any edge or vertex through the array subscript is constant $O(1)$. In comparison, the adjacent linked list requires much less storage space and is more widely used in the industry. For example, Facebook's social graph architecture (Tao/Dragon) adopts the method of adjacent linked lists. Each person is represented as an element in the vertical array (see the right-half data structure of Fig. 6), and the linked list under each vertex represents the person's friends or followers. This design method is easy to understand, but it may encounter hot-spot (supernode) issues. For example, if a vertex has 10,000 neighbors, then the length of the linked list is 10,000 steps, and the time complexity of traversing the linked list is expressed as $O(10,000)$ in Big-O Notation. Operations like insertion, deletion, and modification on the linked list all have similar complexities. More precisely, the average complexity is $O(5000)$ to access any element in a list as long as 10,000. From another point of view, the concurrency capability of the linked list is very bad, you cannot perform concurrent (write) operations on a linked list. Facebook's social graph data structure limits a user's friends to no more than 6000 people, and Tencent's WeChat has a similar limit of 5000 on the total number of friends—have you wondered why?

Now, let us think of an approach, a data structure that balances the following two things.

- Storage space: relatively controllable, occupying a smaller storage space to store a larger amount of data
- Access speed: low access latency, and friendly to concurrent access

In the storage dimension, we should try to avoid using data structures with low utilization rates when facing sparse graphs or networks, because a large amount of empty data occupies a large amount of free space. Taking the adjacent matrix as an example, it is only suitable for graphs with very dense topological structures, such as a fully connected graph (in graph theory, it is called a clique, where any two vertices in the graph are directly connected). If the sample graph with six vertices mentioned above is a clique, there will be at least 30 directed edges ($2 \times 6 \times 5/2 = 15$ outbound edges and 15 inbound edges), if there are self-pointing loops, there will be 36 edges—only in such extreme and synthetic scenarios, matrix is useful.

However, in practical application scenarios, most of the graphs are very sparse. Traditionally in graph theory, the below formula is used to examine a graph's density (and a measurement of graph data complexity):

$$\frac{|E|}{|V|(|V| - 1)} \times 100\%$$

The denominator is the presumably maximal number of directed edges for a fully connected graph (simple graph though). Given such a formula, the density of most graphs would be much lower than 5% which renders matrix-based storage very ineffective without applying some sophisticated compression mechanism. However, many real-world graphs are multigraphs, and matrix data structure alone is not enough to represent multigraph (alternatively, it is possible to use a condensed matrix as the first layer of data structure, and some other data structures on the second and third layer to represent multiple edges between each pair of vertices and attributes of vertices and edges, respectively).

The adjacent linked list saves a lot of storage space. However, the linked list data structure has problems such as high access latency and unfriendly concurrent access. Therefore the breakthrough point should be how to design a data structure that can support high concurrency and low-latency access. Here, we try to design and adopt a new data structure, which has the following characteristics:

- The lowest time complexity of accessing any vertex in the graph is O(1)
- The lowest time complexity to access any edge in the graph is O(1)

The above time complexity assumptions can be realized by some kind of hash function. The simplest way is to access specific vertex and edge elements through the array subscript corresponding to the UUID of the vertex or edge. In C++, the simplest implementation of a hashing data structure oriented to the above characteristics is to use a Vector of Vectors to allow for easy chaining and express dynamically changing vertices and edges:

```
vector < vector <int> > v_of_v[n]; // Vector of vectors
```

```
vector < vector <int> > v_of_v[n];      // Vector of vectors
```

Vector-based data structures can achieve extremely low access latency and have a small waste of storage space, but they still face the following two problems, which we will gradually address next:

- Concurrent access support (while maintaining data consistency)
- Additional cost when data is deleted (such as backfilling of empty storage space, etc.)

In the industry, typical implementations of high-performance hash tables include Google's SparseHash library, which implements a hash table called *dense_hash_map.* In C++11, *unordered_map* is implemented, which is a chaining hash table that sacrifices memory usage for fast lookup performance. However, both implementations do not scale with the number of CPU cores—meaning only one reader or one writer is allowed at the same time.

In a modern, cloud-based high-performance computing environment, superior speed can be (ideally) achieved with parallel computing, which means having a concurrent data structure that utilizes multiple-core CPUs and multiple physically or logically separated machines (computing nodes for instance) to work concurrently against one logically connected data set.

Traditional hash implementations are single-thread/single-task oriented, blocking, a second thread or task that competes for common resources would be blocked. A natural extension would be to implement a single-writer-multiple-reader concurrent hash, which allows multiple readers to access critical(competing) data sections concurrently, however, only one writer is allowed at a given time to the data section (serialization).

Some technical means are usually used in the design and implementation of single-write multiple-read, such as the following.

- Versioning
- RCU (Read-Copy-Update)
- Open-Addressing

Taking RCU as an example, this technology was first used in the kernel of the Linux OS to support multiple reads. In MemC3/Cuckoo hash implementation, open addressing technology is used, as shown in Figs. 7 and 8.

If we are to continue to progress, we certainly hope to realize a real high-concurrency data structure with multiple reads and multiple writes capabilities. However, this vision seems to be against the requirements of ACID (Strong Data Consistency)—in the commercial scenario, multiple tasks or threads writing and reading access to the same data at the same time may cause data inconsistency—but this issue must be solved in a truly scalable system (read

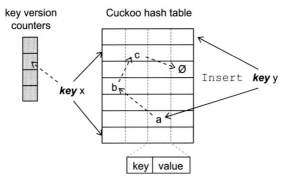

FIG. 7   Hash key mapped to two buckets and one version counter.

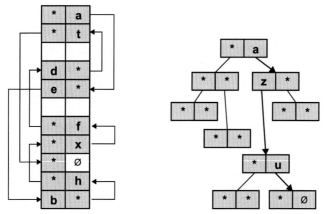

FIG. 8 Random placements vs BFS-based two-way set-associative hashing.

and write access can happen concurrently across multiple instances, multiple tasks/threads).

To realize a scalable-yet-highly-concurrent distributed system, we must evolve the core data structures to be full-concurrency capable. There are a few major hurdles to overcome:

- Blocking → Nonblocking and/or Lock-free
- Coarse Granularity Access Control → Fine-granularity access control

To overcome the two hurdles, both are highly related to the concurrency control mechanism, a few priorities to consider:

- Critical sections:
  ○ Size: Kept minimal
  ○ Execution time: Kept minimal
- Common data access:
  ○ Avoid unnecessary access
  ○ Avoid unintentional access
- Concurrency control
  ○ Fine-grained locking: lock-striping, lightweight-spinlock
  ○ Speculative locking (i.e., Transactional Lock Elision)

A high-concurrency system usually includes at least the following three mechanisms working together to achieve sufficient concurrency. These three are indispensable in the design of a high-performance graph database system:

- Concurrent infrastructure
- Concurrent data structure
- Concurrent algorithm/query implementation

The concurrent infrastructure encompasses both hardware and software architectures. For instance, the Intel TSX (Transactional Synchronization Extensions) which is hardware-level transactional memory support atop 64-bit Intel architecture. On the software front, the program can declare a section of code as a transaction, meaning atomic operation for the transactional code section. Features like TSX provide a speedup of 140% on average. This is a competitive advantage introduced by Intel over other X86 architecture processors. Of course, this hardware function is not completely transparent to the code, and it also increases the complexity of programming and cross-platform migration efforts.

However, with a concurrency-capable data structure, you still need to design your code logic carefully to fully utilize and unleash the underpinning hardware and operating system toolchain's concurrent data processing capabilities. Parallel programming against graph database requires a mindset change for the programmers first—concurrency does not come by default, and most textbooks do not teach us how to do high-concurrency programming, not to mention program against high-dimensional graph data structures where there are more moving factors than traditional databases. Always look to the upside of going through the pain of mastering high-concurrency programming—the performance gain can be in the range of tens to thousands of times compared with traditional mundane sequential programming mode.

Figs. 9 and 10 show the K-hop traversal results against a popular public dataset Twitter that has 1.5 billion vertices and edges in total, and on average each vertex is connected with ~70 other vertices (42 million vertices and 1470 million edges), and the entire dataset has only one connected component (meaning all vertices are connected, a concept and an algorithm we will introduce in the Chapter 4: Graph Algorithms). Note that the starting vertex in Figs. 9 and 10 is a peculiar one (original id = 15738828), unlike most other vertices which may have neighbors only within 6 hops (remember "6-degree of separation"?), its neighborhood structural layout is similar to the green node from Fig. 5, the first 10 hops of it has very few neighbors, but right after that the numbers quickly ramp up (see the steep curve in Fig. 10)—after 16th hop, almost 100% of the graph must be traversed to count the K-hop result—this can be translated to the graph system's capability to traverse ~1.5 billion vertices and edges each second. A similar result can be repeatedly realized on every vertex. This will not be made possible without a native graph computing system architecture that is, real-time oriented with high-concurrency data structures.

In commercial scenarios, the size of the graph is usually in the order of millions, tens of millions, or billions, while traditionally the graph data sets used for research and publishing papers in academia are often orders of magnitude smaller. The difference, when projected to graph queries and algorithms, is both quantitative and qualitative—meaning how you would design your graph system, particularly its architecture, data structure, and code-engineering to ensure varied tasks can be performed with robustness, accuracy, and satisfactory flexibility.

Taking Dijkstra's shortest path algorithm as an example, the original algorithm was completely serial, which may be suitable to work on small graphs (<10,000 vertices and edges), but unfeasible (super slow) on larger graphs (>1,000,000 vertices and edges). Similarly, the original implementation of the Louvain community detection algorithm was implemented in a serial mode,

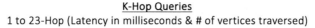

FIG. 9 Real-time ultra deep graph traversal on Twitter dataset.

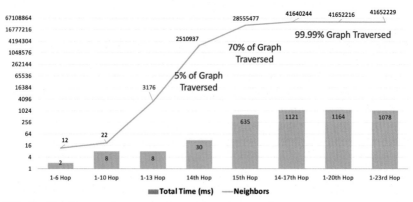

FIG. 10 Near linear performance on deepening graph traversals.

but for a graph data set with a scale of tens of millions of vertices and edges, iterating the entire dataset repeatedly until the modularity stabilizes at 0.001 may take hours or $T + 1$ or longer.

Fig. 11 shows a highly parallel implementation of the Louvain algorithm. Its running time against a densely connected graph dataset of half-million vertices and 4 million edges take <1s, including time for write-back to file system or database. If it were to be processed using NetworkX or Python programs, the running time could easily exceed a couple of hours. The speed-up can be in the range of 10,000 times.

Table 3 is a good illustration of the exponential performance difference in various system implementations, which were presented in 2018 by Turing Award winners David Patterson and John Hennessey, demonstrating that on an 18-core Intel CPU-powered system platform:

- Taking Python implementation as a baseline
- Simply using compiled language like C, a speed-up of 47 times is achieved
- By adding parallel processing to the C program, $366\times$ speed-up
- With added memory optimization, the speed-up can reach 6727 times

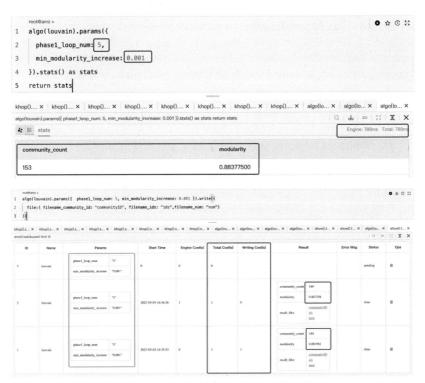

FIG. 11 Real-time Louvain community detection (on a dataset with 4 million vertices and edges).

TABLE 3 Opportunities for exponential speed-up in different system implementations.

System implementation	Speed-up	Optimization items
Python	1	N/A
C	47	Compiled languages are used
Parallel C	366	Concurrent processing
Memory-optimized, parallel C	6727	Concurrent processing, memory access
Intel AVX Instruction set	62,806	Domain-specific hardware is used

- If special lower-level CPU instruction-set optimization is leveraged, a whopping 62,806× speed-up is achieved

High-density parallel graph computing concerns both data structure and system architecture. It is achieved using vector-based native graph data structures (not linked lists or adjacency matrix, which are either too slow for data traversal or too limited to handle path filtering) for metadata (nodes and edges) representation, meanwhile, all operations against the metadata are parallelly executed, and in a recursive fashion ("high-density"). Fig. 12 illustrates the steps for K-hop query concurrency to be performed with high-density parallelization and vectors:

1. Locating the starting vertex (the green node), the time complexity is O(1) via UUID, and determining how many unique adjacent vertices it connects with, again the time complexity here is O(1), if $K = 1$, return directly with the number of adjacent nodes, otherwise, proceed to step 2.
2. $K \geq 2$, determine how many system resources are available, such as allocatable CPU cores and threads as tasks for concurrent computing, divide the nodes from step 1 into each task and continue to step 3.
3. Each task further divides and conquers by calculating adjacent neighbors of the node in question and proceeds recursively until there are no further nodes to work on.

Based on the above algorithm execution logic, let us compare the performance and behavior of two graph databases: System-N and System-U working on the AMZ dataset (400 K vertices and 3.4 million edges). Given the facts and claims that System-N slightly applies parallelism when conducting queries, while System-U claims to go fully concurrent whenever possible, let us see how much difference there is between the two systems. The results are captured in Table 4 and Fig. 13.

Here are our observations:

1. System-U on average is 10–100× faster than System-N (On a larger dataset, the difference is enlarged).

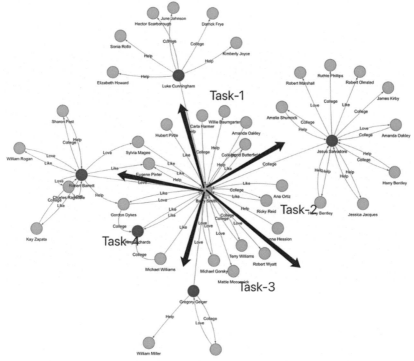

FIG. 12   K-hop query parallelization illustrated.

TABLE 4 K-hop query comparison (Neo4j vs Ultipa).

Comparison matrix	# of neighbors	System-N (ms)	System-U (ms)
1st Hop	21	202	1
2nd Hop	209	247	2
3rd Hop	7884	317	3
4th Hop	48,113	488	10
5th Hop	167,002	1017	19
6th Hop	137,403	2669	19
7th Hop	35,803	3835	33
10th Hop	228	4265	34

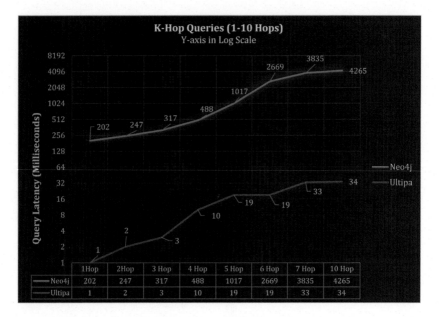

The following data is shown in the figure:

	1 Hop	2 Hop	3 Hop	4 Hop	5 Hop	6 Hop	7 Hop	10 Hop
Neo4j	202	247	317	488	1017	2669	3835	4265
Ultipa	1	2	3	10	19	19	33	34

FIG. 13 K-hop comparison between System-N and System-U.

2. System-N becomes significantly slower starting from the 5th hop (not considered real-time anymore).
3. System-U stabilizes after the 7th hop because that is, when 99% of the dataset is traversed.
4. The testing bed is a 16 vCPU virtual machine, for each query, the max concurrency recorded for System-N is 80%, and 1600% for System-U (this helps explain why the performance difference can reach 100×).
5. The two systems use GQL dialects (the code snippets below show the difference).
6. System-N uses a bespoke graph algorithm (or stored procedure) to support a basic K-hop query—you might feel weird calling it an algorithm.

```
/* System-U GQL/uQL */

khop().src({_id == 1234}).depth(7).boost() as nodes

return(count(nodes))

/* System-N GQL/Cypher */

MATCH (p:Nodx {ultipaId: "U80000000000004D2"})

CALL apoc.neighbors.athop.count(p, "Edgx", 7)

YIELD value

RETURN value
```

On larger datasets, such as the notoriously famous Twitter with 1.5 billion vertices and edges, the latency results can go out of control—If a graph system takes ~1 s on average to complete a query on a small dataset of only 4 million nodes+edges, a 400-times bigger dataset will be proportionally more challenging (and time-consuming)—which means the system may take at least 400 s to return something, and if you pile up 100 queries, the system may take many hours ($T + 1$) or even crash before it can return successfully. In a 2022 PoC (Proof of Concept) test with one of the world's largest retail and commercial banks, the bank's IT department benchmarked nine graph systems, and when they were examining the K-hop results, they did not expect most systems could not return anything (time-out) for 3-hop and deeper queries, due to the sheer data volume and traversal complexity, besides, some systems return discrepant results for the same query (such as the number of neighbors). To illustrate to and convince the folks from the IT department that there are ways to validate the query results (so to tell who is right and who is wrong), we developed a systemic methodology, which is covered in Chapter 7.

For people who are familiar with graph database benchmarks, this has been a real problem, because different systems tend to spit out different results, and the logic here is simple: for a definitive query (not approximation), some systems are not yielding correct numbers. A lot of things could contribute to such discrepancies, such as:

- De-duplication: If a neighbor belongs to K-hop, it should not belong to any other hop, and it should be counted only once.
- Wrongful traversal logics (or simply bugs in the query program)—such as using DFS to simulate BFS, and the results can be off the mark hugely.
- Not feeding the appropriate parameters when invoking the K-hop queries (even though it sounds weird, it does not happen with some graph systems—such as some graph system's K-hop does not de-dup by default …).
- Parallelization can be very challenging to many programmers, and imprudent implementation can generate wrong results as well.

For the impatient readers, jump to Chapter 7 for an expanded discussion on this topic.

Let's summarize the evolution of graph data structures: higher throughput can be achieved through higher concurrency, and this can run through the entire life cycle of data, such as data conversion, modeling, importing and loading, queries, algorithms, etc. In addition, memory consumption is also a nonnegligible factor, over the years we have started claiming that "memory is the new hard disk," and its performance is exponentially higher than SSD or hard disk, however, it does not come without a cost, so conscious use of memory is necessary. Strategies like effective data modeling, compression, de-duplication, avoidance of too much data expansion and projection, and many other techniques can help make your graph system fly with peace of mind.

2.1.2 Applicable scenarios of graph computing

Graph computing applies to a wide range of scenarios. In the early years, graph computing was limited to senior research institutions in academia and industry. With the development of computer architecture, graph computing has been applied in a wider range of industries and scenarios. The development and scope of application of graph computing can be roughly divided into the following stages.

- 1950s to 1960s: shortest path algorithm, random graph theory research, early trading system IBM IMS.
- 1980s to 1990s: graph labels, logical data models, object databases, relational models of relational databases, etc.
- From the mid-1990s to the first decade of the 21st century: Internet indexing technology, web search engine.
- The second decade of the 21st century: the emergence of the earliest graph database, the vigorous development of big data and NoSQL, the emergence of various graph computing frameworks and multimodal databases, and the outbreak of social networks, and social network analysis.
- The third decade of the 21st century: the application of broader business scenarios and innovative scenarios for graph computing and high-performance graph databases.

The development of graph computing in the past half a century is accompanied by the development of mainstream technologies and constantly iterative. Take the most famous Dijkstra's algorithm as an example. It was developed as a side-product of the construction of the third-generation computer ARMAC by the Mathematical Center in Amsterdam, Netherlands between 1955 and 1956. Dijkstra, the only computer programmer in the institution, took only 20 min to come up with the solution to the shortest path problem between the city of Rotterdam and Groningen (a typical breadth-first algorithm in a directed graph with weights), and programmed and verified the accuracy of his algorithm on ARMAC. The shortest path algorithm is widely used in pathfinding, navigation, resource planning, and other scenarios.

Solving the shortest path problem is not limited to Dijkstra's algorithm, there is also the Bellman-Ford algorithm (1955–59), A* algorithm (1968), Floyd-Warshall algorithm (1962), Johnson's algorithm (1977), etc. Taking the A* algorithm as an example, it is a "pathfinding" algorithm invented by the Stanford Research Institute (SRI International) in 1968 during the development of the world's first "universal" mobile robot (The Shakey Project), which is considered an extension of the Dijkstra's algorithm. What these algorithms solve are typical dynamic programming (DP) problems. In the fields of management science, economics, information biology, aerospace, and other fields, DP is widely used. The gist of DP is to find the "optimal solution" in the process, such as minimal travel time, maximal profit, maximal productivity, etc. The

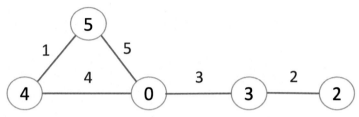

FIG. 14 Vertex labeling and edge labeling.

crown jewel of DP is the Bellman equation[b] (Bellman equation), interested readers can study it in depth.

Graph Labeling is derived from a 1967 paper by Alexander Rosa. Unique identifiers such as integers can be assigned to the vertices and edges in the graph, as shown in Fig. 14—what is interesting about the labels (or values) of the edges is each edge is uniquely identified by the absolute difference between its endpoints, and the value magnitude ranges from 1 to m inclusive, assuming m is the total number of vertices in the graph—such labeling mechanism is called "graceful labeling." It is not hard to imagine that graph labeling preludes and inspires property graphs.

This method of integer assignment is not friendly to humans, because we will soon be unable to distinguish whether an integer refers to vertices or edges, so the labels on the graph are extended to allow and support strings (in the storage engine, string values can be mapped to integer values to achieve better access efficiency), as shown in Fig. 15.

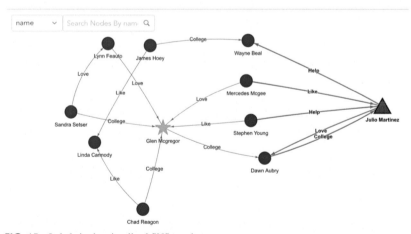

FIG. 15 Labels in the visualized SNS graph.

b. On Dynamic Programming Theory: https://www.ncbi.nlm.nih.gov/pmc/articles/PMC1063639/.

The application scenarios of graph labeling can be seen everywhere with the development of graph computing and graph databases. In the early days, its most well-known application scenario was map coloring. When we abstract different countries in the map into vertices or edges, adjacent vertices use different colors, and adjacent edges use different colors to distinguish them. Other typical scenarios include graph algorithms that use labels to participate in iterations, such as LPA (Label Propagation Algorithm), Louvain (Louvain Community Recognition), and Weighted (Labeled) Degree algorithms. In Chapter 4 of Graph Algorithms, we will go into more detail on graph labels.

An important branch of graph theory is network theory. Euler's falsification of the Seven Bridges of Königsberg is the earliest proof in network theory. The application scenarios of network theory are very extensive, ranging from various network analysis to network optimization scenarios, such as social network analysis, biological network analysis, robustness analysis, centrality analysis, link analysis, etc. Web page ranking algorithms are typical network link analysis, such as PageRank (PR), and various sorting algorithms such as SybilRank in social and financial antifraud scenarios.

Google's PageRank[c] algorithm takes all web pages on the WWW as vertices, and hyperlinks between web pages as edges, and the importance of each web page is expressed by aggregating the weight of all inbound edges and the vertices that are connected with those edges. The aggregation is calculated recursively. The PR algorithm formula is as follows:

$$PR(P_i) = \frac{(d)}{n} + (1 - d) \times \sum_{l_{j,i} \in E} PR(P_j)/\text{Outdegree}(P_j)$$

$$D(\text{damping factor}) = 0.1 - 0.15$$

$$n = |\text{page set}|$$

In some PR algorithm formulas, the damping factor D may also be set to 0.85 (equivalent to taking the remainder of $D = 0.15$ in the formula above). For convenience of calculation, the initial PR value of all web pages is 1, and all web pages participate in the calculation in each iteration. After enough iterations (e.g., five times), the PR value will stabilize (converge), and no further iteration is necessary. Many graph algorithms have similar "converging" characteristics, for instance, the Louvain algorithm, although its computing complexity is exponentially higher than the PR algorithm, the Louvain modularity will converge after a certain number of iterations. Running a minimum number of iterations (and shorter latencies) to reach a state that satisfies business needs is an important feature in measuring many graph algorithms' efficiencies.

After graph computing is gradually integrated into graph databases, its application scenarios have been widely expanded. This section briefly lists

---

c. PageRank Explained and Status: https://seranking.com/blog/pagerank/.

Graph Database Usage Scenarios

FIG. 16   Graph database scenarios.

applications in different industries (Fig. 16). The logic and implementation details of some innovative applications are explained in greater detail in Chapter 6.

- Antifraud: mainly in the financial industry and applicable to all industries.
- Anti Money Laundering (AML): financial regulation, banks, large corporations.
- Risk Control: applicable to all industries.
- Financial Risk Management: credit risk, market risk, liquidity risk, operational risk, risk preference, quantification, technology risk, risk strategy, compliance and audit, risk penetration and early warning, etc.
- Asset and Liability Management (ALM), Liquidity Risk Management (LRM): banks, financial institutions, etc.
- Business Intelligence (BI), Advanced Data Analytics, etc.: applicable to all industries.
- Smart Marketing, Recommendation system, Customer Service Robot: applicable to all industries.
- Accelerated Drug Development: Bio companies, pharmaceuticals.
- Trend Analysis: Energy, Power Grid and Electric Power Sector.
- Network Monitoring: operators and public cloud service providers, etc.
- AI/ML Augmentation: Acceleration of AI and Machine Learning, Improved prediction accuracy, white-box explain ability, LLM/Multimodal Augmentation, etc.
- Smart Enterprise Graph, Knowledge Graph with Deep-Link Analysis.

Taking risk analysis as an example, the essence of risk is the deviation and uncertainty from income or return expectations, and such deviation and uncertainty require quantitative analysis, accurate measurement, attribution analysis, root-cause analysis (to the finest granularity), scenario simulation, and stress testing. Many of these requirements can hardly be satisfied with traditional/previous generations of IT infrastructures and technologies. This is where the graph database has imminent advantages.

2.2 Graph storage

At the functional level, the graph storage engine is primarily responsible for the persistent storage of data in a graph database. Different from traditional databases (RDBMS and many NoSQLs), the graph storage engine offloads many computing efforts to the graph computing engine (server), or things will be running very slowly. Most of the time the storage engine and graph computing engine work together to serve data and analytical results to the end user. As the storage engine is farther away from the user-facing application layer than the graph computing layer (or component), degraded performance by the storage layer will be magnified via the compute layer and application layer, therefore it is important to note that building a performance-oriented storage system architecture is beneficial to everything we talk about concerning a graph database.

2.2.1 Basic concepts of graph storage

Before introducing the principles of Graph Storage Engine, let us first understand the general database storage engine (Fig. 17). There are two mainstream types of database storage engines: B-Tree-based storage engines and LSM-Tree-based storage engines. In addition, to these two categories, there are of course many other types of storage methods.

- File-based: ordered or unordered
- Heap-based (also a kind of file)
- Hash Buckets
- Indexed Sequential Access Method (ISAM)

According to the arrangement of data storage, it can also be divided into row storage, column storage, KV storage, relational storage, and other types.

B-Tree is a self-balancing tree-like, ordered data structure derived from a data structure invented by Boeing Research Labs in the early 1970s, but what exactly B stands for has never been determined (not Binary). B-Tree usually implements the following functions on the database storage engine side:

- Guaranteed key ordering to support in-order traversal
- Use hierarchical indexes to minimize the number of disks read operations
- Accelerate insertion and deletion through block operations
- Index balance is maintained through a recursive algorithm

One advantage of B-Tree is the logarithmic time complexity of data search, sequential access, insertion and deletion, etc. In other words, if the amount of data records is 1 million, the time complexity from the root node to the leaf node in the form of a binary search tree is O(20), because $\log_2 1,000,000 \sim = 20$.

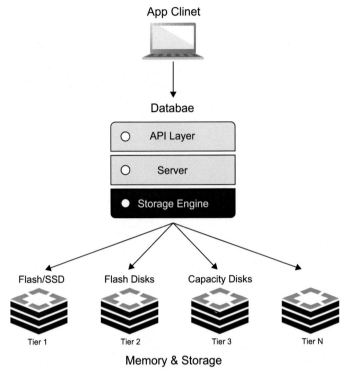

App Clinet

Databae

| API Layer |
| Server |
| Storage Engine |

Flash/SSD   Flash Disks   Capacity Disks

Tier 1   Tier 2   Tier 3   Tier N

Memory & Storage

FIG. 17 General storage engine illustrated.

In database query, data is usually stored in external memory (such as disk) in the form of records, and the disk seek time far exceeds the CPU computing time (the time consumption of the latter can be relatively negligible). Taking a 7200-rpm disk as an example, if the average time for changing tracks and addressing the mechanical arm for reading and writing the disk is 10 ms, and if 20 such kind of operations are required, then the time consumption is 200 ms (0.2 s). In actual situations, some data records may be stored continuously in the B-Tree, which will save some time for disk addressing and track-changing operations.

The B-Tree index usually adopts an auxiliary index to accelerate. The above description is a Binary Search Tree. Through the recursive acceleration of the auxiliary index, the positioning complexity of the above 1 million records can be reduced from $O(20)$ to $O(3) = \log_{100} 1{,}000{,}000$. The index acceleration feature of B-Tree makes it the default index implementation method for almost all relational databases. Of course, a large number of NoSQL databases can also use B-Tree, which is also applicable to the storage engine of graph databases. B-Tree can be seen in many relational or nonrelational databases that we are familiar with, such as Oracle, SQL Server, IBM DB2, MySQL InnoDB, PostgreSQL, SQLite, MongoDB (early edition), Couchbase, etc.

Fig. 18 shows how B-Tree stores sorted data in a flat tree structure, paying attention to the acceleration of the list index addressing leaf nodes. B-Tree is

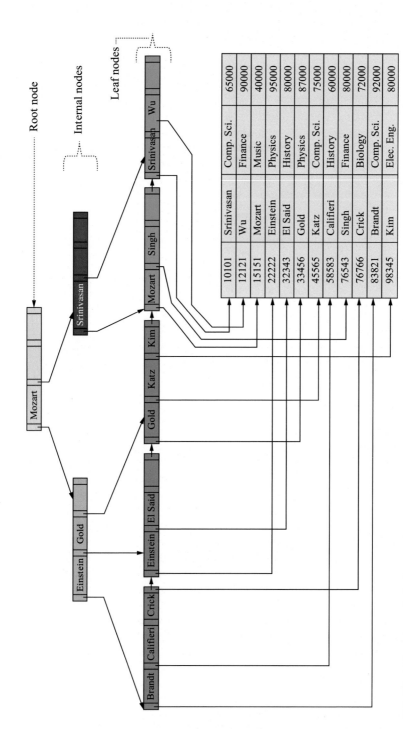

FIG. 18 B-Tree storage logic diagram.

very efficient for sequential reads and writes, but maintaining a dynamically balanced ordered data structure involves a large number of random write operations, and a simple row update operation may require access to the entire disk block where it is located. Read-modify-write operations are not cheap. So, we need to introduce the second tree index architecture: LSM-Tree.

The birth of LSM-Tree (Log-Structured Merge-Tree, hereinafter referred to as LSMT) traces its root to the pre-big-data era. The amount of data is increasing day by day, and the write operation is more frequent than in the previous relational database. Nonrelational databases are rising rapidly, and these emerging databases and big data frameworks are more used for analysis and decision support. LSMT was designed from the beginning to provide high-write performance indexing of disk files. LSMT-related papers were first invented by Patrick O'Neil and others during the database development of DEC Corporation in California in the early 1990s and finally published in 1996. It can be seen in almost all NoSQL database implementations today: Bigtable, HBase, LevelDB, SQLite, RocksDB, Cassandra, InfluxDB, ScyllaDB, etc. The evolution process of LSMT is shown in Fig. 19.

The design concept of LSMT is described in the simplest language by constructing two sets of tree-like data structures of different sizes, and a set of smaller memory data structures C_0, plus another set of data structures $C_1, C_2, C_3 \ldots$ persisted on disk. The new record is inserted into C_0 first, when its size exceeds a certain threshold, a continuous data portion of C_0 will be cleared and merged into C_1, similarly when C_1 is large enough, it will be cropped and merged into C_2, and so on. The schematic diagram is shown in Fig. 20.

The core concept presented by LSMT is **hierarchical storage acceleration**—make full use of memory acceleration, and use hard disk acceleration when the memory space is not enough. Given the continuous development of new storage hardware, such as SSD, NVMe-SSD, and persistent memory (PMEM), the ideal of LSMT is not outdated any time soon.

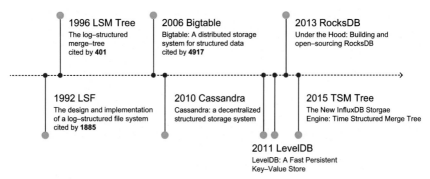

FIG. 19 LSMT from 1992 to now.

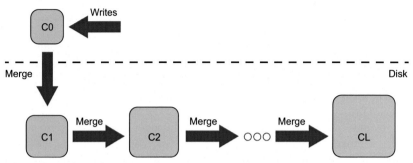

FIG. 20 Illustrated workflow of LSMT.

LSMT is not without its shortcomings. It has two problems compared to B-Tree:

- Reading performance bottleneck (higher CPU resource consumption)
- (Higher) Read and space magnification effects (take up more memory, hard disk space)

In practical applications, LSMT and B-Tree are usually used simultaneously (to get the best of both worlds), and the read performance of LSMT can be greatly improved through Bloom Filter.

The core data structure of the Bloom filter (Fig. 21) is bit-array, which determines that its footprint is very low, and it also means potential speed advantages (if array-subscript is fully utilized). Multiple (k) hash functions are involved in its main operation flow. The actual operation effect of the Bloom filter is directly related to the settings of m and k. (In Fig. 21, $m = 18$, $k = 3$.) On the one hand, the space occupation should not be too high, and on the other hand, do not set too many and too complicated hash functions to ensure index query efficiency and reduce the probability of false positives. The advantages and disadvantages of the Bloom filter coexist. Its time and space advantages are small footprint and fast speed. The disadvantage is that there may be false-positive errors.

Compared with B-Tree, LSMT combined with Bloom filter can be more widely used in distributed system architecture, because its aggregation function is more suitable to be effective under the condition of complete decentralization. Readers who are interested in in-depth research on Bloom filters can read this article.[d]

We spent a lot of time to introduce the storage engine of the "traditional" database. Now let us analyze the storage engine principle of the graph database. As shown in Fig. 22, components such as graph storage engine, computing engine, data management, and operation management are organically integrated to form a relatively complete graph product.

In the previous section, we introduced the computing engine data structure, which is logically derived from the persistent storage data, and maintains

---

d. What are Bloom filters? https://blog.medium.com/what-are-bloom-filters-1ec2a50c68ff.

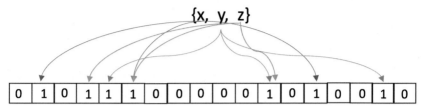

FIG. 21 Bloom filter data structure illustrated.

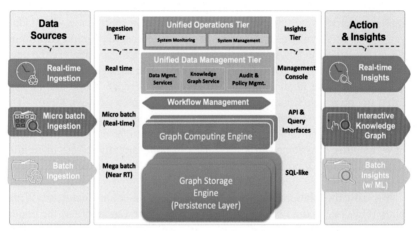

FIG. 22 Logical components within graph database framework.

consistency with the storage engine to achieve overall transactional consistency of the database (i.e., ACID, the first letter of the four English words of Atomicity, Consistency, Isolation and Durability).

The storage of a graph is nothing more than the two most important basic (meta) data structures: vertices and edges. All other data structures are derived from these two, such as various indexes, and intermediate, and temporary data structures. They are used to achieve query and calculation acceleration, as well as data structures that handle heterogeneous returns, such as paths and subgraphs.

Let's analyze the logical data structures of vertices and edges and their suitable storage methods.

- Vertex: Each vertex can be regarded as an array with some regular arrangement of internal elements such as attribute fields, and the combination of multiple vertices is a two-dimensional array. If considering the dynamic changes of vertices (addition, deletion, modification, etc. involving reading, updating, etc.), a vector array is a possible storage container.
- Edge: The data structure of an edge is more complicated than that of a vertex, because an edge not only has a unique ID, but also needs the IDs of the

connecting vertices, the direction of the edge, and other possible attributes, such as weight, timestamp, etc. Two-dimensional vectors can also be used to store edges. The remaining issues we need to focus on are efficiency, storage space occupancy, access efficiency, and index data structure efficiency.

If the data type that can support the combination of vertices and edges is completely static, disk-based files can suffice the storage need. Even if they change dynamically, from the metadata storage and access perspective, the storage engine of traditional databases, such as MySQL's InnoDB, or earlier MyISAM (a variant of ISAM) engine, can handle these basic needs. However, in graph data processing, efficiency is indispensable in many cases—metadata access is only a starting point, as most graph queries and graph algorithms go way beyond metadata, and deeply penetrate, correlate, and analyze the dataset. The traditional/relational database's way of data storage is not enough.

Therefore there is an important conceptual difference in the architecture design of the graph storage engine: Nonnative Graph vs Native Graph. The so-called nonnative graph means that its storage and computing are carried out in the traditional table structure (tabular or columnar database); while the native graph is constructed in a way that can more directly reflect interentity relationship, and therefore there will be better storage and computing efficiency.

If you use the relational database MySQL, the wide-column database (Wide-column) HBase, or the two-dimensional KV database Cassandra as the underlying storage engine for graph data, you can store the vertex and edge data in the form of a table (or list) which is nonnative graph form of storage. Fig. 23 shows the logic of data access flow when a query is performed—how many departments an employee belongs to, and returns information such as the employee's

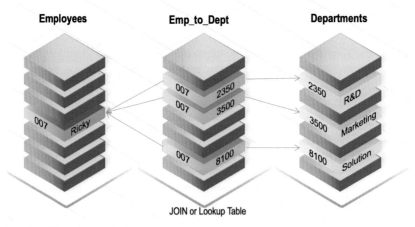

FIG. 23 Nonnative graph storage query mode.

name, employee number, department name, and department number. This simple query involves the SQL-style joins between three tables: the employee table, department table, and employee ID-department ID comparison table.

The entire query process is divided into the following steps:

① In the employee's table, locate employee No. 007
② In the employee-to-department mapping table, locate all department IDs corresponding to the employee ID: 007
③ In the departments table, locate the department names corresponding to all the department IDs in step 2
④ Assemble all the information in the above steps ①–③, and return.

The time complexity of each step in an SQL-style query is shown in Table 5.

The above query (time) complexity does not take into account the physical delay of any hard disk operation or the time of locating and addressing the file system. The actual time complexity in such a simple query operation, if the amount of data is more than tens of millions, can take seconds. If it is a more complex query involving complex associations between multiple tables, it may cause multiple table scan operations. Just imagine the complexity and delay of such operations on the hard disk, even though there are tons of techniques using various in-memory indexes and caches to help accelerate, the overall complexity to deal with is still formidable, and this helps explains why there are so many endlessly running batch processes based on SQL-style queries.

TABLE 5 SQL query complexity.

Step number	Step description	Minimum complexity	Remarks
①	Locate an employee	$O(LogN)+O(1)$	N is the number of rows or columns in the tables; constructing a flattened tree index with a depth of ∼4 layers, that is, $O(4)+O(1)$
②	Location mapping IDs	$3*(O(LogN)+O(1))$	Locating $O(LogN)$, assuming that the minimum complexity of locating records based on employee ID is $O(1)$; three times
③	Locating departments	$3*(O(LogN)+O(1))$	Same logic as above
④	Data assembly	–	–
Total	–	$>O(35)$	

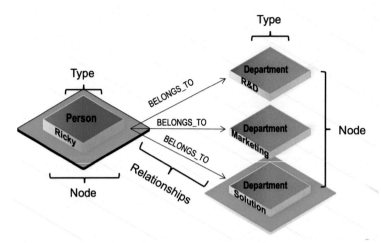

FIG. 24 Core logics of native graph query.

Let's see how the above query can be performed in a native-graph fashion—which we will illustrate the concept of "Index Free Adjacency" as shown in Fig. 24.

The query steps on the native graph are broken down as follows:

① Locate employees in the vertex storage data structure.

② Starting from the employee vertex, find the department it belongs to through the employee-department relationship (this is a typical K-hop query with filtering by edge attribute).

③ Return employee name, employee number, department, and department number.

The time complexity of the above first step is the same as that of the nonnative graph (SQL), but the second step will be significantly faster. Because of the index-free adjacency data structure, the employee vertex is directly linked to the three departments through three edges. If the optimal complexity of the SQL query method is O(35), the native graph can achieve O(8), and the decomposition is shown in Table 6.

The above examples show that there is a large performance gap between native graphs and nonnative graphs in terms of computational complexity. Taking the simple one-hop (1-hop) query as an example, there is a 338% performance improvement. If it is a more complex and deeper query, it will produce an exponential effect, that is, as the depth increases, the performance difference will increase exponentially, as shown in Table 7.

2.2.2 Graph storage data structure and modeling

In practical application scenarios, SQL-type databases are rarely used for querying on 2 hops or deeper, which is determined by its storage structure and

TABLE 6 Query complexity of native graph.

Step number	Step description	Minimum complexity	Remarks
①	Locate an employee	O(LogN)+O(1)	Assuming the index logic is the same as SQL, it is O(5) here. If accessed by UUID, it is O(1)
②	Locate departments	3* O(1)	Since the edge (association relationship) from employee to department adopts an index-free adjacency structure, the complexity of locating each department is O(1)
③	Data assembly	–	Assembling
Total	–	>O(8)	

TABLE 7 Performance gap between nonnative graphs and native graph.

Query depth	Nonnative	Native	Performance gap
0 Hop	O(5)	O(5)	0%
1 Hop	The(35)	O(8)	330%
2 Hop	>>O(100)	–	1000% (10 times) The complexity growth is linear (instead of exponential)
3 Hop	>>O(330)	–	3300% (33 times)
5 Hop	Unable to return	–	33,000% (330 times)
8 Hop	Unable to return	–	More than 10,000 times
10 Hop	Unable to return	–	More than 100,000 times
20 Hop	Unable to return	–	10 billion times

computing (query) model. For example, in a business graph, start from a certain enterprise entity, query for its investors (upstream) recursively, find all shareholders with a shareholding ratio greater than 0.1%, search for 5 hops, and return all shareholding paths (and the complete subgraph).

If this problem is solved with SQL, it will be very complicated, the amount of code will be large, and the code readability will be low due to the problem of "recursive penetration." Of course, the biggest problem is that the computational complexity is high, and the timeliness will become poor (the performance will be exponentially worse than the graph computing system based on native graphs).

However, SQL is very bad at dealing with heterogeneous data, such as shareholding paths. There are vertices, edges, and different types of vertices (companies and natural persons) on a path. Only by encapsulating data structures such as XML and JSON, can such complex query logic (business logic) be supported.

The 70+ lines of SQL code shown in Fig. 25 take 38s to perform a five-hop query in a table with only 10,000 shareholding relationships. In contrast, it takes 7 ms to complete corresponding operations on a dataset that is, 50,000 times larger, with 200 million entities, 300 million relationships, with a real-time graph database, and the latency gap between the two systems is more than 5000 times. In actual scenarios, SQL cannot operate on data sets to the order of hundreds of millions to penetrate more than three layers in seconds (e.g., within 1000 s), let alone much deeper queries.

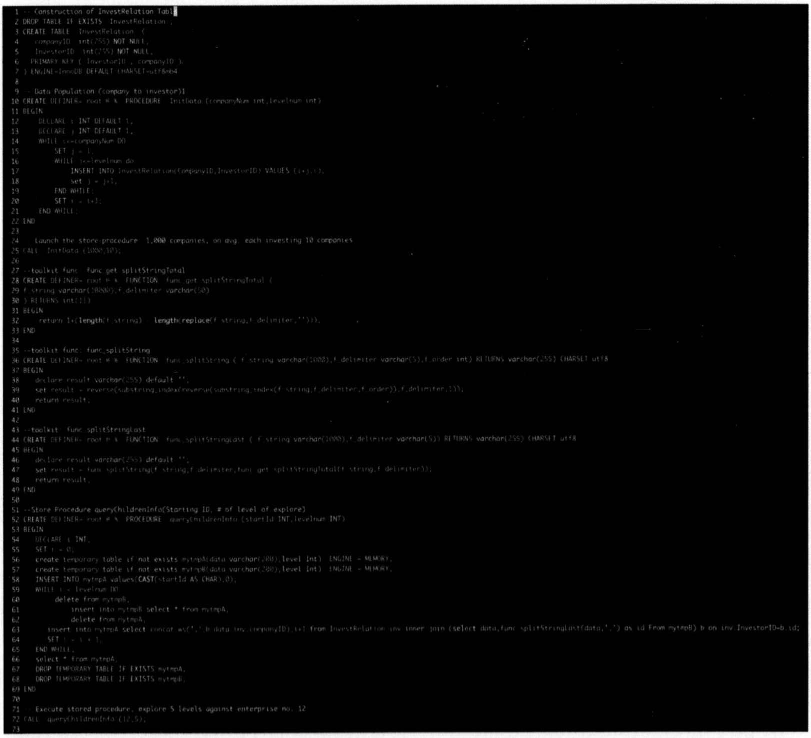

```
-- Construction of InvestRelation Tabl
DROP TABLE IF EXISTS InvestRelation ;
CREATE TABLE InvestRelation (
  companyID int(255) NOT NULL,
  InvestorID int(255) NOT NULL,
 PRIMARY KEY ( InvestorID , companyID ),
) ENGINE=InnoDB DEFAULT CHARSET=utf8mb4

-- Data Population (company to investor)1
CREATE DEFINER= root # % PROCEDURE InitData (companyNum int,levelnum int)
BEGIN
    DECLARE i INT DEFAULT 1,
    DECLARE j INT DEFAULT 1,
    WHILE i<=companyNum DO
        SET j = 1,
        WHILE j<=levelnum do
            INSERT INTO InvestRelation(CompanyID,InvestorID) VALUES (i+j,i),
            set j = j+1,
        END WHILE;
        SET i = i+1;
    END WHILE;
END

-- Launch the store-procedure 1,000 companies, on avg. each investing 10 companies
CALL InitData (1000,10);

--toolkit func func_get_splitStringTotal
CREATE DEFINER= root # % FUNCTION func_get_splitStringTotal (
f_string varchar(10000),f_delimiter varchar(50)
) RETURNS int(11)
BEGIN
    return 1+(length(f_string) - length(replace(f_string,f_delimiter,''))),
END

--toolkit func: func_splitString
CREATE DEFINER= root # % FUNCTION func_splitString ( f_string varchar(1000),f_delimiter varchar(5),f_order int) RETURNS varchar(255) CHARSET utf8
BEGIN
  declare result varchar(255) default '';
  set result = reverse(substring_index(reverse(substring_index(f_string,f_delimiter,f_order)),f_delimiter,1));
  return result;
END

--toolkit func_splitStringLast
CREATE DEFINER= root # % FUNCTION func_splitStringLast ( f_string varchar(1000),f_delimiter varchar(5)) RETURNS varchar(255) CHARSET utf8
BEGIN
  declare result varchar(255) default '';
  set result = func_splitString(f_string,f_delimiter,func_get_splitStringTotal(f_string,f_delimiter));
  return result,
END

--Store Procedure queryChildrenInfo(Starting ID, # of level of explore)
CREATE DEFINER= root # % PROCEDURE queryChildrenInfo (startId INT,levelnum INT)
BEGIN
  DECLARE i INT,
  SET i = 0;
  create temporary table if not exists mytmpA(data varchar(200),level Int) ENGINE = MEMORY,
  create temporary table if not exists mytmpB(data varchar(200),level Int) ENGINE = MEMORY,
  INSERT INTO mytmpA values(CAST(startId AS CHAR),0);
  WHILE i < levelnum DO
      delete from mytmpB,
          insert into mytmpB select * from mytmpA,
          delete from mytmpA,
     insert into mytmpA select concat_ws(',',b.data,inv.companyID),i+1 from InvestRelation inv inner join (select data,func_splitStringLast(data,',') as id From mytmpB) b on inv.InvestorID=b.id;
     SET i = i + 1;
  END WHILE,
  select * from mytmpA,
  DROP TEMPORARY TABLE IF EXISTS mytmpA,
  DROP TEMPORARY TABLE IF EXISTS mytmpB,
END

-- Execute stored procedure, explore 5 levels against enterprise no. 12
CALL queryChildrenInfo (12,5);
```

FIG. 25 Using SQL for deep penetration query.

The above shareholding path query scenario can be summarized as a typical "business relationship graph," which has a wide range of application scenarios, such as KYC (Know Your Customer), Customer-360, UBO (Ultimate Beneficiary Owners), Investment Research, and other due-diligence or auditing scenarios when a customer is opening an account or applying for a loan.

A major feature of graph databases that distinguishes them from SQL databases is the processing of heterogeneous data. The primary data that native graph storage focuses on is metadata, followed by secondary data (auxiliary data), and thirdly derived data or composite data (combo-data). Their detailed descriptions are as follows:

(1) Metadata
  o Entity data (vertex)
  o Associational (relationship) data (edges)
(2) Secondary data (auxiliary data)
  o Entity attribute data
  o Relationship attribute data
(3) Combo data (derived data)
  o Multivertex combination
  o Multiedge combination
  o Path: single path, multipath, loop, etc.
  o Subgraph, forest, network, etc.
  o Combo data across subgraphs, etc.

From the perspective of storage complexity and computing (penetration or aggregation) capabilities, traditional storage engines are suitable for metadata and auxiliary data management, both are considered discrete and two-dimensional, but they will face many challenges to deal with combo data, which are connected and high-dimensional. Such challenges need the help of a "native graph computing engine" which works on top of the storage engine to conduct accelerated networked queries and deep analytics. As illustrated in Figs. 26 and 27, if you like the results to be processed via one real-time query and rendered for intuitive visual presentation, you will need the help of a capable graph engine. More use cases will be demonstrated next in this section.

For the storage of metadata and auxiliary data, the core appeal is to allow vertices, edges, and their attributes to be placed on the disk as soon as possible (persistence), and to provide sufficient efficiency when reading (query) or updating. This process is easy to understand by the most familiar row-based storage method, which can be dissected into two parts: "vertex storage + vertex attribute storage" and "edge storage + edge attribute storage." They respectively have the following data structures, see Tables 8 and 9.

We are using a map-like data structure to illustrate the vertex storage structure in Table 8 for easy comprehension. In real-world systems, many optimization techniques can be applied to allow for efficient storage usage, distributed data access concurrent data manipulation, etc. For instance, the attribute fields

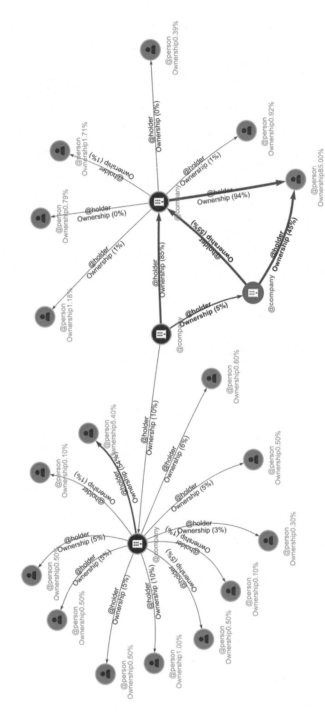

FIG. 26 Deep query and visualization with native graph.

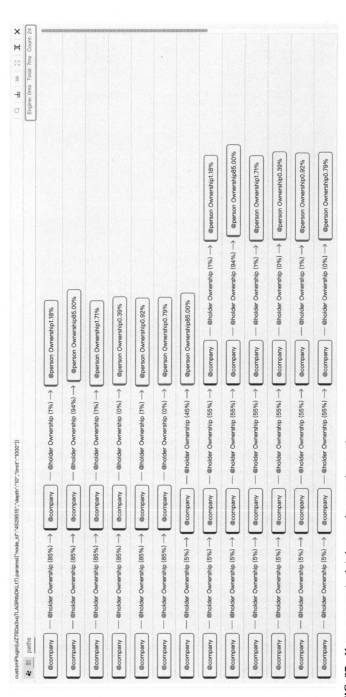

FIG. 27 Heterogeneous (paths) data results on native graph.

TABLE 8 Vertex storage structure.

Vertices	Attribute-1	Attribute-2	Attribute-3	...
Vertex-1	Values	Values	Values	...
Vertex-2	Values	...	...	...
Vertex-3	Values	...	...	...
...	...	...	...	...
Vertex-V	Values	...	...	...

TABLE 9 Edge storage structure.

Edges	Attribute-1	Attribute-2	Attribute-3	...
Edge-1	Values	Values	Values	...
Edge-2	Values	Values	...	...
Edge-3	Values	...	...	...
...	...	...	...	...
Edge-E	Values	...	...	...

can be stored in a separate data structure which can be physically located on separate instances (a form of partitioning or sharing—which allows for improved concurrency). Note the advantages and disadvantages associated with each optimization, especially around distributed data structures, once the data is separated into different instances, the network latency becomes an unignorable factor to consider over the life cycle of a query, while zero network accesses are required on a holistically designed vertex storage structure—which howder could not handle superlarge graphs with many billions of vertices. In Chapter 5, we will talk about issues and concerns around distributed and scalable graph database system design.

Similar logic can be used for edges and their attributes, whether stored as a whole data structure or separately and distributively. Edge storage is inherently more complicated than vertex storage in that edge attribute design is more complicated, for example, we may need to consider the following points:

- Do edges need directions?
- How to express the direction of the edge?

- How to express the start and end of an edge?
- Can an edge relate to multiple origins or multiple endpoints?
- Do edges need additional properties?
- Should multiple edges be allowed between a pair of vertices?

There is no single standard answer to the above questions. We have touched on the answers to some of these questions in the previous section and chapter. For example, regarding the direction of the edge, we can express the direction of the edge by placing the starting vertex ID and the ending vertex ID in the edge record continuously in row-store mode (see Table 9, put Starting Vertex in Attribute 1 and Ending Vertex in Attribute 2). But this solution will soon trigger another question, how to express the concept of reverse edge? Each edge deserves a reverse edge, otherwise, how do you traverse reversely from the ending vertex to the starting vertex? This is a very important concept in graph database storage and computing. Suppose we store an edge in the record as follows:

Edge ID	Starting V-ID	Ending V-ID	Other Attributes	...

When we find the ID of the edge or the ID of the starting vertex through the index-accelerated data structure, we can read the ID of the ending vertex sequentially, and then continue the traversal query in the graph. However, if we first find the ID of the ending vertex, how can we reversely read the ID of the starting vertex to perform the same traversal query?

There is more than one answer to this question. We can design different types of data structures to solve it. For example, set a direction identification attribute in the edge record, and then each edge record will be stored in the forward and reverse directions:

Edge ID-X	Vertex A	Vertex B	Edge orientation: Forward	Other properties ...
Edge ID-X'	Vertex B	Vertex A	Edge orientation: Reverse	Other properties ...

There are many other solutions, such as vertex-centric storage, including the attributes of the vertex itself, as well as all connecting vertices and attributes associated with each edge, this method can also be regarded as an index-free adjacency storage structure, and it is no longer necessary to set up a separate edge data structure. The advantages and disadvantages of this method do not need to be discussed here, and interested readers can conduct independent extended analysis.

Conversely, we can also design storage data structures centered on edges. This structure is at least very common in academia and social network graph analysis. For example, the follower-relationship network between Twitter users can be expressed using only one edge file, and each row in the file has only two

TABLE 10 Edge-centric storage structure (Twitter Dataset).

user 1	user 2
user 1	user 3
user 1	user 10
user 2	user 3
user 2	user 5
user 3	user 7
user 5	user 1
user 10	user 6
......	......

columns, of which one column is the starting vertex, the second column is the ending vertex, and the record of each row is expressed as the user in the second column following the users in the first column, see Table 10.

Based on Table 10, the storage logic of each row record can be greatly expanded, such as adding the unique ID of the edge for global index positioning, adding more attributes of the edge, adding the direction of the edge, etc., to better reflect real-world scenarios. Implicitly, when loading these edge records, each edge should be stored with a matching reverse edge, otherwise, graph traversal query results can be hugely mistaken—in Chapter 7, we will expand on this. Such problems can be identified easily if you know how to validate a system's storage mechanism and its accuracy.

In Table 8 (Vertex Storage), careful readers will ask a question, How to store heterogeneous types of entity data? In traditional databases, different types of entities are aggregated in the form of different tables, such as the employee table, department table, company entity table, and the relationship mapping table between different entities that we discussed earlier. In the storage logic of graph databases, heterogeneous data, such as entities, can be integrated into one graphic table (logically speaking, not physically), which is why graphs are called high-dimensional databases. We take financial transaction data in the financial industry as an example to illustrate how heterogeneous data is stored and queried, and graph data modeling is handled. Table 11 shows the record structure of the three core tables of the card transaction design as expressed in a relational database.

If the entities and association relationships in the above relational table are modeled by a graph database and visualized, the effect is shown in Fig. 28.

The financial transactions captured in Table 11 can be transformed and migrated into a graph database. To do this we first need to do graph modeling, and there are multiple ways of doing so. What is illustrated in Fig. 28 are two typical graph modeling mechanisms:

TABLE 11 Relational tables financial transaction scenarios (three tables).

Accounts table

PID	Name	Gender	Category	
P001	Ricky Sun	M	1	
P002	Monica Liu	F	2	
P003	Abraham L.	M	3	
...				
...				

Cards table

CID	Type	Level	Open_Branch	Phone	Pid
c001	debit	regular	Saint Francis	415xxxxxxx	p001
c002	credit	/	Shenzhen	139xxxxxxxx	p002
c003	debit	vip	Dubai	868xxxxxx	p003
...					
...					

Transactions table

TrxID	Pay_card_id	Recv_card_id	Amt	DeviceID	
t001	c001	c00x	12000	uuid-123	
t002	c002	c01x	24000	uuid-234	
t003	c003	c02x	3600	uuid-4321	
...					
...					
...					

- Simple graph: Two types of entities and two types of relationships
- Multigraph: Four (or five) types of entities and two types of relationships

Let's use a simple mechanism and zoom in to see entity and relationship definitions:

- Entities
  - Account
  - Card: Contains a subtype (Phone)
  - Device

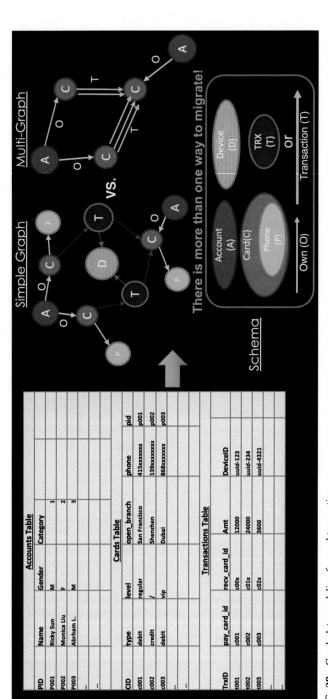

FIG. 28 Graph data modeling for card transaction.

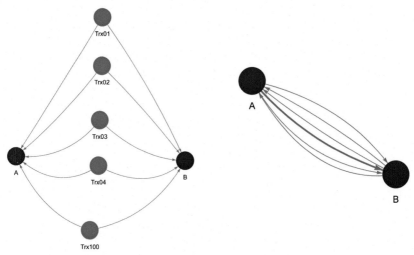

FIG. 29   Simple graph (left) vs multigraph (right).

      ○ Transaction
      ○ Phone: A subtype of Card (the main entity type)
  ● Relationships
      ○ Transaction:
      ○ Ownership: Card belongs to Account (this is a hierarchical relationship)

In multigraph data modeling, as illustrated in Fig. 29, the implementation is more streamlined, for example, there are only two types of entities and relationships respectively.

● Entities: Cards and Accounts (that own the cards)
● Relationship: Transaction (attributes include transaction time, equipment, environment information, etc.) and card-to-account linkage relationship

When we compare multigraph to simple graph, there are a few key differences:

● In multigraph, there are fewer types of entities needed. It is even possible to use just one type of entity, card only, which means the owning account information can be consolidated into the card attribute field, and doing so will also reduce the card-to-account edge relationship.
● The edges in multigraph must be able to handle (store) more attributes than in simple graph. Taking card transactions as an example, in a simple graph, transactions are modeled as entities that connect with cards, the edges in a simple graph contain no material information other than directions, while the transactional edge in a multigraph contains everything that a transaction entity must contain in a simple graph setup.

- Implicitly, the storage engine design for simple graphs and multigraphs can be very different. As there are more vertices and edges in a simple graph, the storage need is three times as much as in a multigraph.
- Compute engine query complexity between simple graph and multigraph can also be different—specifically, assuming you are trying to traverse the graph to find all the downstream cards that had linkage with card A for as far as three legs away, in multigraph, it is a 3-hop query, but in simple graph, it would be a 6-hop query. And the time complexity is exponentially higher than in multigraph.
- Note that multigraph is backward compatible with simple graph, and this is considered a clear advantage of multigraph. The opposite is not true though.

However, in some scenarios, a simple graph way of data modeling has its advantages. For example, it is common in antifraud scenarios to find out whether two credit card applications or loan applications use the same phone number and address. Even though both the phone number and address can be modeled as attributes of the application entity (as you would in a multigraph setup), it is much easier to model both as separate entities (see Fig. 30), so that patterns can be formed to help with fraud detection.

When we judge whether an application is potentially fraudulent, we first check in the graph whether the application is reachable with any other applications, such as a "loop pattern" formed between two applications. For instance, in Fig. 30, if we start traversing from any application, in 4 hops, we will be able to come back to our starting point, and visually, we can see there is a loop, which can also be interpreted as high-similarity between the two applications. When there are many loops in between any two applications or strikingly high similarity, we know the application is likely to be fraudulent.

This section provides some real-world scenarios and possible graph data modeling methods. There is no so-called solution in the process of designing any graph database. The graph database is very close to the business, and graph data modeling directly reflects the business logic. A storage engine is located at

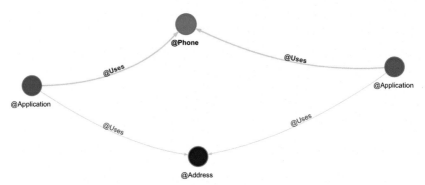

FIG. 30 Loop pattern in the graph.

the bottom of the graph database; the efficiency and flexibility of the storage engine also affect the efficiency and flexibility of computing and analytics running on it. Later chapters will conduct a more in-depth discussion and analysis of these topics.

2.3 Evolution of graph query language

After understanding the computing and storage principles of graph databases, everyone will ask a question, How to operate and query graph databases? We know that relational databases use SQL, which is also the first international standard in the database field, and the only international standard for database query languages before the widespread development of big data frameworks and NoSQL databases. In the past 10 years, as big data and NoSQL were surging, most of them chose to continue to be compatible with SQL, and common to operate through API. Compared with SQL, the API/SDK calling method is very simple. For example, the KV key-value store can only be called through a simple API, so many people do not think that KV Store is a complete database. Though it is possible to continue this status quo—using API to operate a graph database, and even use SQL to do the same, the innate complexity and strength of a graph database system would be drastically confined. That is why a new international standard is being drafted, by the time this book is published (early 2024), the GQL (Graph Query Language) standard should have been published.

This section introduces the basic concept of GQL and the differences between GQL and SQL.

2.3.1 Basic concepts of database query language

We are living in an era of big data. The rapid development of the Internet and mobile Internet has brought about a huge increase in the rate of data generation. Every day and every moment, there are billions of devices (some people predict that there will be more than 100 billion connected sensor devices before 2030) generating huge volumes of data. Databases were created to address this growing data challenge. There are other similar concepts, tools, and solutions, such as data warehouses, data marts, data lakes, or even lake houses, etc., to solve a series of tasks such as data storage, transformation, intelligent analysis, and reporting that we face every day. Compared with the database, the scale of these relatively new "tracks" is far from being able to match the database track (database is a single track close to 100 billion US dollars, taking the industry leader Oracle as an example, its annual revenue reaches 40 billion US dollars). What makes databases so important that we cannot let go of them? The author believes that two things are the core "competence" of the database: performance and query language.

In the world of modern business, performance has always been a first-class citizen. A database can be called a database manifest that it has the processing

| Data | → | Big Data or Fast Data | → | Deep Data |

FIG. 31 The evolution of data to big data to fast data to deep data.

power against data with small, and usually minimal and business-sense-making, latencies. This is in contrast with many BI-era data warehouse solutions or Hadoop-school solutions, they may be highly distributed and scalable, but their performances are quite worrisome. And this is exactly why we are seeing the dying of Hadoop (or replacement with Spark and other newer architectures) in recent years.

The trend of moving from Big Data to Fast Data is to ensure that given the growing data volume, the underpinning database frameworks can handle it without sacrificing performance. Fig. 31 illustrates the evolvement of database-centric data processing technologies: in a word, from data to big data, from big data to fast data, and then from fast data to deep data (or graph data).

Database Query Language is one of the best things invented since the advent of computer programming languages. We hope that computer programs can have the same ability as human beings to intelligent data, filter in-depth, and dynamically correlate information, but the appeal of such artificial general intelligence has never been truly realized so far—the recent boom of LLM and GPT seems to be opening up a new pathway for AGI, but as we have demonstrated in the previous chapter, there still is a very long way to go. Smart programmers and linguists use computer programming languages to help realize human intentions and instructions, and database query language is a kind of "semiintelligent" data processing facility to fulfill our needs for "intelligent data."

Before entering the well-known SQL world, let us take an overview of some features of query languages supported by nonrelational databases (such as NoSQL, etc.).

First, let us take a look at the key-value database (KV Store). Common key-value store implementations include Berkeley DB, Level DB, etc. Technically speaking, the key-value library does not support or use a specific query language, because the operations supported on it are too simple, and a simple API is enough. A typical key-value store supports three operations: Insert, Get, and Delete.

You may argue that Cassandra does have CQL (Cassandra Query Language). No problem with that! Cassandra is a wide-column store, which can be seen as a two-dimensional KV store, and for the complexity of data operations in and out of Cassandra's highly distributed infrastructure, it does make sense to provide a layer of abstraction to ease the programmer's load on memorizing all those burdensome API calling parameters. By the way, CQL uses similar concepts as regular SQL, such as tables, rows, and columns, we will get to SQL-specific discussions next.

A criticism of Apache Cassandra, and perhaps of NoSQL databases in general, is that CQL does not support the common JOIN operations in SQL. This critical thinking is deeply rooted in the SQL way of thinking. JOIN is a double-edged sword. While it solves one thing, it also brings a problem, such as (huge) performance loss. Cassandra can support JOIN operations, there are two solutions:

- SPARK SQL: Apache Spark's Spark SQL with Cassandra (Fig. 32)
- ODBC driver: Cassandra with ODBC driver

However, these solutions come at a price, for example, ODBC performance is an issue for large data sets or clustered operations. However, Spark SQL adds a new layer of system complexity. It is a challenge for multisystem maintenance and system stability, and many programmers hate to deal with new systems and learn new knowledge.

A few extra words about Apache Spark and Spark SQL. Spark was designed in light of the inefficiencies of Apache Hadoop, which borrows two key concepts from Google's GFS and MapReduce and develops into Hadoop's HDFS and MapReduce. Spark is $100\times$ faster than Hadoop mainly due to its distributed shared-memory architecture design. (To draw an analogy, by moving data to be stored and computed in memory, there is $100\times$ performance gain over disk-based operations, period.) Spark SQL was invented to process Spark's structured data by providing a SQL-compatible query language interface. It is an advancement compared to Spark's RDD API. As explained earlier, API can be less efficient when it is getting overly complicated, and at that point, a query language is more desirable and practical (and flexible and powerful).

When the author worked for Yahoo! Strategic Data Department (SDS) about 20 years ago, it was the period when Yahoo! was incubating the Hadoop project. Compared with other high-performance, distributed massive data processing frameworks developed by SDS, Hadoop's performance was unsatisfactory—yet what is impressive is that many common sorting and searching tools in Linux have been rewritten to handle massive data processing with extremely high performance. For example, the performance of the sort command has been increased by more than 100 times to deal with GB to TB level data sets. Perhaps

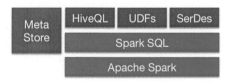

FIG. 32   Spark SQL and ODBC/JDBC.

this can help explain why Hadoop was later donated to the Apache open-source community by Yahoo!, while other projects with obvious advantages were not open-sourced. Perhaps readers should think: Is open source the best?

Referring to Fig. 32, the evolution of Data → Big Data → Fast data → Deep data, the corresponding underlying data processing technology will evolve from the era dominated by relational databases to nonrelational databases (such as NoSQL, Spark, etc.), NewSQL and eventually graph database frameworks. According to this trend and evolution path, we can boldly predict that the future core data processing infrastructure must at least include or be dominated by graph architecture.

The evolution of database technology in the past half-century may be summarized in Fig. 33—from the navigational database before the 1970s to the relational database that began to emerge in the 1970s and 1980s, to the rise of the SQL programming language in the 1990s, to the various NoSQLs that appeared after the first decade of the 21st century—perhaps Fig. 33 leaves us with a myth—will graph query language be the ultimate query language?

Ask a question from another angle: What is the ultimate database for human beings? Perhaps you will not object to the answer: the human brains! What kind of database technology is the human brain built with? The author thinks it is a graph database, at least in terms of probability, it is much higher than relational databases, column databases, key-value databases, or any document database. Maybe a database we have not invented yet. But graph database is the closest to the ultimate form of database, without a doubt.

If the reader has an understanding of the evolution of the SQL language, it directly promoted the rise of relational databases (if you can recall how clumsy and painful to use databases before the emergence of the SQL language). The rise of the Internet has led to the birth and rise of NoSQL. One of the big factors is that relational databases cannot cope with the demands of data processing speed and data modeling flexibility. NoSQL databases are generally divided into the following four or five categories, each with its characteristics.

- Key-Value: performance and simplicity
- Wide-Column: volume and performance
- Document: data diversity
- Graph/Vector: deep data + fast data
- (Optional) Time Series: IOT data, time series data

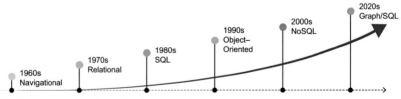

FIG. 33   Progressive view of database evolution.

From the perspective of query language (or API) complexity, there is also an evolutionary path for data processing capabilities. Fig. 34 vividly expresses the comparison between NoSQL databases and SQL-centric relational databases.

- Primitive key-value library/API → Ordered Key-Value Store
- Ordered key-value library → Big Table (A typical Wide Column Store)
- Big Table → Document Database (such as MongoDB), with full-text search capability (search engine)
- Document Database → Graph Database

SQL has been around for nearly 50 years and has iterated many versions (on average, one major iteration every 3–4 years), among which the most well-known ones are SQL-92 and SQL-99. For example, in the SQL-92 version, the FROM statement added a subquery function; in the SQL-99 version, CTE (Common Table Expression) functionality was added. These features greatly increase the flexibility of relational databases, as shown in Fig. 35.

However, there is a "weakness" in relational databases, and that is the support for recursive data structures. The so-called recursive data structure refers to the function realization of the directed relationship graph. Ironically, although the name of a relational database includes relations, it is difficult to support relational queries from the beginning of its design. To implement relational queries, relational databases have to rely on table join operations—every table join means potential table scan operations, and the subsequent exponential decline in performance, as well as the exponentially rising complexity of SQL code.

The performance loss of table join operations is directly derived from the basic design ideas of relational databases:

- Data normalization
- Fixed/Predefined Schema

If we look at NoSQLs, the primary data-modeling design concept of most NoSQLs is data de-normalization. Data de-normalization is basically to trade time with space, meaning data may be stored in a multitude of copies so that they can be accessed in a storage-in-close-proximity-to-compute way, this is in contrast to SQL's just-one-copy all-data-normalized design, which may save some space, but tends to yield worse performance when dreadful SQL operations like table joins are necessary.

Predefined schema is another major difference between SQL and NoSQL, and this can be hard to conceptualize. One way to digest this is to think:

Schema First, Data Second vs Data First, Schema Second

In a relational database, the system administrator (DBA) needs to define the table structure (Schema) before loading the first row of data into the database. It is impossible for him to dynamically change the table structure. This rigidity may not be a big problem with fixed schemas, immutable data structures, and

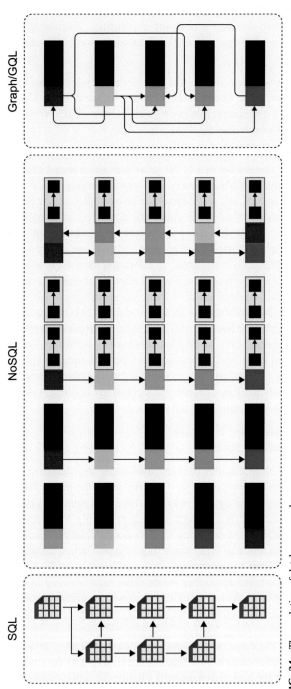

FIG. 34 The evolution of database query languages.

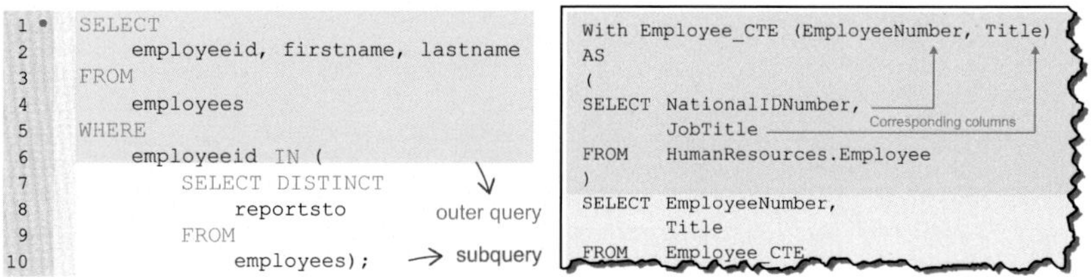

```
1 •  SELECT                                      With Employee_CTE (EmployeeNumber, Title)
2        employeeid, firstname, lastname         AS
3    FROM                                         (
4        employees                               SELECT  NationalIDNumber,
5    WHERE                                                JobTitle
6        employeeid IN (                         FROM    HumanResources.Employee
7            SELECT DISTINCT                      )
8                reportsto         outer query    SELECT  EmployeeNumber,
9            FROM                                         Title
10               employees);      → subquery      FROM    Employee_CTE
```

Corresponding columns

FIG. 35   Subquery per SQL-92 and CTE per SQL-99.

never-changing business requirements. But let us imagine if the data schema could adjust dynamically based on the incoming data, which gives us tremendous flexibility. For people with a strong SQL background, this is hard to imagine. But what if we could temporarily ditch our rigid, limiting mindset and replace it with a growth mindset? The goal we want to achieve is a Schema-Free or Schema-less data model, that is, there is no need to preset the data model, and the relationship between data does not need to be defined and understood in advance. As the data flows in, they will naturally form a certain associative relationship. What the database needs to do is to improvise.

In the past few decades, database programmers have been trained to understand the data model first, whether it is a relational table structure or an entity E-R schema diagram. Knowing the data model certainly has its advantages, but it also complicates and slows down the development process. Programmer readers, do you remember how long was the development cycle of the last turnkey solution you participated in? A quarter, half a year, a year or longer? In an Oracle database with 8000 tables, no DBA can fully grasp the relationship between all tables—how much different if we compare this fragile system to a timed bomb, and all your businesses are bound to it.

Schema-free was first introduced in document databases or wide-column databases, although they have somewhat similar design concepts to graph databases, the query models are different, and the intuitiveness of a query can best be reflected in a graph query model powered with a native graph database. In the following section, we will use some specific examples of graph databases to help readers understand schema-free.

2.3.2   Graph query language

In a graph database, logically there are only two types of basic data types: vertices and edges. A vertex has its ID and attributes (label, category, etc.), and an edge has similar bells and whistles except that it is usually determined by the order of the two vertices (the so-called directed graph refers to each edge corresponding to an initial vertex pointing toward a terminal vertex, plus other attributes, such as edge direction, label, weight, etc.). Apart from these basic data structures, graph databases do not require any predefined schema or table structure, and a gigantic networked data structure can be formed simply

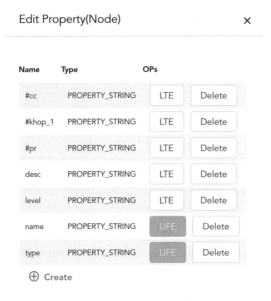

FIG. 36 Vertex attributes in graph dataset.

by expanding on the vertices and edges. This extremely simplified concept is very similar to how humans think and store information—after all, humans usually do not set table structures in their minds (except for those dedicated SQL programmers).

Now, let us look at some graph database implementations in real-world scenarios. For example, the attribute definition of a vertex in a typical graph data set in Fig. 36 contains the definitions of the first few fields, such as desc, level, name, and type, but there are also some dynamically generated and expanded fields, such as #cc, #pr, #khop_1, etc. Compared with relational databases, the structure of the entire "table" capturing vertex attributes is dynamically adjustable.

Note that the properties of the Name and Type fields in Fig. 36 are of type STRING, which can maximize compatibility with a broad spectrum of data types, thereby providing great flexibility. However, flexibility does not come without a tradeoff for performance—processing against string type can be much slower than integer type. To achieve both flexibility and performance, optimizations must be done. The usual practice is to achieve this through indexing and storage-tiering. We covered indexing techniques in previous sections, now we can start covering the tiered-storage approach. To achieve excellent computing performance, data can be dynamically loaded into memory for speedy computing and unloaded as per needs. This dual process is called

LTE vs UFE (Load-to-Engine vs Unload-from-Engine). Certainly, in-memory computing and low-latency data structures that support concurrent computing are also essential for boosting performance.

Key-value stores can be seen as pre-SQL and nonrelational, graph databases on the other hand are post-SQL and support truly recursive data structures. Lots of today's NoSQL databases support SQL to be backward compatible, however, SQL is very much table-confined, and this is, in our humble opinions, counter-intuitive and inefficient. By saying table-confined, we mean it is two-dimensional thinking, and doing table join is like three-dimensional thinking, but as the foundation in the data model is two-dimensional based, the SQL framework is fundamentally restrictive!

Graph, on the other hand, is high-dimensional. Operations on graphs are also high dimensional, many of which are natively recursive, for instance, breadth-first-search (BFS) or depth-first-search (DFS). Moreover, graphs can be backward compatible, that is to support SQL-like operations just as NoSQLs do—but the point is why would we bother doing that?

Let's examine a few graph queries next, and please bear in mind and think about how SQL or other NoSQL would accomplish the same.

Task 1: Start from a node, finding all its neighbors and the connectivity of these neighbors recursively up to K layers or 8000 edges, and return the resulting dataset (subgraph).

To achieve this, we must understand that this is not a regular K-hop query, because K-hop only concerns neighboring entities in the result set (not the connections between neighbors), and this is also NOT a regular DFS or BFS query—it is a BFS-DFS-hybrid query.

```
/* System-U's GQL */
spread().src({_id == 1159790032}).depth(6).limit(8000) as paths
return paths{*}
```

The GQL code itself is self-explainable, starting with the source code (ID = 1159790032), expanding for up to 6 hops, and returning up to 8000 paths (and edges, implicitly). The resulting subgraph consists of the entire 8000 edges and connecting vertices, therefore we are calling it: a holistic view of the graph.

The spread() operation is equivalent to finding the connected subgraph starting from any vertex—the shape (topology) of its neighbor network can be directly calculated and displayed intuitively through the visual interface. In this way, the spatial characteristics of hotspots and aggregation areas formed by vertices and edges in the generated subgraph can also be seen, without the need for E-R model diagrams in traditional databases.

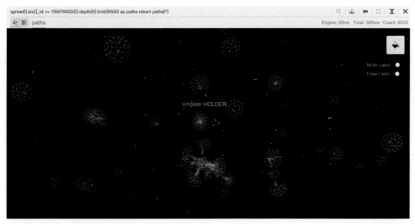

FIG. 37 3D holistic view of spreading results.

FIG. 38 K-hop and count of results.

If you were ever interested in knowing how many neighbors in total within 6 hops from the vertex from Fig. 37, you can easily achieve this with K-hop count as shown in Fig. 38:

```
/* System-U GQL */
khop().src({_id == 1159790032}).depth(6).boost() as nodes
return count(nodes)
```

Task 2: Given multiple vertices, automatically form a network (network of inter-connected vertices).

This query may be too complicated for readers who are familiar with traditional databases, and this networking function may not be realized with SQL. However, for the human brain, this is a very natural appeal—when you want to form a network of associations between Account A, Account B, Account C, and many others, you have already begun to draw the following in your mind—refer to Fig. 39.

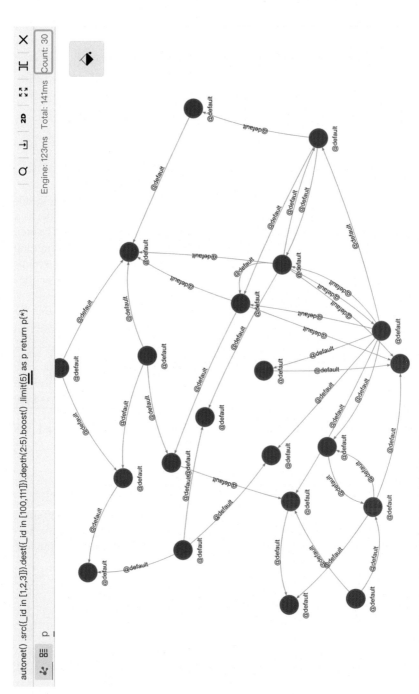

FIG. 39 Automatic networking (subgraph formation) in Ultipa graph.

If we were to achieve this operation with a one-liner, we could do this:

```
/* System-U GQL */
/* Inter-group Networking */
autonet()
    .src({_id in [1,2,3]}).dest({_id in [100,111]}).depth(2:5).boost()
    .limit(5) as p
return p{*}

/* Intra-group Networking */
autonet()
    .src({_id in [1,2,3,100,111]}).depth(2:5).boost()
    .limit(5) as p
return p{*}
```

autoNet() is the function to invoke, the starting nodes is a set with multiple vertices (as many as you desire), the search depth between any pair of nodes is a range between 2- and 5-hop, and the number of paths between any two is limited to five. There are two modes to form a network: intergroup vs intragroup (refer to Fig. 40)—meaning the network is constructed between two groups of entities (M*N) or automatically formed amongst just one group of N entities (N * (N−1)/2).

Let's straighten out this type of networking query mathematically so that we can have a better idea of how complex it is. The computational complexity is:

Paths to Return	3 * 2 * 5 = 30 paths (Intergroup)
	5 * (5-1)/2 * 5 = 50 paths (Intragroup)
Max Query Complexity	30 paths * (2 * E/V)^5 = 30 * 16^5 = 31,457,280
	50 paths * (2 * E/V)^5 = 50 * 16^5 = 52,428,800

Assuming E/V is the ratio of vertex-to-edge, the graph we used has $|E|/|V| \sim = 8$.

$$\frac{|E|}{|V|(|V|-1)} \quad \longrightarrow \quad \left(2 \times \frac{|E|}{|V|}\right)^k$$

Traditionally in graph theory, the left formula is used to examine a graph's density (a measurement of graph data complexity), but this formula is counterintuitive in assessing time complexities. We propose to use the right formula to

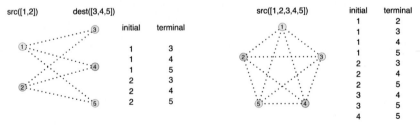

FIG. 40 Intergroup networking (left) and intragroup networking (right).

indicate the average complexity to traverse K-hop from any vertex (node) in the graph, where E stands for the total number of edges, and V stands for the number of vertices in the graph. In our testing dataset (AMZ 0601), there are 0.4 million vertices and 3.4 million edges, on average, each vertex is connected to $\sim 2\times$ $(3.3M/0.4M) = 16$ neighboring vertices, and traversing 2-hop's complexity is 16^2, 3-hop 16^3, and so on.

This query makes a lot of sense in real-world applications. For example, law enforcement agencies will use the telephone company's call records to track the characteristics of the deep network formed by the calls of multiple suspects to determine whether other suspects are associated with them. They can look into the characteristics of the resulting call networks for hints such as how close or distant any two suspects are and how they operate as gangs (crime rings).

Traditionally, these operations are extremely time-consuming on any big-data framework, and practically impossible with relational databases. The reason is computational complexity, assuming there are 1000 suspects/gangsters/nodes to query against, if it is queried in an intragroup mode, there are $(1000 * 999)/2 =$ 500 K possible paths and an extremely large number of possibilities if the query depth is 6-hop and the dataset is densely populated! Apache Spark with GraphX may take days to finish the query, but on a real-time graph database like Ultipa Graph, it can be completed in real-time or near real-time, possibly with a turnaround time within minutes or even seconds. Fig. 41 illustrates the real-time deep (6-hop) network formation of many suspects (red dots) on a large telecom phone call dataset. When you are fighting crimes, latency matters!

For real-time high-concurrency graph databases, performance is a "first-class citizen," but this does not make us regard the simplicity, intuition, and

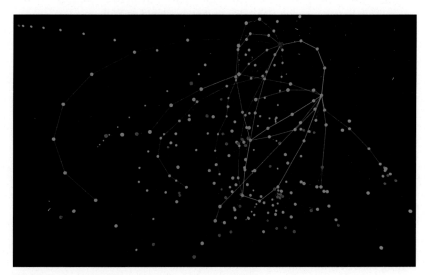

FIG. 41 Large scale deep network formation in real-time.

understandability of the query language as a second-class citizen. The dialectic GQL illustrated above and below adopts chain-query practice, it is not unfamiliar to programmers who have experience with the popular document database—MongoDB, a typical simple graph chain-query is like this.

In Fig. 42, paths between two vertices (card accounts as encapsulated in src and dest) with a depth of four are queried, ten paths are returned, and all attributes (property fields, schema, labels, directions, etc.) of nodes and edges along the paths are returned.

Let's look at a slightly more complicated example—template query:

```
/* Ultipa GQL */
ab().src({_id == "CA001"}).dest({_id == "CA002"}).depth(:6)
  .path_ascend(@transfer.time)
  .node_filter({@card}).edge_filter({@transfer}) as p
limit 10
return p(*)
```

The template path query tries to find paths between two card accounts according to custom-defined templates, with the following filtering requirements:

- The returned path must be in ascending order per card-transfer time
- Filter by card type
- Filter by edge type
- Return the found paths

Like SQL, aliases can be set in GQL, which can be quite convenient. But note the returning paths which contain heterogeneous types of data—a set of path results corresponding to the entire template search, collection of a group of vertices and edges organized in a way satisfying the template filtering requirements. This heterogeneous flexibility is not available in SQL.

In SQL, this kind of flexibility is difficult to achieve without writing user-defined functions, and the search depth is also prohibitive for relational databases.

Task 3: Mathematical statistics type queries, such as count(), sum(), min(), collect(), etc.

Example 1: Find out the sum of the transferred amount from a card. This example is no stranger to SQL programming enthusiasts.

```
n({_id == "CA001"}).re({@transfer} as pay).n({@card})
return sum(pay.amount)
```

The above one-liner GQL is to find all outbound 1-hop transfer paths from the starting card account and aggregate by edge attribute "amount."

In a small SQL table, this operation is stress-free, but in a large table (hundreds of millions of rows of transactions), perhaps this SQL operation

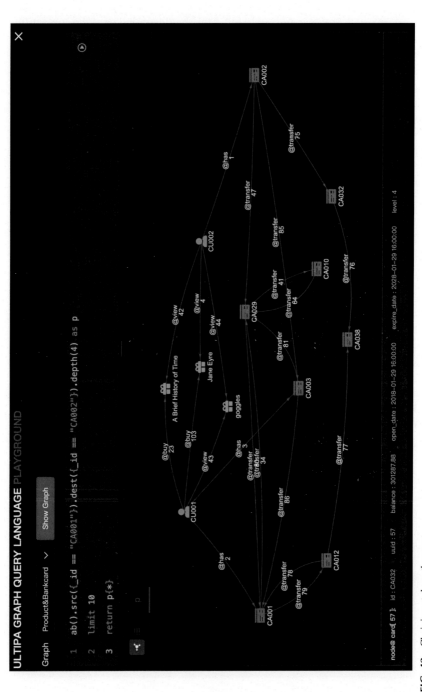

FIG. 42 Chaining-style path query.

will become slow due to table/index scanning. If this were to find 3-hop deep transfer paths, SQL would simply fail. This should remind us of the competitive advantages of index-free adjacency (as illustrated in Fig. 24) in doing such kind of traversal queries. In native graph queries, the possible neighboring vertices and edges that can be reached in each hop are hardly affected by the entire size of the data set, which in turn can make the query execution time both fast and predictable.

Example 2: In a financial transaction network, find all transactional counter-parties of a certain account that are within 3 hops, and return ALL paths with accounts lined up in each path.

The example is not difficult to depict and understand in natural language, but to do it with a programming query language can be quite a challenge. There are three key things to handle:

1. Identify the requirement is a path query (template-based)
2. The desired returned results are a collection of paths
3. Each path consists of nodes that had transactions consecutively (and forming a path).

With these key things in mind, we can come up with a simple GQL solution in Ultipa (Fig. 43):

```
n({_id == "CA001"}).e()[3].n() as p
with pnodes(p) as a
return collect(a)
```

The two examples above illustrate the capabilities and convenience of GQL. For very basic operations, you could do the same with SQL, but when the networked analysis goes deep, GQL is your best friend.

Task 4: Powerful template-based full-text search.

If a database system cannot support full-text search, it is difficult to call it a complete database. Supporting full-text search in graph databases is not a new

FIG. 43 Aggregation of vertices into path results.

thing. Some graph databases have chosen to integrate with third party full-text search frameworks like ES (Elastic Search) or the engines powering ES (Solr and Lucene). However, this way of integrating open-source frameworks has a side effect, which is the gap between performance expectations and actual (query) operations—graph queries often focus on multilayer, deep paths or K-neighbor queries, while a full-text search is only the first step in graph queries if the full-text matching takes a long time, the overall query execution would take much longer.

Let's use a series of GQLs to illustrate how full-text search can be integrated seamlessly into a graph database:

```
/* Ultipa GQL */
/* Step-1: Take a look at some sample nodes */
find().nodes({@`default`}) as nodes
return nodes{name, type, fullName} limit 10

/* Step-2: Create full-text index for node property */
create().node_fulltext(@default.name, "ft_name")

/* Step-3: Check the status of full-text index(es) */
show().node_fulltext()

/* Step-4: Use full-text index to search meta-data */
find().nodes({~ft_name contains 'mary'}) as nodes
return nodes{name, type, fullName} limit 10

/* Step-5: Use full-text to search path template */
n({~ft_name contains "Mary*"}).e()[:5].n({~ft_name contains "Dav*"}) as
paths return paths{*} limit 10
```

The five steps from above are self-explanatory but note that full-text search should not be confined to metadata only. It can also be used to empower deep and fuzzy search. The 5th step illustrates finding paths between starting nodes and ending nodes, both are the results of fuzzy match, as illustrated in Fig. 44. The integration of full-text search can be expanded to allow for fuzziness equivalently on both nodes and edges—this would be very challenging if the full-text search capabilities are not native to the graph database because the recursive traversal (and search) nature of graph queries would require the full-text search to seamlessly integrate with graph computing engine, instead of as a 3rd party library.

The beauty of an advanced (database) query language is not how complex it is, but how concise it is. It should have some generalities: easy to learn and understand; high performance, of course, this depends on the underlying database engine; the underlying complexity of the system should not be exposed to the language interface level. Especially the last point, if readers have lingering fears about the complex nested logic in SQL, Gremlin, Cypher, or GraphSQL, you will understand this metaphor better: when Atlas, the Titan in ancient Greek mythology, carrying the whole world (earth) on his shoulders Citizens of the

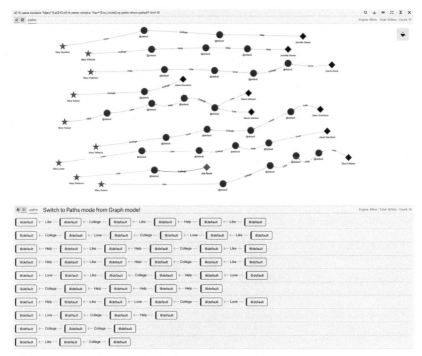

FIG. 44 Real-time full-text path search (graph vs path mode).

world (database users) do not need to perceive how heavy the world is (how sophisticated the underpinning database is).

Task 5: Complex graph algorithms.

Compared with other databases, one of the obvious advantages of graph databases is the integrated algorithm function support. There are many algorithms on the graph, such as in-out degree, centrality, sorting, propagation, connectivity, community recognition, graph embedding, graph neuron network, etc. With the increase of graph-powered commercial scenarios, it is believed that more algorithms will be ported to or invented in the graph.

Take the Louvain Community Recognition algorithm as an example. This algorithm has only been around for more than 10 years. It is named after its birthplace—Louvain University in the French-speaking part of Belgium. It was originally invented to traverse the nodes and edges in a large graph composed of social relationship attributes through complex multiple recursions to find the clustered communities (such as people, products, and things), closely related vertices will be in the same community. In the field of Internet and financial technology, the Louvain algorithm has received considerable attention.

FIG. 45   Real-time Louvain community detection w/statistics.

The following query statement invokes the algorithm and returns with real-time statistics (Fig. 45):

```
/* Ultipa GQL */
/* Real-time Statistics */
algo(louvain).params({
  phase1_loop_num: 5,
  min_modularity_increase: 0.001
}).stats() as stats
return stats
```

In a graph database, calling an algorithm is like executing an API call, both which require providing some required parameters. In the above example, the user only needs to provide a minimum of two parameters to execute Louvain. A more complex set of parameters can be set to fine-tune the Louvain algorithm. In Chapter 4, there will be a comprehensive and in-depth analysis of the Louvain algorithm.

Note: The implementation of the original Louvain community recognition algorithm is serial, it needs to start from all vertices in the whole graph, one by one, and to perform repeated operations. Just imagine that in a large graph (more than 10 million vertices), the time complexity must be measured in $T+1$. For example, in Python's NetworkX library, it takes tens of hours to perform Louvain on a multimillion scale graphs datasets. By transforming the algorithms to run in a highly concurrent way, the time can be drastically shortened to a millisecond-second level. Here, what we are talking about is not 10–100 times performance improvement, but tens of thousands of times performance improvement. If readers feel that the case given by the author is a fantasy, maybe you should re-examine your understanding of system architecture, data structure, algorithm, and their optimal engineering implementation.

Graph query language can support many powerful and intelligent operations. The above five examples are only a modest beginning. The author hopes that they can reveal the simplicity of GQL and arouse readers to think about a question: Writing tens of or even hundreds of lines of SQL code has aggravated cognitive load. We should consider using a more concise, convenient yet more powerful graph query language.

About database query language, the author thinks:

● Database query language should not just be a proprietary tool for data scientists and analysts, any business personnel can (and should) master a query language.

- The query language should be easy to use, and the complexity of the underlying architecture and engineering implementation of all databases should be transparent to upper-level users.
- Graph databases have great potential and will gradually (quickly) replace SQL loads in the next 5-to-10 years. Some top companies in the industry, such as Microsoft and Amazon, have estimated that 40%–50% of SQL loads will be migrated to GQL by 2028.

Some people including quite a few well-known investment institutions and industry "experts" think that Relational databases and SQL will never be replaced. The author thinks that this kind of view cannot withstand scrutiny. If we look back at the not-too-distant history, we will find that the relational database began to replace the navigational database in the 1980s, and it has dominated the industry for nearly half a century. If history has taught us anything, it is that obsession with anything old never lasts long.

Chapter 3

Graph database architecture design

3.1 High-performance graph storage architecture

In the era of digital transformation, distributed architecture concepts can be seen everywhere, high-performance systems are at least familiar to readers of this book, and they seem to be always within reach. However, for a long time, the Apache Hadoop system was also considered an "artifact" for data processing. Many practitioners even believed that the Hadoop system is also high-performance, but the performance of all systems can only be determined in real-world scenarios—which will help distinguish the performance, stability, scalability, and ease of use of each system.

Like all other types of databases, designing a high-performance graph database requires attention to at least three major links: computing, storage, and query analysis and optimization.

The problem to be solved by the computing layer (or computing component) of the database is to return the data stored in the database to the request initiator after performing necessary (calculation) processing according to the query instructions. It is broken down into specific steps as follows:

(1) The client initiates a query request for the database server.
(2) The server receives the query request, parses, and optimizes the query command.
(3) Read data from the storage engine (some data may need to enter memory).
(4) The central processing unit performs corresponding calculations (mathematical operations such as various aggregations, sorting, filtering).
(5) Return the corresponding result.

Combined with Fig. 1, the core pillars of the above five steps of a high-performance system are steps (3) and (4), namely the storage layer and the computing layer. This section mainly introduces the design gist of graph storage component, and the computing, query analysis, and optimization will be introduced in the remaining sections.

The Essential Criteria of Graph Databases. https://doi.org/10.1016/B978-0-443-14162-1.00008-8

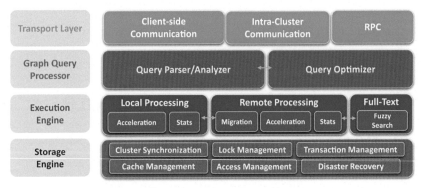

FIG. 1 Layered diagram of database management system architecture.

3.1.1 Key features of high-performance storage systems

The three characteristics of a high-performance storage system are efficient storage, efficient access, and efficient update.

Storage efficiency mainly refers to two aspects: On the one hand, the data ingestion efficiency is high. In the traditional sense, ingestion can be equivalent to writing to disk (persisted on the hard disk file system). However, with the emergence of new technologies such as persistent memory, the concept of "writing to disk" is not accurate any more, and readers need to pay attention to the distinction. However, writing efficiency can always be reflected in trans-actions per second (TPS), that is, the number of records (or transactions) written per second; on the other hand, how much additional overhead will be incurred to store these data records (payload).

Regarding storage efficiency, two questions are often asked in the industry: space for time and cost performance.

Trading space for time is a common practice to improve storage efficiency, not only in the narrow sense of storage operations, but also in various operations such as access and update. Many NoSQL database design concepts adopt a strat-egy to improve the timeliness of the storage engine by using a larger storage space, such as using some intermediate data structures to achieve higher con-current write performance, that is, higher TPS; there is also the case of storage amplification, for example, there will be multiple copies of data, which can improve access efficiency (avoid I/O pressure caused by real-time migration) while satisfying data security; some databases use Copy-on-Write method to achieve concurrent read and write operations by using doubled storage space.

Access efficiency refers to locating and reading the required data records with the minimum time consumption (or the minimum number of operation steps, and the minimum algorithm complexity). The various data structures we introduced in the previous chapter will be the focus of this chapter. The efficiency of the update operation means that when records need to be updated or incremental data is written or old records are deleted, the operation complexity at the storage engine level is the lowest (minimum scale update, and minimum number of update steps).

In commercial scenarios, the cost-effectiveness of storage systems is also of major concerns. In scenarios that are not sensitive to delay, low-cost hardware and software solutions will be used. Otherwise, high-performance and high-cost solutions will be designed according to business needs.

The complexity of the storage engine is different on varied storage hardware levels. The design complexity of the memory-level data structure is lower than that of the hard disk-level data structure; and the data structure design based on the solid-state drive (SSD) at the hard disk level is also very different from the data structure design based on the disk (HDD). It can be said that no storage engine design optimized for a certain type of storage medium can be universally applicable and generalized to all other types of storage media. Usually, some read or write performance will be greatly reduced due to changes in storage media. This is also one of the areas where the architectural design of the storage engine is complicated.

To better explain the diversity and complexity of storage engine design, we use the schematic diagram of the seven-layer model in Fig. 2 to help analyze

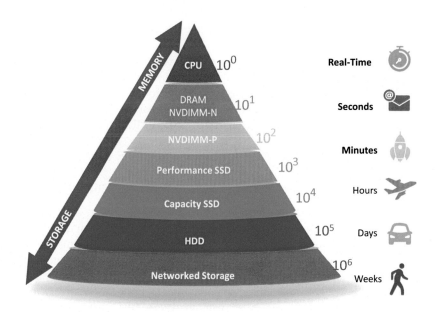

FIG. 2   Seven layers of database storage engine.

what kind of storage architecture and data structure combination to achieve high-performance storage (and computing). At present, the industry has set off a trend of "separation of storage and computing" since around 2015. This statement was first proposed by the global storage giant EMC. This formulation logically refers to the development of big data and cloud computing. Storage should not be limited to the local storage of a machine, but should realize the logical separation of the storage layer and the network layer, that is to say, storage resources can be relatively independent and (horizontally) expansive. Obviously, the separation of storage and computing implies that storage is far (moving farther) away from computing.

When the storage is far away from the computing, that is, when the data needs to be calculated, it needs to go through a migration path before it can be processed by the CPU (and/or GPU), and the traveling of data along this migration path has an exponential performance drop effect—there is a million times performance gap in terms of data processing throughput between the data in the network storage module and the top-level central processing unit. If any database query operation needs to trigger the operation of a group of network storage layers so far away from the central processing unit, it is conceivable that this operation takes a lot of time. On the other hand, it is impossible for us to stack all the data in the cache layer(s) of the CPU (although this is the operation with the lowest latency in theory), nor is it always possible to compress all the data in the dynamic memory.

The logic of hierarchical storage gives us a good inspiration; in fact, this is the storage engine design logic used in all databases:

- The full amount of data is persisted on the "as fast as possible" storage media.
- Accelerated data structures such as indexes and caches use memory and disk.
- When memory is overloaded, data overflows to the persistence layer and makes full use of the storage medium access characteristics of the persistence layer (distinguish between HDD, SSD, PMEM, etc.) to accelerate access.
- Make use of the multicore and multithread concurrency capabilities of central processing as much as possible.
- Make use of patterns of database queries to optimize multilevel cache utilization (hit rate).

The last two points above are important design ideas for graph computing acceleration (the traditional database storage engine part contains computing logic by default, but in graph databases, it is necessary to introduce the graph computing engine part as an independent function module, given the general trend of separation of storage and computing, the computing layer has its relatively independent characteristics), which we will analyze in the next section. This section focuses on the first three points.

3.1.2 High-performance storage architecture design ideas

The design ideas of storage architecture and core data structure are usually carried out around the following four dimensions:

- Storage and memory usage **ratio** (role allocation and allocation ratio).
- Whether to use **cache**, how to optimize cache.
- Whether to sort data or records, and how to **sort** them.
- Whether and how **changes** to data or records are allowed (mutability).

(1) Storage and memory occupation ratio

A database that only uses 100% external memory is extremely rare, and it can even be said that it has little to do with high-performance databases. Similarly, databases that use 100% memory are also very rare. Even if there is a category of "memory database", usually as a subclass of relational database or key-value store, data will still be persisted on external memory when needed. The reason is nothing more than the "volatility" of memory.

In the architecture design of most modern databases, the data structure characteristics of external storage and memory are used to achieve some kind of storage acceleration. In addition to caching, sorting, and data variability, for graph databases, because of its special topological structure, the memory storage (RAM) layer is used to store some key data (such as metadata, mapping data, temporary algorithmic data, indexes, attribute data of metadata that needs real-time calculation), and the external storage layer is used to store massive remaining data that does not need real-time network-analysis type of processing—combining such two storage layers will subjugate a decent increase in terms of system throughput.

There is no fixed formula for the proportion of internal and external memory, but in many traditional IT architectures, the proportion of memory resources of servers is relatively low, and this situation is often intensified (worsened) in the process of cloudification. It is a kind of retrogression—in database application scenarios, the volume of the memory should not be lower than 1/8 (12.5%) of the external memory, that is to say, if there is 1 TB of external memory, there should be at least 128 GB of memory. In high-performance computing scenarios, the ratio of internal and external memory can be more than 1/4 (25%). If there is 2 TB of external memory for persistent layer data, the memory volume can be at least 512 GB. In some private cloud deployment scenarios, large memory cannot be allocated, and virtual machines with large memory cannot be managed. This kind of cloud virtualization itself runs counter to the principle of high-performance computing.

In addition to the ratio of internal and external memory, another important issue is what type of hardware is used for external memory. There are usually two types: mechanical disk and solid-state drive. Disk operations involve relatively expensive (slow) physical operations. The robotic arm needs to be moved, and the disk needs to be rotated and positioned before reading and writing operations can begin. Although many people think that disks have advantages in continuous read and write operations, and disks seem to be more reliable than solid-state drives, solid-state drives have balanced read and write and high-efficiency performance advantages. It is only a matter of time before the traditional disk architecture is fully replaced, at least in database applications.

It should be noted that although the database data structure and code logic optimized for disk access can be run directly on the solid-state drive, there is a lot of room for optimization and adjustment. These details are also the most complicated part of the database architecture design. If we introduce the hardware solution of persistent memory, it will make the problem more complicated. Obviously, it is unrealistic for a set of code logic and data structures to be generalized and adapted to all hardware architectures, and to achieve the best performance. In the actual database architecture design, we always make trade-offs (balancing), including making corresponding adaptations and compromises for the underlying hardware. We believe that the design scheme that ignores the underlying hardware differences (hardware-agnostic) is not feasible—although the underlying file system and operating system of the database have done a lot of abstraction and adaptation work, a high-performance database still requires precise data structures for specific hardware environments, architecture and performance tuning.

(2) **Cache**

The purpose of caching is essentially to reduce the number of I/Os and improve system throughput. Almost all disk-persistent data structures utilize caching in some way for acceleration (to help locate, read, or write back more quickly). In addition, cache design, data mutability, and sorting often need to be considered comprehensively, because whether data is allowed to be mutated, or stored after sorting directly affects the design logic and implementation of the cache, and vice versa.

The storage engine of graph database seems to make more use of cache than other types of databases. Let's list the operations that may use caching: data writing, data reading, metadata index, full-text search index, neighborhood data, intermediate operation results, and other situations that may use caching, such as high concurrency, multiple versions maintenance scenarios.

It should be noted that whether it is a mechanical hard disk or a solid-state drive, the smallest unit of reading and writing is a block. Mechanical hard disks are usually represented by sectors at the hardware level, as shown in Fig. 3, but at the operating system level, they are uniformly represented by blocks like solid-state drives. A block can usually be set to a fixed size, such as 4 KB which may contain multiple sectors. In the Linux

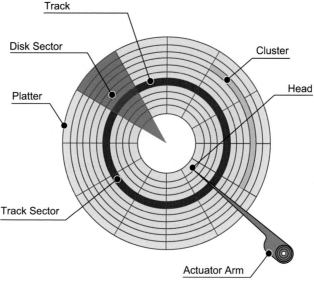

FIG. 3 Storage components of mechanical hard disk (HDD).

operating system, the following two commands can help find out the size of Block and Sector:

```
fdisk -l | grep "Sector size"
blockdev -getbsz /dev/sda
```

On mechanical hard disks, the addressing of moving parts and disk rotation take up more than 90% of the time of each operation, while in solid-state hard disks (and persistent memory, dynamic memory), there is nothing that consumes time due to the movement of physical parts, thus exponentially shortens the time consumption of each operation (as shown in Fig. 4), and the time consumption is reduced from a few milliseconds to tens to hundreds of microseconds (a performance improvement of about 40 times or more).

In solid-state drives, the smallest granular storage unit is Cell, followed by String, Array, Page, Block, Plane, Die, and Solid-State Drive (SSD) with all parts and units organized in a nested fashion. The last (and the largest) part gave the commercial name SSD (Fig. 5). The smallest unit that can be programmed (written) and readable is Page, and the smallest unit of storage that can be erased is Block. Each block is usually composed of 64 to 512 Pages.

Whether it is a mechanical hard disk or a solid-state drive, the most granular unit of reading and writing interfacing with operation system, file system, or database storage is not bits nor bytes. This design at the

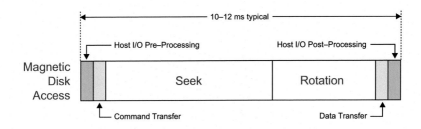

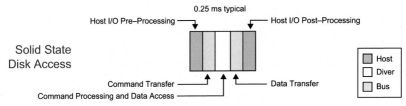

FIG. 4 Access delay decomposition comparison of HDD and SSD.

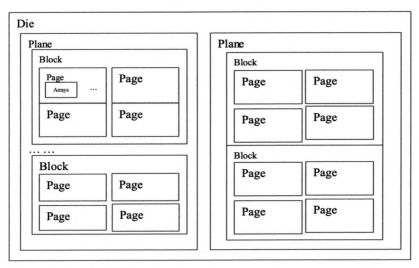

FIG. 5 Storage units of SSD.

hardware level is to improve storage efficiency. For example, if a person is 20 km from home to the company, and it takes 1 h to take a taxi, he needs to bring 10 books from home to the company. He can choose to deliver 10 books to the company at one time, or divide them into multiple deliveries. In terms of cost and time cost, it is obvious that one-time delivery is the most cost-effective way. Similarly, when facing storage media, cache

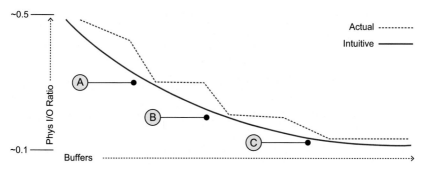

FIG. 6 "Inverse ratio" between cache and I/O.

design is this "one-time" batch delivery (access) technical means. In a distributed architecture design, if each operation involves a network request and data migration and transmission process, it is an optimal solution to combine multiple operations reasonably and then perform a unified transmission.

Fig. 6 shows the "inverse ratio" relationship between cache and physical I/O. In theory, the more cache, the lower the I/O, but in fact, after the cache reaches point C, the I/O decreases (performance improvement) effect is not obvious any more (flattened curve). And this point C is a "high specification" goal when we design the storage cache, while B or A can be considered as the "low-to-medium specification" goal when low-cost budget is set.

All in all, caching is a core technology that runs through the entire development of computer hardware and software. Even in the most extreme cases—for example, if you do not actively use any caching technology in your database or application software design, the underlying file system, operating system, hardware such as hard disk, CPU, etc. will still achieve access and calculation through caching acceleration. The emergence of the cache makes the design of the system architecture more complicated, because the system has more components to manage, such as improving hit rate, resetting, and replenishing.

(3) **Sort (or not)**

Data sorting is an increasingly common storage optimization technique as the number of data records grows. Whether to sort depends on the specific business scenario. If the data is sorted and inserted, then in the continuous storage type data structure, you can directly use the address subscript to access a record with the lowest delay, or quickly read a data record stored in an address space. If combined with the cache acceleration strategy, the continuously stored data records can be more effectively accessed, but if a record needs to be inserted or deleted in the middle, the operation cost of this will be more expensive.

In Chapter 2, we introduced the "index-free adjacency" storage data structure, which allows us to locate any vertex's neighboring vertices (neighbors within one hop) with O(1) time complexity. The access efficiency of this data structure design itself is already higher than that of indexes. Without considering complex situations such as concurrent access, the efficiency of incremental writing, reading, and updating can all be O(1), but the complexity of deletion will be greater. For continuous storage data structures such as arrays and vector arrays, deleting records from the middle will cause the displacement of the entire subsequent data records. At this time, if the data records do not need to be sorted by Key or ID, you can consider switching the last record with the to-be-deleted record, and then delete the last record after the replacement. This operation is called Switch-on-Delete, and its complexity can still be equal to O(1) (the actual complexity is 7–10 times that of the lowest delay O(1)). The schematic diagram of the operation is shown in Fig. 7.

The characteristic of non-sequential storage is that the data structure can be designed to delete a single record, which may be more efficient, such as O(Log N) (usually this value is not greater than 5), but the cost is high for reading multiple consecutive records, especially for range queries, for example, the time consumption of reading K records is $K * O(Log N)$.

Range queries are heavily used in many application scenarios; however, it seems to conflict with nonsorted storage. In Chapter 2, we introduced LSMT. Here we introduce WiscKey which is based on LSMT optimization (Wisc is the abbreviation of University of Wisconsin-Madison, and Key refers to a Key-Value Store storage optimization scheme). WiscKey data structure can take into account the efficiency of writing and the space utilization of nonsorted storage.

First, for hard disks, the characteristics of LSMT are as follows:
- Ideal for batch and sequential write.
- High-throughput sequential access (the efficiency of sequential access is hundreds of times that of random access).

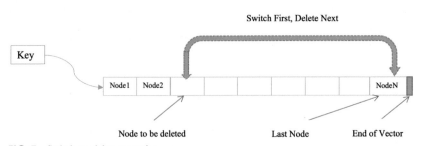

FIG. 7 Switch-on-delete operation.

- Read and write amplification effect (it is not optimized for SSD, and record read and write will be magnified many times).

  Second, the characteristics of SSDs are as follows:
- High-performance random read.
- Ability to operate in parallel.

  The core design concept of WiscKey is key-value separation (only keys need to be sorted), and it includes the following optimization strategies for solid-state drives (Figs. 8 and 9):
- Split sort and GC (garbage collection).
- Take advantage of the concurrent capabilities of SSDs in range queries.
- Online lightweight GC.
- Minimize I/O amplification.
- Crash consistency.

  Fig. 10 shows the sequential and random access performance on the WiscKey-optimized SSD. We can see that when the size of the accessed data is in the range of 64 K to 256 KB, the throughput rate of concurrent access by 32 threads is comparable to that of sequential access.

  WiscKey is optimized based on the core code of LevelDB. After optimization, WiscKey shows a certain degree of performance improvement in random access and sequential access, especially after the value corresponding to the key occupies more than 4 KB (Fig. 11). Of course, how many key values in the actual business require more than 4 KB of storage space is a question that needs to be discussed in detail. According to the author's experience, the actual key value is generally within 1 KB. At this time, the performance of WiscKey is almost the same as that of conventional LSMT.

LSM-tree   LSM-tree   Value Log

FIG. 8   LSMT to WiscKey.

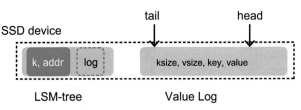

LSM-tree   Value Log

FIG. 9   LSMT and vLog are separated in WiscKey.

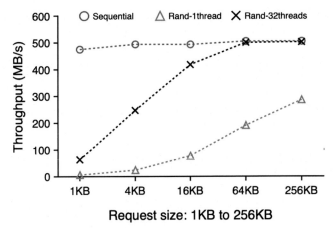

FIG. 10 WiscKey random range query performance comparison.

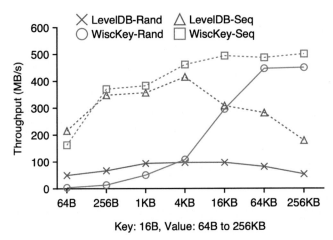

FIG. 11 Query performance comparison between WiscKey and LevelDB.

(4) Data can be mutated (or not)

Whether the data is mutable (mutability) refers to whether the update of the data record adopts the append mode (append-only) or allows local (in-place) changes. Different modes are suitable for different scenarios and thus have evolved different "acceleration" models. For example, the Copy-on-Write mode (referred to as COW for short) was first implemented in the Linux kernel design, it does not change the original data, but uses the method of copying the page first and then changing it. The difference between LSM and B-Tree is also usually summarized as the difference between immutable and in-place modification cache acceleration modes.

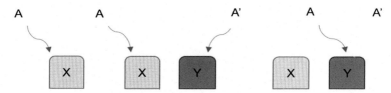

FIG. 12   Lock-free variable (record) update.

One technique COW employs is called lock-free record updating, as shown in Fig. 12. The specific operation steps are as follows:

(1) Locate variable (record) A (content is X).

(2) Copy A to A' (non-deep copy, pointer-level copy), and create the content as Y.

(3) In an atomic operation (compare-and-swap), A points to Y, cutting off the relationship between A' and the variable, and the previous X records are GCed (garbage collected).

If the above operations are placed in a binary tree, the same logic applies. The core of COW is to avoid deep copy as much as possible, and use pointers to implement shallow copy (copy-by-reference) through lightweight operations, and perform deep copy and precise record update operations only when necessary. In the kernel of the Linux operating system, when the parent process needs to spawn a child process, this COW technology is used to save memory and CPU clock consumption.

Although the B-Tree-type data structure realizes the function of updating records in place (in-place update), it brings the problem of write amplification, and the design complexity of dealing with locks and concurrency is relatively high. Bw-Tree came into being from this. Its full name is Buzzword-Tree, which is a memory-level, variable page size, high-concurrency, lock-free data structure.

Bw-Tree[a] has four core design parts (Fig. 13):

- Base nodes and delta chains.
- Mapping relationship table.
- Merge and garbage collection.
- Structural change (SMO): To achieve lock-free.

To be precise, these four designs are all for achieving lock-free high concurrency, nevertheless at the cost of using more memory space. However, its demonstrational significance is that the design ideas of high-performance systems are nothing more than the following points:

(1) Use and adapt to faster storage devices (including memory, including tiered storage).

---

a. A B-tree for new hardware platforms: https://www.microsoft.com/en-us/research/wp-content/uploads/2016/02/bw-tree-icde2013-final.pdf.

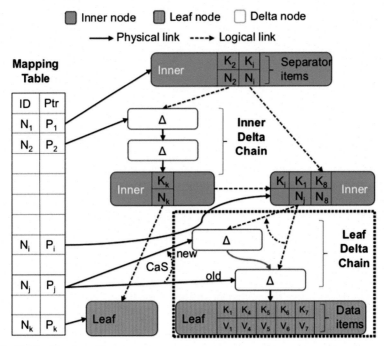

FIG. 13 Bw-Tree internals.

(2) Achieve higher concurrency as much as possible (with data structure support).

(3) Avoid using too many locks with too coarse granularity.

(4) Avoid expensive operations such as frequent copying of data or deep-copying (data utilization rate is too low).

(5) Be close to the underlying hardware, avoiding too much virtualization and too many intermediate links or layers.

3.2 High-performance graph computing architecture

In the traditional type of database architecture design, the computing architecture is usually not introduced separately, and everything revolves around the storage engine. After all, the storage architecture is the foundation, especially in the traditional disk storage-based database architecture design. Similarly, in the design of graph database architectures, some products are developed around the way of storage. For example, the open-source Titan project (suspended in 2016) and the subsequent JanusGraph project (started in 2017) are all based on the storage engine of third-party NoSQL database(s). The fundamental working mechanism of these graph database projects is calling the interface provided by the underlying storage to complete any graph computing request, and there are many intermediate links (refer to the last bullet point in the previous section)—it is possible for such projects to handle extremely large amount of data storage-wise, but not performant computing-wise. The core challenge to be solved by graph databases is not storage but computing. In other words, the business and technical challenges that cannot be fulfilled by traditional databases (including NoSQL databases) are concentrated in the computational efficiency of in-depth penetration and analysis of networked data. Although these challenges are also related to storage, they are more about computing efficiency.

3.2.1 Real-time graph computing system architecture

Regarding the design of database computing architecture, there is no so-called single correct answer. However, the author believes that among the many alternatives, the most important thing is common sense, and sometimes it also includes some reverse thinking.

(1) Common sense
- RAM is much faster than external storage.
- The CPU's L3 cache is much faster than the memory.
- Data caching is much faster in memory than in external storage.
- Relying on certain language's memory management is dangerous (i.e., Java).
- Never ignore time complexity: the search time complexity of linked list data structure is $O(N)$, the search time complexity of tree data structure (index) is $O(\text{Log } N)$, and the complexity of hash data structure is $O(1)$.

(2) Reverse thinking
- SQL and RDBMS are the best combination in the world in the past 40 years. But this is due to change dramatically, as the continuous innovation of business scenarios warrants new architectures to meet business needs. In the Internet business, a large number of multiinstance concurrency is used to satisfy massive user requests (such as lightning-deal or flash-sales scenarios). In these high-concurrency scenarios, the processing mode belongs to the "short-chain transaction mode" and the "long-chain transaction mode" led by the financial industry is very different—the main difference between the two is that the computational complexity of long-chain transactions increases exponentially, while the data communication or synchronization amount between nodes of distributed systems in short-chain transactions is very small and the logic is much simpler (why it's called short chain).
- The perception that large tech giants such as MAGA or BAT must be the most powerful provider of graph technology—if this idea is followed, they may never have had the chance to grow from a small startup to today's giant over the past 20 years. Every time a new huge IT upgrade opportunity arises, there will be some small companies outperforming the rest of the market.

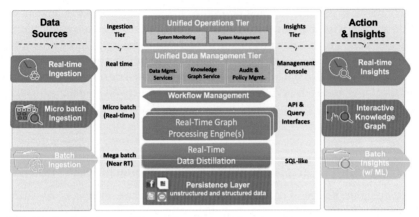

FIG. 14 Graph database architecture (data flows and feature stack).

Fig. 14 is the overall architecture design idea of a real-time graph database.

The core components of the real-time-capable graph database are as follows:

- Graph storage engine: data persistence layer.
- Graph computing engine: real-time graph computing and analysis layer.
- Internal workflow, algorithm flow management, and optimization components.
- Database and data warehouse docking components (data import and export).
- Graph query language parser and optimizer components.
- Graph management, visualization, and other upper management components.

The schematic diagram of the internal structure of the graph database is shown in Fig. 15.

In Fig. 15, we use three colors to identify the three workflows in the graph database: data workflow (orange), management workflow (blue), and query computational workflow (green). Data worker and management workflow should be familiar to readers who have designed and implemented software-defined storage systems. Logically, they represent the separation of data transmission and system management instructions, which can be seen as the transmission of graphs on two separate planes (data plane vs control plane). The query workflow can be regarded as a special graph data workflow, which is the flow of data and instructions flowing through the graph computing engine.

We use concrete examples to illustrate the differences between computational workflows and data workflows. Figs. 16 and 17 are the visual representations of the returned results of the two real-time path query commands. The difference between them is whether vertex (entity) and edge (relationship) attributes are returned. In Fig. 16, no vertex and edge attributes are returned; in Fig. 17, all attributes are returned.

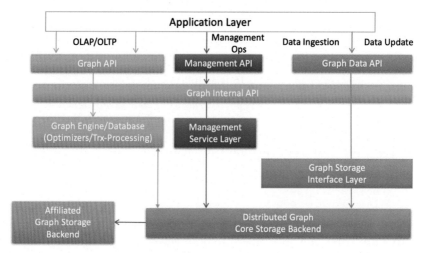

FIG. 15 Functional components and internal data flow of graph databases.

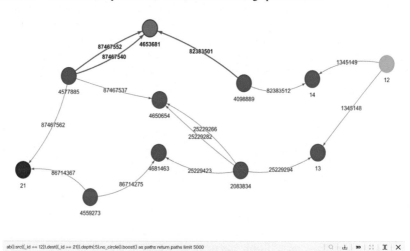

FIG. 16 No attribute returned in path query.

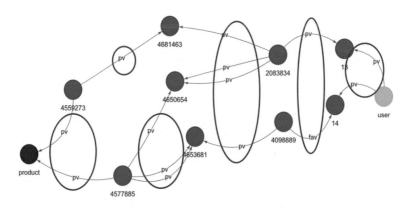

FIG. 17 Return with (all or part of) attributes in graph query.

The query command in Fig. 16 is as follows:

```
/* Path Query, No Circucles, No Attributes Returned */
ab().src({_id == 12}).dest({_id == 21}).depth(:5)
.no_circle().boost() as paths return paths limit 5
```

The query command in Fig. 17 is as follows:

```
/* Path Query, No Circles, All Attributes Returned */
ab().src({_id==12}).dest({_id==21}).depth(:5)
.no_circle().boost() as paths return paths{*} limit 5
```

The above two path query statements both want to query for acyclic paths with a max depth of 5 between two vertices, and only five paths are needed to return. The difference is that the latter requires returning all attributes of the vertices and edges along the paths. The difference between the underlying graph computing and storage engines at this semantic level is that if no attributes are returned, the graph computing engine can use serialized IDs to perform queries and calculations—which is obviously more memory-efficient and performant. If we need to return all (or some) attributes, this work can be assigned to the storage engine to find the attributes of each vertex and edge in the storage persistence layer and return them. The query latencies can be hugely different if we are to return many paths, the storage engine will be busy working on retrieving those attributes—in no-attribute mode, finding 5000 paths may take just 100 milliseconds, while full-attribute 5000 paths may take exponentially longer time, say, 1000 milliseconds—such differences can be seen in Figs. 16 and 17's engine time-cost and total time-cost.

Of course, through optimization, especially the implementation of caching and other functions, the storage engine can also return a large amount of attribute data at the millisecond to second level (especially the acceleration effect is obvious when there are multiple queries). However, the crux of the problem is that for deep path queries like the one above, the efficiency of a computing engine based on memory acceleration will be hundreds or even thousands of times faster that of a traditional storage engine, and with the increase of the query depth, the performance gap increases further, possibly exponentially.

3.2.2 Graph database schema and data model

Graph databases are generally considered to use a schema-free (schema-less) approach to data processing, and therefore have higher flexibility and the ability to deal with dynamically changing data. This is also an important difference between graph databases and traditional relational databases.

The schema of a relational database describes and defines database objects and the relationship between objects, which plays an outlining role. Fig. 18 shows 19 tables and their data types, and the association relationship formed between the tables through primary and foreign keys. However, databases from different manufacturers have very different definitions of schemas. The schema

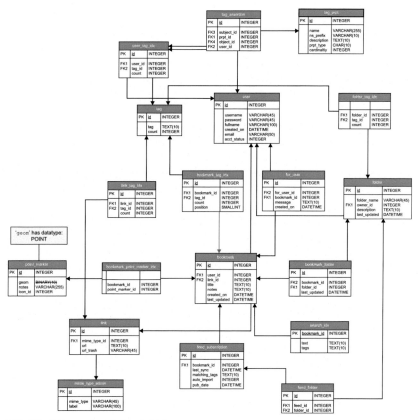

FIG. 18   Schemas for relational databases.

in MySQL can be equivalent to the database itself, while Oracle uses the schema as a part of the database, or make the schema to belong to each user. In SQL server, after the schema is created, information such as users and objects can be attached to it.

Schemas in relational databases are precreated, and once created and the database is running, it is very difficult to make dynamic changes. However, in a graph database, being able to define the skeleton of the graph data objects and the relationships between objects in an outline vastly improve its ability to handle things in a more dynamic way. There are two solutions to this challenge: one is schema-free; the second is dynamic schema. In fact, there is a third solution, which is to continue the static schema of traditional databases, but this is contrary to the goal of building a graph database that is more flexible than traditional databases, and is beyond the scope of this book.

The schema-free scheme, as the name implies, does not need to explicitly define the schema of the relationship between database objects up front, or the schema between objects is self-evident or implicit but clear. For example, in social network graph, the relationship between entities concerns following and being-followed by default. This simple-graph relationship does not require the intervention of schema definitions. Even in a more complex relationship network, such as a financial transaction network, there may be multiple fund-transfer relationships between two accounts (vertices) (multitransaction = multiedge), this kind of graph belongs to the multigraph. In the existence of multiple edges of the same type between vertices of the same type (isomorphism), there is no necessity to define a specific schema to distinguish them. It would be a different story if there are heterogeneous types of relationship between such two accounts, which, ideally, would require schemas to be defined, on-the-fly, to differentiate different types of edges for instance.

The dynamic-schema mode is relative to the static mode. Obviously, in the above example, if there are multiple types of vertices (accounts, merchants, POS machines, debit cards, credit cards, etc.), and multiple types of edges (transfer, remittance, credit card, repayment, etc.), defining schemas can more clearly describe (differentiate) the objects and relationships between objects. If the graph data is changing dynamically, for example, after new types of data (vertices, edges) appear, it is necessary to dynamically define a new schema or change the existing schema to better describe and process the data, without the need to going through a series of complex operations such as stopping and restarting the database as with traditional fixed-schema databases.

Because there are no concepts such as tables, primary keys, or foreign keys, graph databases can be largely simplified to only include the following components.

- Node: also known as a vertex (plurality: Nodes and Vertices).
- Edge: Often called relationships, each edge usually connects a pair of vertices (there are also complex edge patterns, such as hyperedge that connects more than 2 nodes, but this is very rare and confusing, and can be realized with just a bunch of simple edges. To avoid complications, this book does not deal with hyperedges).
- Edge direction: For each pair of nodes connected by an edge, the direction is meaningful. For example: $A \rightarrow$ father (is $\rightarrow$ B; user $A \rightarrow$ (owns) account $\rightarrow$ account A.
- Attributes of nodes: Attributes attached to each node, each attribute is expressed by a key-value pair, for example: <Reference Book> − <Conversations with God>, the key before the dash is the name of the attribute field, and the value after the dash is the actual value of the attribute.
- Edge attributes: Edge attributes may include many things, such as relationship type, time span, geographic location, description information, and key-value pairs of decorative edges.
- Schemas: In graph databases, the definition of schemas can occur after graph database objects are created, and the objects contained in each schema can be dynamically adjusted. Vertex and edge can respectively define their own multigroup schemas.

- Index and query acceleration data structure: The disk-based index part is the same as the traditional database, but graph query acceleration is a unique innovation of the graph database which has some similarities with in-memory database, but more advanced.
- Labels: Labels can be either regarded as a special attribute to vertex or edge, or a special kind of index (for query acceleration). Some graph databases use labels to simulate the effect of schemas, but there are major differences between the two. In schema-free graph databases, label can only be regarded as a simplified but incomplete implementation of schema. To get the best of both worlds (schema-free and schematic concurrently), some serious and fundamental system architecture design must be done to allow that to happen, which we'll get to in the following sections and chapters.

Table 1 lists the possible data types in the data source of a graph database.

In addition to the data types listed in Table 1, there are many other types of data, such as those semistructured, quasistructured, and unstructured data types. But when the graph computing engine is running, these rich data types do not necessarily need to be loaded into the engine, because this kind of operation may be neither meaningful nor important. In other words, they should be loaded

TABLE 1 Typical graph data types (partial list).

Data type	Descriptions
Integer	4 bytes
Long integer (Long)	8 byte integer (in large graph)
Character type (String)	variable
Boolean	1 or 0 (true or false)
Byte	1 byte
Short integer (Short)	2 bytes
Single-precision floating-point type (Float)	4 bytes of floating point precision
Double-precision floating-point type (Double)	8 bytes of floating point precision
Unique Identifier (UUID)	Graph database internal/unique IDs
Original ID	Original and external/unique IDs (to be ingested into graph database)
Date (Date) or timestamp	Dates, Timestamps and Subtypes
Other types	Too many types to list them all

into the persistent storage layer or auxiliary database (such as document data store) for processing to meet the extended query needs of customers.

3.2.3 How the core engine handles different data types

One thing to keep in mind is that certain types of data are not memory friendly, such as string types of data, these data types tend to bloat in memory, and if not processed and optimized (data distillation or compression), there will be a problem of insufficient memory.

For example, if a graph has 1 billion vertices and 5 billion edges and a unique identification code (UUID) is used to represent the edge and node, each edge takes two connecting nodes, each UUID is a 64-byte string, then the formula for calculating the unoptimized memory usage of this graph is as follows:

$$64 \times 2 \times 5,000,000,000 = 640 \text{ Billion Bytes} = 640 \text{ GB RAM}$$

If stored as an undirected or bidirectional graph (allowing reverse traversal), requiring each edge to be reversely stored (a common technique in graph traversal), the memory consumption using UUIDs doubles to 1280 GB, and this does not include any Edge ID, vertex-edge attribute data, regardless of any caching, runtime dynamic memory needs, etc.

In order to reduce memory usage and improve performance, we need to do the following things:

- Serialization: Convert the UUID types of data into integer, and establish a mapping table (such as Unordered Map or some kind of hash-mapping table) between the source data and the serialized data. The establishment of the mapping table for 1 billion vertices and 5 billion edges to integer IDs (8-byte) can save 80% memory.
- Index-free Adjacency: Use appropriate data structures to achieve vertex-edge-vertex traversal with minimal algorithmic complexity. Readers can refer to the IFA data structure design introduced in the previous chapter to build their own efficient and low-latency graph computing data structure.
- Finding a way to effectively saturate the CPUs and run them at full speed is the ultimate goal of every high-performance system, because memory is still 100 times slower than the CPU. Of course, this also has to do with how multithreaded code organizes and manipulates worker threads to be productive as much as possible. This is closely related to the data structure used—if you have used heap in Java, it must not be as effective as map or vector, so is linked-list to array, etc.

On the other hand, not all data should be directly loaded into the graph core computing engine. It is important to note here that the number of attributes of vertices and edges are multiplied by the number of vertices and edges respectively, and the result may be far greater than the number of vertices and edges.

Many people like to use the number of fields in traditional databases to count the size of graph databases, for 1 billion vertices and 5 billion edges, and a hypothetical vertex has 20 attribute fields, the edge has 10 attribute fields, its (field) size $= 10 \times 20 + 50 \times 10 = 70$ billion $\sim = 100$ billion. Many so-called trillion-scale graphs may be far smaller than the real situation.

The essence of graph computing is to query, calculate, and analyze the topological structure of the graph (the skeleton of the graph) and involving only the necessary and limited vertex and edge attributes. Therefore, in most graph computing and query processes, a large number of vertices, edges, and attributes are irrelevant to the content of the subject query. This also means that they do not need to occupy valuable computing engine resources during real-time graph computing, nor do they need to be on the critical path of real-time processing.

To better illustrate this point, some examples are given below:

- In order to calculate the friends of friends of friends of a person, it is equivalent to the 3-hop (3-hop) operation in graph computing, which is a typical breadth-first (BFS) query operation.
- To understand all relationships within 4 hops between 2 people, it is a typical AB path query (BFS or DFS or both), involving a lot of nodes in the middle, which may be people, accounts, addresses, telephones, social security number, IP address, company, etc.
- Regulators look at the phone records of a company's executives, such as their incoming and outgoing calls in the past 30 days, and how these callers or recipients made further calls, understand their extended call network within 5 hops, and check the relationship between them.

These three examples are typical network analysis or antifraud scenarios, and the query depth and complexity gradually increase. Different graph systems will take different approaches to this problem, but the key points are:

- How to do data modeling? The pros and cons of data modeling will affect the efficiency of the job and the consumption of storage and computing resources, and even create a lot of noise. These noises will not only slow down the query job, but also make it difficult to guarantee the accuracy of the job. For example, isomorphic graph, heterogeneous graph, simple graph, and multigraph. Different data modeling logics may affect the final query efficiency.
- How to get the job done as fast as possible? Returning results in real time is always better as long as the cost is acceptable.
- Price/performance is always a consideration, and if the same operation is going to be run over and over again, it makes sense to create a dedicated graph for it. In addition, under the condition that the data is not updated frequently, the cached results can also achieve the same effect.
- How to optimize the query? This work is done by the graph traversal optimizer with or without human intervention.

The graph traversal optimizer works in two ways: one is to filter after traversing all nodes and edges; the other is to filter while traversing. The core differences between these two traversal optimizations are:

- In the first method, the traversal is conducted against the original graph dataset, first obtain a subgraph, and then filter and retain the possible path set according to the filtering rules.
- In the second method, the filtering rules are implemented in the form of dynamic pruning during the graph traversal process, filtering while traversing.
- These two approaches can be quite different in real business scenarios, as it affects how attributes (for nodes/edges) are handled. The second approach requires attributes to coexist with each node/edge, while the first approach can allow attributes to be stored separately.
- The implication here is: The first method is more Analytical Processing (AP) oriented, while the second one is Transactional Processing (TP) oriented.

Different graph data modeling mechanisms and different query logic may make one method run faster than another. Supporting both methods and running two instances at the same time can help optimize and find the more suitable method for graph data modeling (and schema) through comparison.

3.2.4   Data structure in graph computing engine

In the storage engine of the graph database, persistent storage solutions can be divided into three categories: row storage, column storage, and KV storage. At the graph computing engine layer, although we are eager for high-dimensional computing models, the data structure level is still divided into two categories: vertex data structure and edge data structure. Each data structure also includes fields such as attributes, etc.

Note: In some graph computing frameworks, because nodes and edges have no attributes, only edge data structures can exist, and no vertex data structures are needed, because each edge is an ordered pair of integers composed of a start vertex and an end vertex, and the edge direction is implicit.

Possible graph-computing data structure schemes are as follows.

- Vertices and edges are stored separately: vertices, edges, and their respective attribute fields are expressed in two logical sets of data structures.
- Vertex and edge combined storage: the vertex data structure contains edges, or the edge data structure contains vertices.
- Vertices, edges, and their respective attribute fields are stored separately: it may be expressed by 4 or more sets of data structures.

The graph data structure shown in Fig. 19 is an "integrated" data structure design scheme of vertices, edges, and attributes all merged into one. The first vertical column is the list of vertices, and on each row followed by the attributes

FIG. 19   Graph computing data structure (integrated mode).

of the vertex, next the connecting edge and edge attributes. This data storage model is very similar to Google's distributed storage system (Bigtable)—a typical wide column storage. The advantages of this type of data model design are:

- For graph traversal, this is an edge-first data model with high traversal speed.
- Data models can be optimized with the most appropriate data structures for best traversal performance.
- It will make partitioning (or sharding) easier.
- This data structure is also applicable to the persistent storage layer (row storage mode).

Weaknesses are:

- Data structures that use contiguous storage (memory) space cannot update (delete) data quickly.
- If you use aligned field boundaries, it will inevitably lead to waste of space (the so-called space for time tradeoff).

Expanding on the idea of Fig. 19, readers can figure out ways to design more efficient graph computing data structures.

3.2.5   How to partition (shard) a large graph

Partitioning (sharding) a large graph can be a daunting task. Traditionally, the graph sharding problem is an NP-hard problem, and the complexity of solving it is at least as difficult as that of an NP-complete (NP-Complete) problem, which is very challenging. Even if a graph is partitioned, due to the complex dependencies among the subgraphs formed after partitioning, the performance of the entire cluster system may be reduced exponentially.

Graph partitioning (also known as graph cutting) is an interesting topic in graph theory research. There are generally two ways to cut graphs, namely edge-cut and vertex-cut, as shown in Fig. 20.

Carefully observe Fig. 20, we will find that in the two new subgraphs formed after edge cutting, not only the vertices (A, C, E) are stored repeatedly, but also the edges (AC, AE) are repeatedly stored; after that, the vertex (A) is replicated. The logic of cutting edges is more sophisticated than cutting on vertices, so more graph computing frameworks adopt the vertex-cutting mode. For example, the GraphX graph computing framework based on the Apache Spark

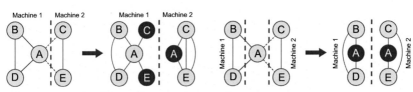

FIG. 20   Illustrated difference between edge-cut and vertex-cut.

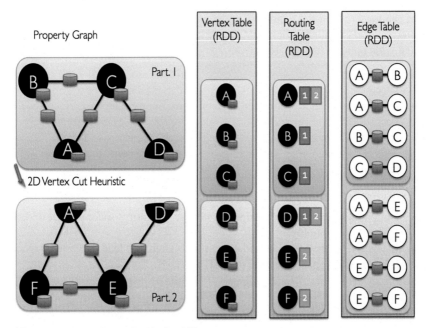

FIG. 21   Vertex-cutting method in GraphX.

system in Fig. 21 adopts a typical vertex-cutting method. After the two vertices A and D in the graph are split, GraphX uses three sets of distributed data structures to express the complete original graph: vertex table, routing table, and edge table.

Among them, the two split vertex tables use a completely nonoverlapping method, that is, any vertex will only appear in one vertex table and will not be duplicated (in both tables), so it will appear that although A and D are both cut, but A is in only found in the above table, D is in the following table; the routing table indicates which edge table(s) each vertex corresponds to (every cut vertex will appear in at least 2 edge tables); the edge table is responsible for describing the start vertex, edge attribute, and end vertex of each edge in each subgraph formed after partitioning.

There are other schemes and variants of the logic and specific data structure design of the vertex-cutting mode, such as the PowerGraph graph computing framework from Carnegie Mellon University (CMU)—identifies other graph computing framework's inefficiency in dealing with hotspot supernodes, optimizes on vertex-cut subgraphs, and the computational efficiency of some graph algorithms (such as the famous PageRank algorithm).

Due to historical reasons, most of the datasets worked on by the graph computing frameworks are synthetic data sets (such as the Graph500 data set in supercomputing), while the data sets of the real world generally come from the Internet, social networks, or road network class, such as Twitter, LiveJournal, and others. These datasets generally have several shared characteristics:

- Simple graph
- No attributes (vertices and edges have no attributes)
- Existent of super nodes (hotspots)

Perhaps because of these commonalities, the graph analysis and algorithm research on the graph computing framework are generally concentrated in the following types of algorithms: PageRank (webpage sorting), triangle calculation (statistics by vertex), shortest path or its variants, connected components (weakly connected Component), map coloring problem.

A common characteristic of these algorithms is that the query depth is relatively shallow. Except for the shortest path algorithm, other algorithms can be simply understood as having only 1–2 hops of query depth. However, as more industries (such as the financial industry) are paying attention to graph technology, the limitations of the (academia-focused) graph computing frameworks are beginning to be exposed, especially these common requirements require support at the architectural level:

- Multigraph: There are multiple edges between vertices, such as multiple transfer transactions, credit card transactions, and call records.
- Multiple attributes: There may be multiple attribute fields on vertices and edges.
- Dynamics: The data on the graph, vertices and edges and their attributes may be dynamically added, deleted or updated.
- Real-time: In some scenarios, it is necessary to perform calculation and analysis in real time.
- Query depth: queries with a traversal depth of more than 3 hops.
- Time series: Most graph data in the industry tends to have timing attributes which is relevant to how graph can be partitioned logically.

Contemplating the more comprehensive general characteristics of graph data above, there are at least the following four methods to segment a graph.

1. Brute-force Partitioning: Brute-force partitioning is very similar to random partitioning, which basically ignores the inherent topology of the graph (from a human perspective, a large graph may contain subgraphs that form communities, making some subgraphs very dense, while others very loose, thus giving the opportunity to partition the graph at the loosest points), and simply cutting between arbitrary (e.g., equally partitioned) vertices or

edges. At present, the partitioning logic of most graph computing frameworks adopts the vertex-cut method, and a small part adopts the edge-cut method.

2. Time-series-based Partitioning: This caters to scenarios where the data model contains time-series data, for example, there are 1 billion accounts, recording 10 billion transactions per week, if you look up in the past 12 months account groups, it makes sense to create a graph with 52 subgraphs. Each compute-n-storage instance in the cluster captures 1 week of transactional data. Each query can be sent to all 52 subgraphs for execution, and finally merged (aggregated) to a designated (extra) computing and management node for final result aggregation. The logic of this partitioning mechanism is very clear, and it also conforms to the concept of dividing and conquering big data, which can solve the needs of many real-world scenarios.

3. User-specified partitioning: Similar to the logic of time-series partitioning, it's up to the DBA/user to define how to segment the graph data set. This graph-slicing mode can be regarded as a superset of time-series graphs. This method normally involves understanding of business logics as well how business data is generated.

4. Smart Partitioning: Smart graph cutting is still an NP-hard ($\geq$ NP-complete) problem in theory. However, in fact, because the connectivity of most real-world graphs belongs to sparse graphs, that is, far from reaching the density of complete connectivity (clique), it is logically possible to find the smallest set of vertices with the sparsest connectivity in a large graph, and create a split that balances the size of the subgraph as much as possible. Of course, one possible problem with this kind of partitioning is that if the graph data continues to change dynamically after partitioning, it may cause problems such as load imbalance among the split subgraphs. Therefore, how to solve the dynamic balance cutting problem is a problem that the industry, academia, and research circles are paying attention to.

Fig. 22 shows a possible graph cutting method at the data structure level, but upon careful observation, this method does not seem to be a typical vertex-cutting or edge-cutting method, because this data structure itself is difficult to complete cross-subgraph operations (Because the terminal vertex that an edge points to may belong to another subgraph, but the data structure itself does not have a record of another subgraph). Therefore, similar to the logic in Fig. 21, we need a routing table or a mapping table to query the subgraph affiliation of any vertex, so as to complete cross-subgraph query. Obviously, the more graphs are cut, the heavier the burden of cross-subgraph communication.

Some studies have shown that after a large graph is cut into 10 subgraphs, 90% of the vertices will fall into multiple subgraphs, and if it is cut into 100 subgraphs, 99% of the vertices will be cut—running complex graph algorithms such as path query, K-neighbor query, random walk, or community identification on the network will be a huge challenge, and the performance (timeliness)

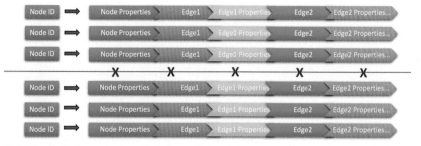

FIG. 22   Typical vertex-cut graph data structure.

after cutting the graph may have a drop of tens of thousands of times. (Refer to the following 2 papers for more information: Scalability! But at what COST?,[b] and Designing Highly Distributed Graph Databases without Exponential Performance Degradation.[c])

3.2.6   High availability and scalability

Any enterprise-level commercial system must have high availability, scalability, and fast failure recovery capabilities.

(1)   High availability

Like mainstream databases, data warehouses, or big data systems, there are many ways for graph systems to provide and achieve high availability, as follows:

- Two-instance cluster system (HA): The simple and standard HA setup is very simple, the primary server and the secondary (backup) server each have a 100% graph system. They are kept in sync, and any modification/ addition/deletion operations are performed through the internal synchronization mechanism of the dual-instance cluster system. Once the primary server is shut down or offline, the backup server is activated immediately (usually with a delay of milliseconds). This dual-machine cluster system (HA) is also called a hot standby system.
- Master-slave server cluster (MSS): A master server and multiple slave servers form a cluster consisting of multiple (≥ 3) servers, all of which have read permissions, but only the master server can perform additions, deletions, and modifications. It is very similar to the configuration of the master-slave dual-machine cluster system of the classic MySQL database. One of the setting modes of the master-slave server is distributed consensus (RAFT)—A cluster, usually consisting of 3 instances, is

---

b. Scalability at what cost: https://www.usenix.org/system/files/conference/hotos15/hotos15-paper-mcsherry.pdf.

c. Designing highly-scalable graph database systems without exponential performance degradation https://dl.acm.org/doi/abs/10.1145/3579142.3594293.

based on a distributed consensus algorithm called RAFT, in order to simplify cluster data synchronization, usually one server is responsible for writing operations, and all servers can participate in cluster load balancing. As long as more than 50% of the nodes in the cluster are online, the entire cluster can work normally. When an instance is down, and back online again, the cluster can re-elect for master node (the leader). In some naïve RAFT-based graph database implementations, the backup nodes (follower or learner nodes) do not participate in the cluster load-balancing. They only recommend a node to become the master node through the election mechanism when the master node is offline, such as some vendor's causal cluster (causality cluster) is a typical hot backup mode based on distributed consensus. Its disadvantage is that it wastes the computing resources of (at least) 2/3 instances in the cluster with no load-balancing.

- Distributed high-availability cluster (DHA-HTAP): The minimum size of a distributed high-availability cluster is usually three instances. At this time, the role of each node is based on the incoming query/operation type—TP or AP. The HTAP-capable system will be able to distinguish between real-time TP-type operations and AP-type operations, and automatically route the operations to appropriate instances for execution—which can be preset or dynamically allocated according to needs. Usually, because AP-type operations (such as algorithmic operations or full-graph traversal) consume a lot of computing resources, it is not suitable to mix with TP operations on the same server instance. It is precisely because of this separation of operations that the hybrid transactional and analytical processing architecture (HTAP) came into being. In larger horizontally distributed clusters (>6 instances), there are usually different types of instances such as management servers, sharded computing and storage servers, name servers, meta servers, etc. This kind of large cluster setup is often called having Grid, Federation, or Shard architecture.

- Disaster recovery distributed high availability (DR-DHA): This kind of high-availability setup is to support disaster recovery of multiple data centers in the same city or in different regions—similar to the above DHA-HTAP mode, but obviously the scale of the cluster is larger (multiplied), and the communication across data centers makes data synchronization more complicated due to the greater delay. For large enterprise users, especially financial industry users, although this relatively complicated solution is necessary, the main cluster is basically the only operating entity, and the disaster recovery cluster will only be activated (manually or semiautomatically) after the main cluster is forced to go offline.

(2) Scalability

    With the rise of commercial PC server architecture, cloud computing, and Internet scenarios, horizontal scalability is more and more popular than vertical scalability. Many developers think that horizontal scaling is universally feasible and ignore the fact that most graph datasets are not large enough to require true horizontal scaling; on the other hand, all modern PC server architecture has high concurrent/parallel processing capabilities, and it is still very meaningful to make full use of the concurrent computing capabilities of each server and each CPU. The computing power of a single server with 40-core is greater than that of five 8-core multicomputers, especially in the context of graph database queries.

    Two approaches to achieve graph database scalability:

- Vertical scalability: Fully considering the processing power of today's computers, a motherboard with 2 CPUs and 20 cores (40 threads) per CPU has much better computing performance than a 5-node cluster (each node has 1 CPU, 8 cores, and 16 threads)—because the former has more intensive (denser) computing resources, and the communication efficiency within a machine is much higher than multiple machines communicating through the network. By simply upgrading your hardware intentionally, you can get better system performance, especially in graph systems. Generally, the larger the memory, the higher the performance, which can greatly improve the performance and capacity of the system. For example, if you have a server with a powerful CPU, but only 64GB of memory, then you can hardly process a dataset of 10 billion plus edges effectively (via memory acceleration), now consider upgrading the memory to 512GB—in fact, the number of threads of the CPU and the number of GB of memory remain at 8–16 ratio is more appropriate. For example, 16 threads correspond to $128\,G \sim 256\,GB$ memory, 32 threads correspond to $256\,G \sim 512\,GB$ memory, and so on. The point here is that while horizontal scalability may sound more enticing, vertical scalability is more effective and cost-productive in many, if not more, use cases.

- Horizontal scalability: There is a layered concept in horizontal expansion. The simplest horizontal expansion is to increase the hardware resources, such as number of servers, to obtain higher system throughput. Referring to the master-slave or distributed consensus algorithm-based architecture introduced above—an HTAP cluster of 3 instances has $3\times$ (or more) throughput than a singular instance system on most read-type operations (but not write-type operations). In a more sophisticated Grid-architecture setup, both the read and write operations efficiency and throughput can be linearly increased as the number of instances are increased. The uniquely complicated part of designing a distributed graph database system is that many people have assumed when a large graph is divided (partitioned onto multiple instances), both reading and writing against the graph dataset become very difficult because any operation may require the access of multiple instances and lots of synchronizations among the cluster instances. As more and more machines participate in the horizontally expansive architecture, the throughput rate of the system will decrease instead, because the query performance of the more extended graph will be lower as the communication between instances becomes more frequent. In view of this, the design and implementation of horizontal distributed graph architecture in the industry usually adopts the multiple graph mode, that is, logically, multiple graph datasets will be stored in a large cluster, and each graph dataset can be self-contained in a small cluster, and multiple clusters can work together via the help of metaserver and name-server. Intercluster communications and large-scale data migrations are limited to help reduce latencies and boost system performance. We'll expand on this topic in Chapter 5.

3.2.7 Failure and recovery

In graph databases, there are two major types of failures.

- Instance failure: It is a serious failure type, which means that the running instance (in the worst case, the entire cluster system) is offline, and usually the instance will automatically restart. If there is a hot standby instance, it will automatically take over from the primary instance. But if not, this issue needs an IT hot ticket and be fixed ASAP.
- Transaction failure: There are many reasons why a transaction can fail, because a transaction is of ACID type, if it fails, some kind of rollback may be needed to ensure that the system wide state is in sync, otherwise the problem will end up being unmanageable. Failures are logged for developer review, and an IT support ticket should be raised. Otherwise, only rerun the transaction, if the problem no longer exists, it may be a false alarm or for other obscure reasons.
- All of our discussions above have roughly assumed that all graph servers are connected to the same high-speed backbone for very low network latency (for state and data synchronization). The big question to be addressed now is: How do you support disaster recovery across data centers? Taking dual data center disaster recovery as an example, there are at least two solutions:
- The scheme in Fig. 23 uses near-real-time synchronization of data from the master (primary) cluster to remote slave (secondary) cluster. When the primary cluster fails, the business can be switched (manually or automatically) to the secondary cluster. In this scheme, logs are used for asynchronous replication (log replication) between two clusters. In extreme cases, temporary

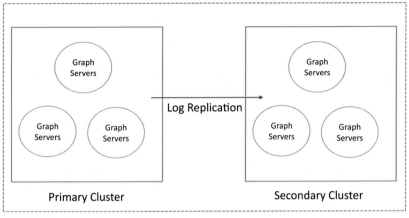

FIG. 23 Dual-center cluster backup scheme.

data inconsistency (temporary loss of data in one cluster, waiting to be synchronized from the other one) may occur in this scheme.

- The solution in Fig. 24 uses a single cluster spanning two centers, and each center has three instances (number of instances may vary). If a disaster occurs in any center, business continuity may be temporarily affected, but because of the strong data consistency protocol which ensures all data having multiple copies, the scale of impact is very limited, data will not be lost, and the business can be quickly restored by reducing the number of copies (manually operated by maintenance personnel)—for example, using the three instances in the second center to pull up the business ($\geq 50\%$ to ensure business continuity).

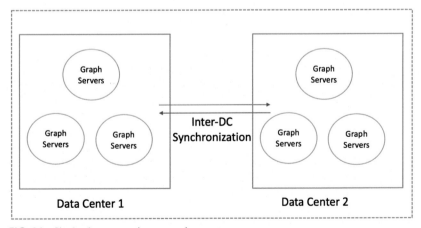

FIG. 24 Single-cluster spanning across data centers.

When building a business continuity plan and disaster recovery plan (BC/DR), the questions (and assumptions) to consider are as follows:

- Design business continuity and disaster recovery plans (BC/DR planning).
- Establish a time frame within which systems and processes must be restored.
- Evaluate IT infrastructure, data center (or cloud) contracts and understand their limitations.
- Evaluate the priority of the graph database in the entire business-oriented IT support system.
- Implement business continuity and disaster recovery plans (BC/DR).
- Test real-world situations through drills and mock-ups.
- Iteratively optimize the above plans and implementation plans.

Graph computing and analysis are applicable to a wide range of scenarios. Gartner released a market research report in May 2021 indicating that by 2025, 80% of innovations in the field of business intelligence will be powered with graph analysis; and in the 2020 research report, its forecast for 2023 was 30%. As more and more industry applications begin to explore graph database-based solutions in the digital-transformation processes, we have reason to believe that more and more innovative design solutions will be rooted with graph databases. The author sorts out the following possible "destructively innovative" architecture designs (in no particular order).

- Persistent memory architecture: large-scale use of large memory architecture, the hard disk does not participate in the real-time graph computing process.
- Distributed grid architecture: In the horizontally distributed architecture, memory computing and storage grids are used to virtualize (containerize) massive heterogeneous types of storage (disk, memory, cache, index, etc.), and perform multiinstance interconnection and data exchange through low-latency networks.
- Low-carbon footprint & High-density parallel computing architecture: Through system architecture design optimization, the underlying hardware concurrent computing capabilities can be fully utilized, and fewer instances can be used to achieve higher system throughput and lower.
- Accelerated architecture that integrates heterogeneous types of processors: such as FPGA, GPU, ASIC, SoC, and other computing acceleration architectures that cooperate with CPUs.
- Intelligent graph sharding and dynamic multigraph load balancing architecture: At the software level, it can intelligently split large graphs and maintain dynamic balance and high efficiency of multiple graphs and multiclusters in the face of dynamically changing data sets.

3.3   Graph query and analysis framework design

After understanding the computing and storage principles of graph databases, everyone will definitely ask a question, how to operate and query graph databases? We know that relational databases use SQL, which is also the first international standard in the database field (before the official release of GQL, SQL has been the only database query language international standard). In the past 10 years when big data and NoSQL are surging, SQL standard has not been seriously challenged, they are either stay compatible with SQL or manipulate the database queries through API/SDKs. Graph database operations via SQL is impractical given all the limitations of SQL. The author believes that the query and analysis of the graph database should be a combination of graph query language and API/SDKs natively supported by the database, and the GQL can add not only a layer of standardization but also white-box explainability to how the graph data is processed, meanwhile, allowing users to decide which method is more practical, efficient, and easy to use.

3.3.1   Graph database query language design ideas

As a graph database with complete functions, the support of graph query language (GQL, generally speaking, not necessarily the standard though) is essential. However, not all graph query languages are created equal, and each query language has its own characteristic, advantages and disadvantages. When the user deploys multiple sets of graph systems of heterogeneous types of query languages (for application development), it's no doubt a disaster. That's why there is a need to develop international standard. On the other hand, only when the industry reaches a consensus that a certain type of product will be widely adopted will there be a rigid need to formulate international standards, which also shows that graph databases will be a mainstream database just like relational databases. Therefore, the development process of SQL, the query language standard of relational database, has reference significance for GQL.

The development of the relational model (shown in Fig. 25) can be traced back to IBM's research lab in Silicon Valley in 1970s—for their research work, including the implementation of the SQL prototype in System R. At the end of the 1970s, commercial relational databases (RDBMS) began to appear. The earliest SQL standards were released in 1986 and 1989 (commonly known as

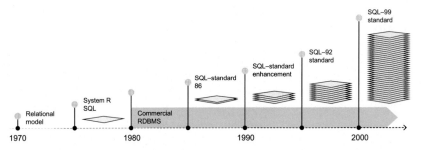

FIG. 25   History of SQL standard formation.

SQLv1), but the relatively mature and fully functional one is SQL-1992 (commonly known as SQLv2), such as join, inner-join, union, intersect, alter table, alter view, and other functions, we also say that SQL-1992 laid the foundation of contemporary SQL. The modern SQL standard version was released in 1999, incorporating functions such as Group By. It should be pointed out that the commercial databases usually implement these beneficial functions first, and then successively incorporated into and become international standards. For the next 20 years, the SQL standard experienced minor variations, this time period also happens to be the period of rapid development and iteration of NoSQL and big data frameworks. Whether there is cause and effect or just pure coincidence is thought-provoking.

From the international standard formulation of SQL, we can sum up keys as follows:

- Time-consuming: It takes a long time to form a standard, and it may take more than 10 years after the commercial product/edition appears.
- Iterations: The earliest standards are usually incomplete and require 1 or 2 major iterations to take shape (and be fully practical).
- Long cycle: Each major version iteration takes about 3 years.
- Promoting development: Once the standard is fixed, the market develops rapidly, but it also means that products will enter homogeneous competition, creating opportunities for the emergence of new products.

The commercialization process of graph databases (native graph databases) in the true sense began in 2011. The representative product is Neo4j's v1.4 database version with Cypher graph query language (precisely v1.0~v1 0.3 version can only be regarded as a server with only computing functions), and the version that supports the label attribute graph (LPG) will have to wait until the v2.0 and v2.1 versions 3 years later.

Around the same time period, the Tinkerpop Gremlin (now Apache Gremlin) query language also began to take shape, its designers seem to prefer Gremlin to become a general-purpose graph traversal language, open to all graph databases. Although Gremlin claims to be a Turing-complete graph language, writing complexity when supporting complex scenarios is a disadvantage of Gremlin.

As for the query language SPARQL in the RDF field, although it was recognized as an international standard by the World Wide Web Consortium in 2008, its latest standard stagnated at the v1.1 version released in 2013. The industry generally believes that the graph of the RDF schema is too complex and inefficient. The query language is also not smart enough.

Different graph query languages have very different styles of accomplishing the same thing, even for a seemingly simple thing. For example, Fig. 26 describes the simplest path starting from a person-type vertex and passing through it to connect with another (type of) vertex which is of occupation type named chef. In knowledge graph and RDF, this kind of vertex-edge-vertex data association is called triplet. In graph databases, how to describe such using graph query languages? We hereby give the following four solutions:

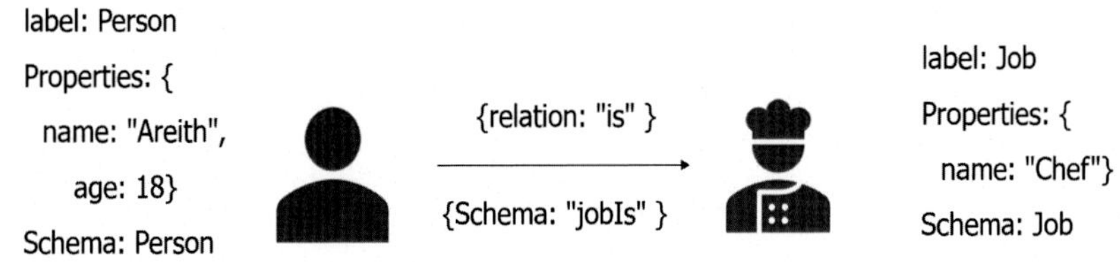

label: Person

Properties: {

    name: "Areith",

    age: 18}

Schema: Person

{relation: "is" }

{Schema: "jobIs" }

label: Job

Properties: {

    name: "Chef"}

Schema: Job

FIG. 26 Schematic property graph.

Option 1: System-N

```
match path = (p:Person) - [{relation:"is"}] - (j:Job)
where p.name = "Areith" && j.name== "Chef"
return path
```

Option 2: System-T

```
CREATE QUERY areithjob(vertex<word> w) for graph test {
    SetAccum<node> @@nodeSet;
    SetAccum<edge> @@edgeSet;
    Start = {persion.*};
    Result = select j from Start::p - (jobIs:e) - job:j
             WHERE  p.name == "Areith" AND j.name ==
        "Chef"
            accum @@nodeSet += p,
            accum @@nodeSet += j,
            accum @@edgeSet + = e;
    print @@nodeSet;
```

Option 3: System-U (Schema-free)

```
n({name=="Areith"}).e({relation=="is"}).
n({name=="Chef"}) as paths
return paths
```

Option 4: System-U (Schematic)

```
n({@person.name == "Areith"}).e({@jobIs})
.n({@job.name == "Chef "}) as paths
return paths
```

Option 1 is Neo4j's Cypher solution, which has good code readability and can clearly express the logic of path query at the semantic level; Option 2 is Tigergraph's GSQL solution, which is complicated in terms of code logic and size, the readability of the code is not high because it is trying to be consistent with the SQL style. The key is to return two separate vertex and edge data sets, which need to be encapsulated on the client side; the third option is a chain query language, using vertex-edge-vertex flow semantics to naturally express path return results; the difference between option 4 and option 3 is whether to support schemas, although these two graph query styles are quite different from the style of SQL, but the code is concise and easy to read.

The differences of the above solutions will be more prominent in more complex scenarios.

Fig. 27 describes a typical heterogeneous graph pattern, that is, there are multiple types of entities and relationships in the graph. Taking the financial industry as an example, heterogeneous fusion graph data may include various types of entity data such as branches, industries, commercial customers, retail customers, accounts, customer groups, customer managers, deposits, and loans; and these entities form multiple types of relationship, such as transfer relationship, management relationship, reporting relationship, and other affiliation relationships. If we need to find the relationship pattern highlighted by the thick green arrow at the bottom right of the figure, we need to conduct a reverse query to analyze which industries the branch's operations are predominantly serving, which financial indicators the industries care about, and which customers' deposit accounts contribute the most to the indicators. This type of "attribution analysis" is highly sought after in the quantitative analysis of the contribution of the core business of the financial industry (or any industry). The traditional data warehouse architecture of commercial banks has almost no way to complete this kind of in-depth penetration and aggregation queries in real or near-real time.

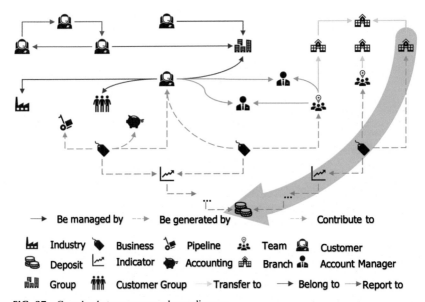

FIG. 27 Complex heterogeneous schema diagram.

If you use the pattern query in Option 4, the effect is as follows. Although the query path is long, it is not difficult to write and read the query statement.

```
// Template Path Query
n({@branch}).le({@generate}).n({@business}).re({@contribute}).
n({@indicator}).n({@deposit})
```

Path query

The ability of path query is one of the most significant differences between graph databases and traditional databases. It is equivalent to allowing users to obtain query results with context (vertex-edge-vertex-edge-vertex), and this context makes content naturally easier to be understood and digested.

There are many different modes of path query, enumerated as follows:

- Shortest Path query
- Loop/circle Path Query
- Template Path query (also known as path query with filter conditions)
- Interstep Filter Query (can be regarded as a special template path query)
- Subgraph query (compound path query)
- Networking Query (Composite Path Query)

Among them, the subgraph query can be seen as is a query consisting of multiple path queries describing a subgraph or a way to match a template. Fig. 28 shows the realization of a credit antifraud scenario composed of three path queries.

A specific implementation of a possible GQL statement is as follows:

```
graph([
  n({@apply as A1).e().n({@email}).e().n({@apply._uuid >
A1._uuid} as A2),
          n(A1).e().n({@phone}).e().n(A2),
          n(A1).e().n({@device}).e().n(A2)
]) return A1, A2
```

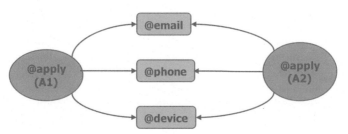

FIG. 28 Multipath subgraph query.

The subgraph query logic is as follows:

(1) Take the credit application A1 as the starting point, pass through the email relationship, and arrive at another credit application A2 with a different ID.

(2) There is another path between A1 and A2 through the phone-type relationship.

(3) A1 and A2 also have a third path associated through device-type relationship.

(4) If the conditions in steps 1–3 are all met, return the vertex pair A1 and A2.

Note: Even though A1 is unique, there may be multiple A1-A2 vertex pairs, because A2 represents all known credit applications with the same email, phone, and device ID as A1.

This subgraph query statement example allows the antifraud business logic to be easily expressed at the graph query language level, and the execution efficiency of this type of query can also be completely real-time. Even in a large graph of billion-scale, it can be done in an online pure real-time manner. In a real business scenario, the new application A1 and its relationship with other vertices are constructed in real time, which is also the core capability of the graph database and the main difference from the graph computing framework that can only process static data.

Let's look at another example of loop query implementation. As shown in Fig. 29, in a transaction network, starting from a certain account (U001), find 10 transfer loops within 6 degrees of separation from the transaction card under the account. The so-called loop here means that the trading card at the starting point is the same as the trading card at the ending point. We can try to describe it in a very concise and templated way. Note that the alias of the trading card is defined in the first line as origin, and its outward (right) transfer step does not exceed 6 steps (degrees of separation), and eventually returning to the same card. Assign the entire path with an alias and return up to 10 paths:

```
// Deep loop query
n({_id == "U001"}).re({@own}).n({@card} as origin).
re({@transfer})[:6].n(origin._uuid) as path
return path limit 10
```

In the loop path query example above, we have limited the return path to 10. If a completely unlimited method is used, it is very likely that the query will contain many paths due to the potentially high connectivity of the transaction network when query depth is deep. Issues like resource exhaustion (e.g., OOM) may take place. After all, the above search conditions are too broad. In a real antimoney laundering query, at least a clearly defined transfer relationship with increasing time and decreasing amount (or other filtering conditions such as specific ranges) is required.

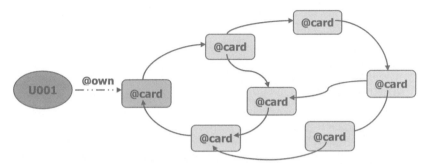

FIG. 29 Loop path query.

So how to realize the function of gradually increasing the transfer time? The answer is: interstep filtering, that is, in the path query, define a keyword *prev_e* reserved by the system, which is used to refer to the previous connected edge, so that we can use the time attribute of the current edge and the time attribute of the previous edge to make comparison, and the final statement looks like:

```
// Inter-step query
n({_id == "U001"}).re({@own}).n({@card} as origin).
re({@transfer.time > prev_e.time})[:6].n(origin._uuid) as
path
return path limit 10
```

Path query is the most distinctive and challenging mode of query of graph database query languages. It can be said that all complex graph query modes are inseparable from path query. The underlying implementation of the path query may adopt either the breadth-first method or the depth-first method, or even a mixture of the two methods. This is difficult for beginners to understand, so it is necessary to explain it. In theory, regardless of the breadth-first or depth-first traversal method, as long as all possible path combinations can be traversed, the same result set will eventually be obtained. However, in actual query operations, different query modes will have different search efficiencies, and different data structure designs will also lead to large differences in the algorithm complexity and time complexity of the specific implementation of the two query modes. For example, if you query the first-degree neighbors (direct neighbors) of a vertex by using the index-free adjacency data structure, the complexity of accessing all the first-degree neighbors is O(1), even for super nodes; However, if the adjacency-list method is used, obtaining all the neighbors of the super node (i.e., 1 million neighbors) may be a rather time-consuming and system resource-intensive operation, because it is necessary to traverse and copy the records of all linked-list nodes. In the loop query shown in Fig. 29, the depth-first search mode is adopted, because the conditions for priority search include attributes such as edge direction and transfer amount, and as long as these conditions are met, the drill-down (depth) mode can be continued, and the operation will continue until the qualifying condition is reached or a qualified loop is found.

K neighbor query

K-neighbor query (commonly known as K-hop query) is another type of graph query in addition to path query. Its purpose is usually to quantitatively analyze the trend of an entity's influence step by step as the traversal depth increases. K-hop queries are usually processed in a breadth-first manner, which is determined by the characteristics of K-hop by definition—to find vertices that meet the filtering conditions with the shortest path distance K from the queried vertex. If the depth-first method is used for K-hop, after traversing all possible path combinations, it is necessary to mark whether each vertex on each path is on the shortest path from the starting vertex—the query complexity is obviously much higher (less inefficient and error prone).

Note: In a graph with high connectivity (that is, a dense graph), the algorithmic complexity of the K-hop query that returns all results may be comparable to a full graph traversal, especially when the query is performed on a super node.

Although the logic of the K-hop query is simple, there are still some details that are easy to be ignored in the specific traversal implementation, which leads to errors in the code logic implementation:

- Pruning problem
- Vertex deduplication problem
- Computational complexity and concurrency challenges

We use Fig. 30 to illustrate the above three problems. Suppose we start from vertex C001 to find all the neighbors whose shortest distance is 4-hop or 5-hop. The following statement can be used to achieve this:

```
// K-hop query with range (depth)
khop().src({_id == "C001"}).depth(4:5) as neighbors
return count(neighbors)
```

In the process of implementing the khop() function query, the pruning problem can be seen everywhere, because K-hop query focuses more on vertices and less on edges, that is, the significance of edges is limited to traversing and measuring whether the path between vertices conforms to the shortest path rule. Once the shortest path is locked, redundant and repeated edges will not affect the traversal results. (This isn't always true, because there are cases where K-hop may be performed with filtering against specific edge attributes).

Deduplication is the most easily overlooked problem in K-hop queries, and it can also lead to errors in the results, and it is difficult to check the accuracy of

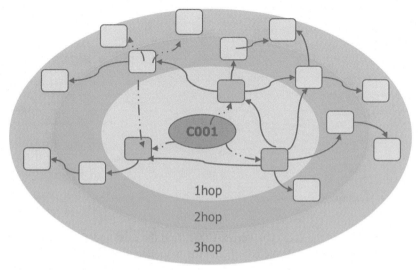

FIG. 30   K-hop query.

results in large graphs. For example, the three nodes in the innermost circle in Fig. 30 are the first-degree neighbors of C001, but because there are connected edges between these three nodes, if there is no correct deduplication mechanism, these nodes may be double-counted in the second-degree (second hop) neighbors of C001. By analogy, when calculating each degree (K) neighbor set, it is necessary to ensure that the set does not contain any neighbors with degrees from K-1 to 0, otherwise the vertex needs to be deleted (deduplicated).

When K-hop queries face super nodes, it is easy to form a situation of traversing a significant portion of the entire graph. At this time, concurrent queries can be very helpful in penetrating the super node and complete the query in a much shorter time in a "divide and conquer" manner. Readers can refer to the schematic diagram of the K-hop concurrency algorithm in Chapter 2's Fig. 12 to consider how to implement their own high-performance K-hop query.

Using K-hop query can also achieve an effect similar to path filtering query. For example, in Fig. 31, we use template K-hop query to realize a typical anti-fraud pattern recognition.

```
// K-hop with filters
khop().n({_id ==
"U001"}).re().n({@card}).le().n({@apply}).
re().n({@phone}).le().n({@apply}).re().n({@card}) as
suspect
return count(suspect)
```

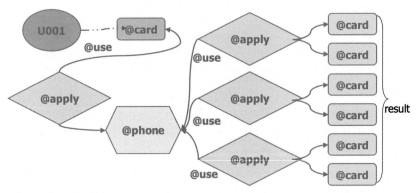

FIG. 31 Template K-hop query.

Starting from the user type vertex U001, find the card under the account to the right (the direction of the edge), and the associated card application which links with a phone number same as that of some other card applications, counting all qualified paths. The counting result is the number of all suspected fraudulent cards associated with the account.

K-neighbor query is more common than path query in graph theory research, especially in the era when simple graphs are more common, and many graph traversal algorithms are essentially performing K-hop queries. However, with the emergence of multigraphs, coupled with the problems caused by logic loopholes in the underlying K-hop query code implementation of many graph computing or graph database systems, many K-hop query results may have accuracy problems. In Chapter 7 of this book, we will use a special section to introduce the correctness verification of graph databases. In many cases, K-hop query can also be considered as a special shortest path query, except that the result it focuses on is a collection of vertices rather than a path (a collection of vertices and edges).

Metadata query

In a graph database, the most basic query is a metadata query, that is, a query for the "skeleton" data (so called metadata) such as vertices, edges, and their attribute fields in the graph. This type of query is similar to the traditional database queries, including CRUD operations (create, read, update, and delete).

Two characteristics of the graph database query statements are evident in Figs. 32 and 33:

- The use of schema plays a key role, which is more efficient than the traditional way of filtering records.
- The returned results can be combined in any way, including the combination of different fields, and aggregation operations on certain fields.

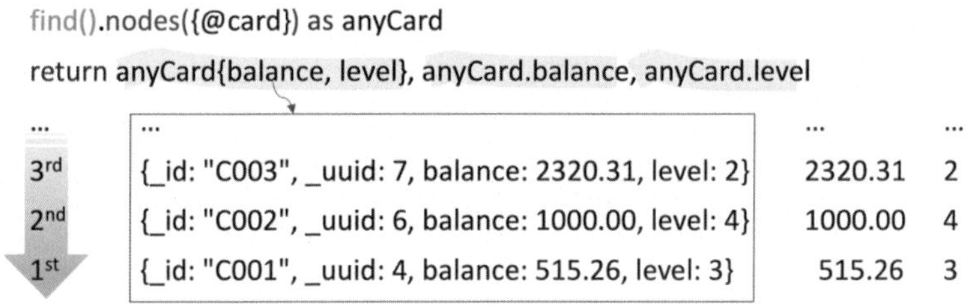

find().nodes({@card}) as anyCard

return anyCard{balance, level}, anyCard.balance, anyCard.level

...	...	...	...
3rd	{_id: "C003", _uuid: 7, balance: 2320.31, level: 2}	2320.31	2
2nd	{_id: "C002", _uuid: 6, balance: 1000.00, level: 4}	1000.00	4
1st	{_id: "C001", _uuid: 4, balance: 515.26, level: 3}	515.26	3

FIG. 32   Vertex data query.

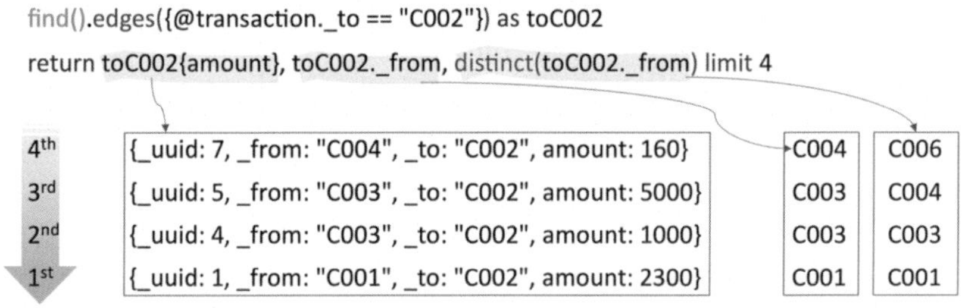

find().edges({@transaction._to == "C002"}) as toC002

return toC002{amount}, toC002._from, distinct(toC002._from) limit 4

4th	{_uuid: 7, _from: "C004", _to: "C002", amount: 160}	C004	C006
3rd	{_uuid: 5, _from: "C003", _to: "C002", amount: 5000}	C003	C004
2nd	{_uuid: 4, _from: "C003", _to: "C002", amount: 1000}	C003	C003
1st	{_uuid: 1, _from: "C001", _to: "C002", amount: 2300}	C001	C001

FIG. 33   Edge data query.

For any combination of query results, from the perspective of data format, support of heterogeneous data types is a must. We can choose lightweight data exchange formats, such as JSON, Binary JSON (BSON) or YAML (can be seen as a superset of JSON, supports embedded annotations) to realize the connection between Web front-end components and graph database servers (some NoSQL databases, especially document databases, such as MongoDB and CouchDB use BSON or JSON as data storage formats). JSON is much more compact and efficient (lightweight) than the more traditional XML format.

Metadata-oriented insert, update, and overwrite operations in graph databases are similar to insert, update, overwrite, or replace operations in traditional databases.

The syntax of the insert statement is very simple. If it is expressed in chain syntax, you only need to specify the schema (if not specified, it will automatically point to the default schema), and link with nodes() or edges():

```
// Insert a vertex by schema
insert().into(@<schema>)
    .nodes([
        {<property1>:<value1>, <property2>:<value2>, ...},
        {<property1>:<value1>, <property2>:<value2>, ...},
        ...
    ])
```

```
// Insert an edge by schema
insert().into(@<schema>)
    .edges([
        {<property1>:<value1>, <property2>:<value2>, ...},
        {<property1>:<value1>, <property2>:<value2>, ...},
        ...
    ])
```

```
//Insert a vertex by account schema and return its _uuid
(system generated)
insert().into(@account)
    .nodes([
        {_id: "U001", name: "Graph"},
        {_id: "U002", name: "Database"}
            ]) as nodes
    return nodes._uuid
```

```
//Insert an edge into transaction schema
insert().into(@transaction)
    .edges({no: "TRX001",
            _from: "C001",
            _to: "C002",
            amount: 100,
            time: "2021-01-01 09:00:00"
    }) as edges
    return edges
```

In the above insertion operation, the _id of the vertex is specified by the user (it can be a string type), and many of the user's original ID data records are strings, such as 32-byte or 64-byte UUIDs (longer string-type UUIDs takes up much storage space, which is contrary with the true meaning of UUID— Universally Unique IDentifier). The graph database storage engine needs to serialize these original IDs to achieve more efficient storage (reduce space occupation), access addressing, and the serialized system internal ID can be called _uuid or _system_id. If several different data structures are listed here, their access efficiency is in the following order from high to low: _uuid > _id > storage system index. That is, when you can use a higher performance index or ID to access it, use it as much as possible.

The insert() operation will fail if the _uuid already exists and conflicts arise, and because of this there are several variants of the insert operation.

- Upsert: update or insert, this operation can be regarded as a smart operation of either insert or update.
- Overwrite: overwrite or insert, the difference from Upsert is that attribute fields that are not specified in overwrite will be automatically cleared

(depending on the specific field type, it can be an empty string, NULL or a value of 0).

Note: Interestingly, the Upsert operation is already well known in relational databases, but the Overwrite operation has not been coined as a new proprietary vocabulary oversert.

Finally, let's analyze the deletion operation of metadata. We can delete by vertex or edge, through the unique ID of the metadata, or through the schema or any attribute field in the schema to delete, or even delete the entire picture.

```
// Delete by ID
delete().nodes({_id == "C003"})
// Delete by _from ID
delete().edges({_from == "C003"})
// Delete by multiple attributes
delete().nodes({@card.level == 5 && @card.balance == 0})
```

```
// To truncate all the nodes and edges in a certain graphset
// in the current Ultipa instance
truncate().graph("<graphset>")
// To truncate all the nodes in a certain graphset
// in the current Ultipa instance
truncate().graph("<graphset>").nodes("*")
// To truncate all the edges in a certain graphset
// in the current Ultipa instance
truncate().graph("<graphset>").edges("*")
```

Note that there is a dependency relationship between nodes and edges. When deleting a vertex, all its connecting edges need to be deleted by default (implicitly); and when deleting an edge, it is not necessary (and not possible) to delete the connecting vertices. In other words, deleting edges may form orphan vertices, but deleting vertices should not form orphan edges, otherwise it will cause memory fragmentation and leaks. In addition, the deletion operation of the super node can be very expensive. If it is stored with bidirectional edges, a vertex with 1 million neighbors needs to delete 2 million edges in both directions.

Graph query language compiler

The query language compiler of the graph database is not fundamentally different from the traditional database in terms of logic and processing flow. Especially with the development of Integrated Development Environment (IDE) which offers better user experience, many grammar analysis tools can make the work of developing database query language compiler easier. For example, software such as ANTLR and Bison can help graph database development engineers get started faster.

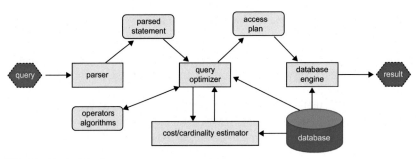

FIG. 34 Typical database query statement parsing process.

The steps of generating, transmitting and processing graph database query statements are as follows (Fig. 34):

- Generate query statements (through the CLI or the front end of Web manager, or send query instructions through API/SDK calls).
- The syntax analyzer (parser) processes the statement to form a syntax tree and the corresponding execution statement.
- The query optimizer (if it exists) optimizes the corresponding execution statement to form an execution plan (instructions).
- The database engine processes the corresponding execution plan and assembles the returned results.

The core components of the parser work as follows:

- Lexical analyzer (lexer, also known as scanner): Convert the string sequence of the query statement into a token sequence. Usually, this process produces an IR (intermediate language or intermediary code), which is convenient for compiler optimization work, while ensuring that the information of the query statement is not lost.
- Constructing a parse tree (or syntax-tree): This process first generates a specific syntax tree CST (Concrete Syntax Tree, also referred to as a parse tree) based on a context-free grammar, and then converts and generates Abstract Syntax Tree (AST).

Fig. 35 shows the effect of the concrete syntax tree constructed by the following query statement. Pay attention to the root of the syntax tree and the pathPatternList on the fourth layer (indicates that the string after MATCH and before RETURN is by default in the syntax logic and defining the path template list), the nodePattern of the 9th layer, and the leaf node of the 12th layer.

```
// Find all vertices and their attributes...
MATCH (*) RETURN *
```

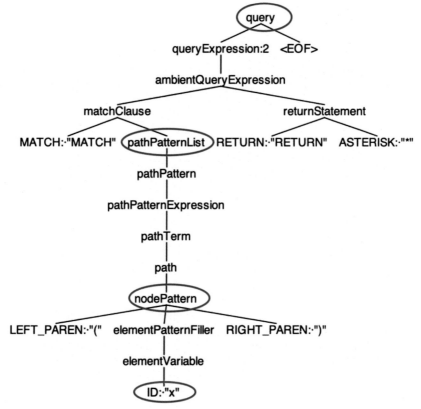

FIG. 35 Schematic diagram of the parse tree (syntax tree).

Note: The difference between the concrete syntax tree and the abstract syntax tree is that the abstract syntax tree implicitly expresses some content (such as brackets, connectors) through the tree structure itself, while the concrete syntax tree restores all the details in the statement in detail, as shown in Fig. 35.

There are many details and optimizations of the parser, given the existence of the parser and optimizer, it cannot be considered that any statement can be completed efficiently. In fact, the query logic of the query statement is still the most important. What kind of query method can be used to obtain the results most quickly depends more on the query and traversal logic used by the user, and which specific query function or algorithm to call, the parser is only responsible for "faithfully" executing the user's will—different query functions obviously have great differences in query timeliness and resource consumption due to differences in internal traversal logic and algorithm complexity, even though they may all return the same and correct result.

In addition, when the parser encounters an input sentence syntax error, it is essential to provide error feedback. For example, in the ad hoc network path

FIG. 36 Prompts for query statement errors in the web manager.

query in Fig. 36, because there is an illegal character "A" in the input parameter of the function that limits the return result, it will be intuitively and accurately prompted in the Web manager. Note that the error message shows that the parameter is abnormal, and at the error position what is expected is a closing parenthesis. This is a very typical error that is located by traversing the syntax tree.

3.3.2 Graph visualization

The operation interface of the traditional database is mainly the command line interface (CLI), and the visual interface appears with the advancement of the display technology. With the advent of the Internet, browsers, C/S and B/S mode database visualization management components are becoming more and more common, especially the Web-CLI, given its convenient development process and rapid iteration, is gradually replacing traditional desktop windows management programs. Due to the high dimensionality and complexity of heterogeneous data in graph databases, the demand for visual presentation of graph data is particularly prominent.

The main users of databases and big data frameworks in the traditional sense are IT technology developers and data scientists. The aesthetics, rationality, and user experience of the interface are not the first priority, which directly leads to business personnel being far apart from these underlying technologies. Technicians largely act as translators between business language and program execution commands—whether generating a report, running a batch program, or implementing a business function, business personnel need to rely on the technicians to complete the operation which may take a few days to a few weeks, and the efficiency is very low.

In the era of graph databases, all this need to be changed. The graph database is closer to the business. The modeling of the graph data directly reflects the demands of the business. The logic and method of the graph query language are 100% mapping of the business language and logic. Therefore, it is possible for a graph database to serve business scenarios in a way that is close to the business, and the visualized, intuitive, and easy-to-use graph database management component (Web-CLI management front end) or the customized application constructed based on this component is aimed at full-spectrum user groups, including both technical and business personnel.

The main data presentation method of a traditional relational database is list, which can be viewed as discrete data clustering presentation. Data in a graph database may be aggregated and presented in the following (much enriched) ways.

(1) List mode.
  o Vertex
  o Edge
  o Attributes
  o Path
  o Aggregated data
(2) High visualization method.
  o 2D plane
  o 3D Stereo
  o 2D/3D interactive
  o Programmable (secondary development) graph

Graph data visualization has many ways and genres in terms of layout strategy, for example, Force-directed Layout (becoming mainstream in recent years), Spectral Layout, Diagonal Layout, Tree Layout, Hierarchical Layout, Arc graph, Circular Path Layout, Dominance Layout.

For the layout of networked data and associated data, there are usually seven topology visualization methods as shown in Fig. 37:

• Point-to-point
• Bus
• Ring
• Star
• Tree
• Mesh
• Hybrid, etc.

From the perspective of designing a general-purpose graph database front-end visualization component, the force-directed graph is obviously the most

Network Topology Types

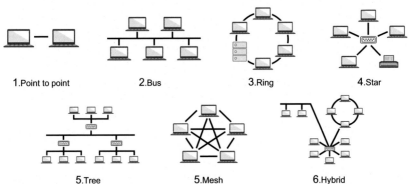

1.Point to point   2.Bus   3.Ring   4.Star

5.Tree   5.Mesh   6.Hybrid

FIG. 37   Seven topological structures and visualization of networked data.

universal, and relatively speaking, it is most suitable for dynamic data and massive data visualization.

In terms of functional modules, the visual management components include the following categories:

- User management, login management, privilege management.
- Graph dataset management, schema management, metadata management.
- Query management, result management.
- Graph algorithm management, task management, state management.
- Data import and export management.
- Secondary development, plug-in management, user-customizable function management.
- System monitoring module, log management, etc.

Fig. 38 shows some user-customizable functional modules (shortcuts) that can be quickly accessed for:

- Metadata query.
- Path query, K-neighbor query.
- Automatic spreading and networking.
- Template query, full-text (fuzzy) search.
- System monitoring and management.
- Customized business modules and plug-in management, etc.

All the above functional modules are integratable with the Web-CLI by calling the API/SDK of the graph database. By exposing the predefined interface, users can easily complete plug-in management and secondary development and truly achieve independent and controllable functionality and application development.

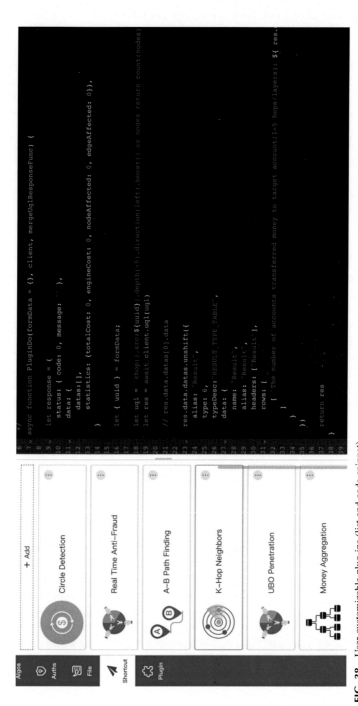

FIG. 38 User-customizable plug-ins (list and code snippet).

Schema management and metadata display

Schemas give wings to data flexibility in graph databases, allowing users to manage data more conveniently. Multiple schemas can be defined on each graph, and users can outline the overall location of the data through the schemas, and conveniently describe the features of type as well as relationship between the data and directly reflect the business logic. However, this flexibility also poses challenges to schema management, and refined management is necessary.

The left side of Fig. 39 shows the definition of attribute fields in vertex and edge schemas. When the attribute is defined (under a schema), you can search for relevant metadata by referring to the schema, or search for data globally (without referring to any schema), and the data in different schemas will be aggregated and displayed per category.

Various display methods of query results

The biggest difference between a graph database and a relational database at the level of visual dimensions is the high-dimensionality of the returned result set. The data returned by graph databases can contain paths composed of vertices and edges, networks (subgraph), or a combination of heterogeneous types of data (metadata, aggregated data, etc.). There are three modes for network visualization: list mode (Fig. 40), 2D mode (Fig. 41), and 3D mode (Fig. 42).

Usually, the 2D mode can handle the display layout and interaction fluency of the network view composed of "vertices + edges" with a scale of less than 5000 (the force-directed graph shown in Fig. 41). However, if the number of vertices and edges is larger, the 2D layout will be sluggish, because the 2D plane rendering with many edges crossing each other takes lots of computing power of the front-end Web browser. At this time, the 3D visualization effect is better, and the browser's WebGL can be used to accelerate the realization of 100,000-level graph network visualization.

The drawing and rendering complexity of the force-directed graph is a problem that needs to be paid attention to during the design and implementation process. In theory, its complexity is $O(N^3)$, that is, when there are N vertices to be rendered, iterate N times, each time each vertex needs to perform N square rendering calculations with all other vertices. However, this theoretical assumption is the case of a fully connected graph (that is, all nodes are interconnected). In actual situations, the degree of association of each node can be Log N, and the complexity of rendering can be reduced to $N * $ Log N. We can further reduce the computational complexity of "big graph" rendering by taking a layered rendering approach (as user interactively zoom in or out of the graph results), thereby improving the timeliness of rendering and the fluency of interaction.

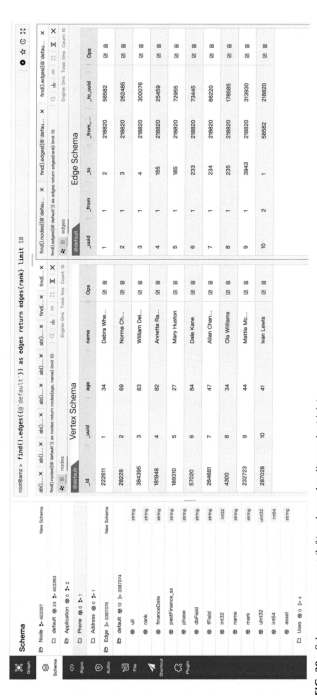

FIG. 39 Schema management (left) and metadata list results (right).

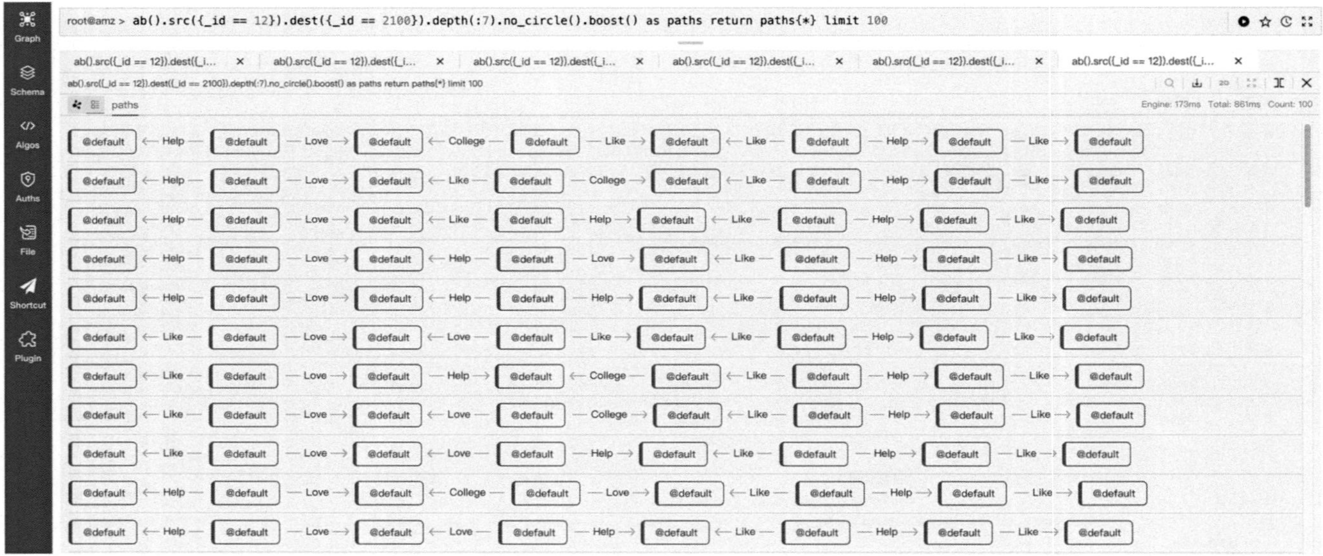

FIG. 40 List display of path query results.

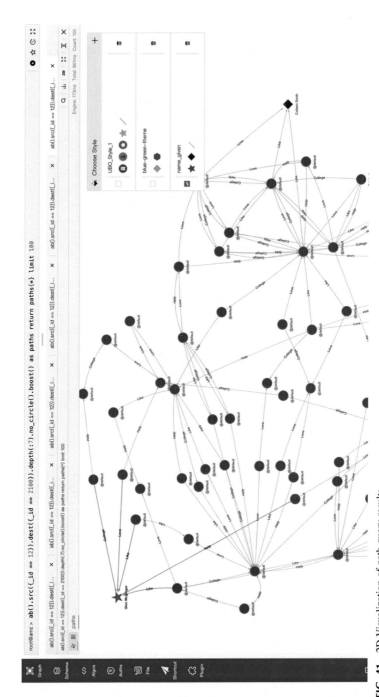

FIG. 41 2D Visualization of path query results.

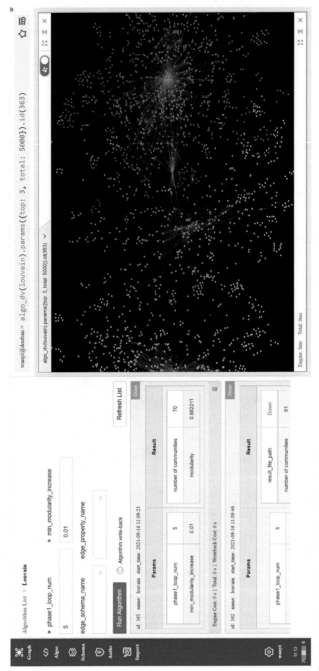

FIG. 42 Graph algorithm result list and visualization.

Graph algorithm visualization

Graph algorithm visualization is a new field of experiment. Given the complexity of many algorithms, having visualizations on the results churned out by the algorithms would help with explaining things in a white-box fashion. There are two types of graph algorithms if measured by the inherent logics of how data is processed—local data oriented algorithms vs global data oriented algorithms. The former type includes degree calculation and similarity calculation against specific vertices, and the other type includes algorithms iterate through all graph data such as PageRank, Connected Component, Triangle Calculation, Louvain, etc.

For the former, visualization can clearly display the results of the algorithm and help users clarify problems; for the latter, on large data sets (above a million vertices and edges), it is only suitable to visualize the data by sampling, as shown in Fig. 42, the visualization of the Louvain algorithm which samples 5000 nodes from the top three communities in the result set to form a 3D network showing how the vertices are relating with each other within their own communities (distinguished by different colors), and how the three communities interact each other on a larger scale.

Chapter 4

Graph algorithms

Graph algorithms can be regarded as special kinds of graph queries that are often times convoluted. If measured in terms of computational complexity, some graph algorithms are aimed at individual nodes, and some are aimed at the full amount of data therefore required to run in batch processing fashion on large graphs (over tens of millions of nodes and edges). Some graph algorithms seek exact solutions, while others seek approximated approach at a lower cost (exponential saving on computing resource consumption and reducing time consumption).

About exact and scientific classifications of graph algorithms, the industry does not have a strict consensus. Some graph computing frameworks that focus on academic research even include breadth-first or depth-first searches as separate algorithms, although they are more suitable as a graph traversal mode. There are also shortest path algorithms. In the first 30 years of computer development in the 20th century, the shortest path algorithm was continuously revamped, which promoted the development of computer architecture. In this book, the shortest path algorithm is placed in the graph query section and not considered a genuine graph algorithm.

This book synthesizes the latest developments in the field of graph computing in academia and industry and divides graph algorithms into the following nine categories.

- **Degree**: such as vertex in-degree and out-degree, full graph in/out-degree.
- **Centrality**: such as closeness centrality, betweenness centrality, graph centrality.
- **Similarity**: such as Jaccard similarity, cosine similarity.
- **Connectivity**: such as strongly and weakly connected components, triangle calculation, induced subgraph, bipartite graph, MST, hyperANF.
- **Ranking**: such as classic PageRank, SybilRank, ArticleRank, and other algorithms.
- **Propagation**: such as label propagation LPA, HANP algorithm.
- **Community** Detection: such as Louvain algorithm, KNN, k-Means.
- Graph **Embedding**: such as random walk, FastRP, Node2Vec, Struc2Vec, GraphSage, LINE.
- Advanced/Complex Graph Algorithms: such as HyperANF, kNN, K-Core, MST.

The Essential Criteria of Graph Databases. https://doi.org/10.1016/B978-0-443-14162-1.00003-9

There are also some classifications starting from other dimensions, such as loop analysis, minimum spanning tree class, shortest path class, maximum flow, and other algorithms.

It should be pointed out that although classification helps us sort out knowledge, it is not static or exclusive. Some algorithms may span multiple classifications. For example, minimum spanning tree (MST) is both a connectivity algorithm and a complex graph algorithm. The collection of graph algorithms will continue to evolve rather quickly, and new ones are being introduced as you read this book.

Some algorithms made some assumptions when they were invented, but as the data changes, those assumptions are no longer suitable, such as the MST algorithm, its initial goal is to start from a vertex and use the edge with the smallest weight to connect all nodes associated with it. The algorithm assumes that the whole graph is connected (that is, there is only one connected component), but in many real scenarios, there are a large number of solitary vertices and multiple connected components; therefore, the algorithm must adapt to such situations—the algorithm call interface and parameters need to support multiple IDs, specify the attribute field corresponding to the weight, limit the number of returned result sets, and so on.

This chapter introduces the implementation logic of representative graph algorithms in the first eight categories and focuses on the explainability of graph algorithms in the last section.

4.1 Degree

Degree refers to the total number of edges connecting with a single vertex or batch of vertices on a graph data set, or the sum of a certain type of attribute on an edge.

Degree calculation has many subdivision scenarios, for example: out-degree calculation, in-degree calculation, in-n-out-degree calculation, weighted degree calculation; graph-wide in or out-degree calculation, etc.

Degree computing plays a vital role in scientific computing and feature extraction. The degree calculation for a vertex depends on the design and implementation of the specific graph database or graph computing framework and may be executed in real-time or batch processing; while the degree calculation for all vertices of the entire graph is mostly done in the form of tasks (that can be scheduled, pipelined or terminated). In graph computing, it is generally considered that degree computing is the simplest type of graph algorithm, and it is also an algorithm that is easy to implement in a distributed system to improve efficiency through horizontal expansion. After all, it is metadata-oriented shallow (≤ 1 layer) computing.

In addition to being returned to the client in real time, the calculation results can also be processed by updating the database in real time or writing back files to disk. Typical application scenarios include running degree calculations on the full graph of data, and writing back databases or files to pave the way for subsequent operations.

The interface design of degree algorithm is as follows:

- Write back to the database, with attributes that may be created or overwritten, such as #degree_right, #degree_left, #degree_bidirection or #degree_all.
- Algorithm call, the command is algo(degree).
- Parameter configuration, as shown in Table 1.

Example 1: Calculate a vertex's (UUID = 1) out-degree:

```
algo(degree).params({
    ids: [1],
    direction: "out"
}) as degree
return degree
```

Example 2: Calculate the degree centrality of a group of vertices, which is the sum of in-degree and out-degree, return results in descending order with statistical data (Fig. 1):

```
algo(degree).params({
    ids: [1,2,3,4,5],
    order: "desc"
}) as degree, stats
return degree, stats
```

TABLE 1 Degree algorithm parameters and specifications.

Name	Type	Spec	Description
ids / uuids	[]_id / []_uuid	/	ID/UUID of the nodes to calculate, calculate for all nodes if not set
edge_schema_property	[]@<schema>?.<property>	Must LTE	Edge properties to use for weighted degree
direction	string	in, out	in for in-degree, out for out-degree
limit	int	≥–1	Number of results to return, -1 to return all results
order	string	asc, desc	Sort nodes by the size of degree

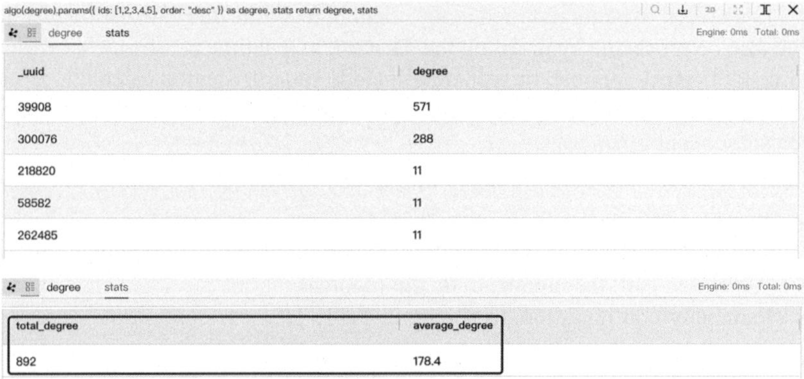

FIG. 1 Degree algorithm with results and stats.

Example 3: Calculate the degrees of all graph and write back to database property field degree which will be created automatically if not existent:

```
algo(degree).params({
}).write({
   db:{
      property: "degree"
   }
})
```

Example 4: Find 2-hop neighbors of the node with the highest degree (across the entire graph), return the name and age of those neighbors (Fig. 2):

```
algo(degree).params({
   order: "desc",
   limit: 1
}).stream() as results
khop().src({_uuid == results._uuid}).depth(2) as neighbors
return neighbors{name, age}
```

FIG. 2 Degree algorithm + K-Hop.

One important feature about white-box explainability in graph databases is the ability to validate the correctness of graph algorithm outputs. The following demonstrates how to verify the accuracy of the degree algorithm in a white-box manner through various methods.

Taking vertex (_id 222611) as an example, we can use template path query to explore how many 1-hop paths are connected with the vertex. This is done intuitively in Fig. 3; a total of 10 neighbors and 18 edges are connected. In the above Example 2, we have run the degree algorithm, and the degree for the vertex should have been written back into its property field—Fig. 4 shows that the vertex's degree field is populated with number 18.

We can also use the K-Hop query to validate the results. As shown in Fig. 5, we use both k-hop template query and regular k-hop to count the # of immediate neighbors of the vertex, and find both return number 10.

Fig. 6 shows the degree algorithm has been run three times, as the graph has not been changed, the total and average degrees stay put. In our case, the graph data set is Amazon 0601 (half-million vertices and 3.4 million edges), the whole-graph degree can be run in real-time as a task, and most of the latencies

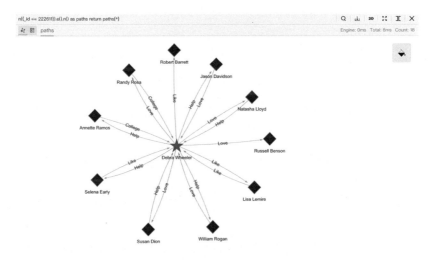

FIG. 3   Template path query to validate Degree Algo results.

_id	_uuid	degree	Ops
222611	1	18	☑ 🗑

FIG. 4   Find a vertex and its attributes (after running Degree).

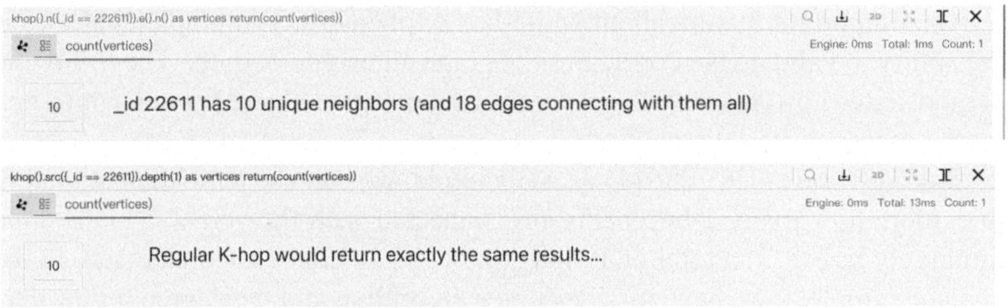
khop().n({_id == 22611}).e().n() as vertices return(count(vertices))
count(vertices)
Engine: 0ms Total: 1ms Count: 1
10
_id 22611 has 10 unique neighbors (and 18 edges connecting with them all)
khop().src({_id == 22611}).depth(1) as vertices return(count(vertices))
count(vertices)
Engine: 0ms Total: 13ms Count: 1
10
Regular K-hop would return exactly the same results...

FIG. 5 K-hop queries (template and regular) to validate Degree Algo results.

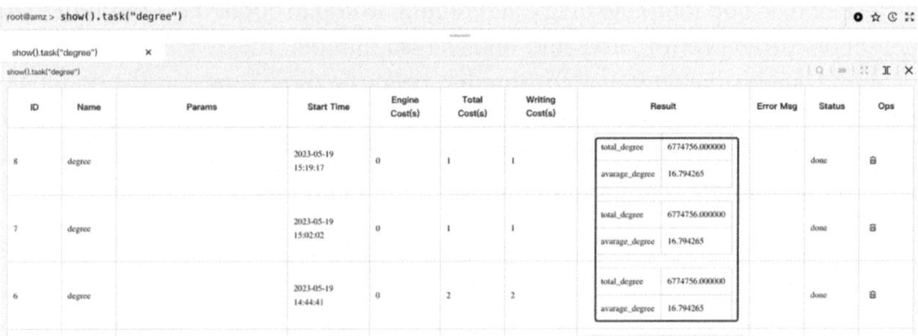

root@amz > show().task("degree")

ID	Name	Params	Start Time	Engine Cost(s)	Total Cost(s)	Writing Cost(s)	Result	Error Msg	Status	Ops
8	degree		2023-05-19 15:19:17	0	1	1	total_degree 6774756.000000 avarage_degree 16.794265		done	
7	degree		2023-05-19 15:02:02	0	1	1	total_degree 6774756.000000 avarage_degree 16.794265		done	
6	degree		2023-05-19 14:44:41	0	2	2	total_degree 6774756.000000 avarage_degree 16.794265		done	

FIG. 6 Degree algorithms run as tasks.

are on writing back to the database (disk-level write back). On larger graphs, the write back time would be proportionally longer.

Note that the degree algorithm is relatively easy to run in a highly parallel fashion as well as in a highly distributed environment given its simplistic logics and low computational complexity. The time complexity of degree algorithm against the entire graph is $\mathbf{O(|E|)}$, assuming E is the total number of edges in the graph. For running against a particular vertex, the complexity is O(1) assuming index-free adjacency type of data structure is used.

4.2 Centrality

Centrality computing refers to a large class of graph algorithms. In essence, centrality is to answer a key question—what are the characteristics of an important vertex? In order to better answer this question, the centrality algorithm performs quantitative calculations on vertices in the graph and sorts out vertices that are more "important". Most centrality calculations will measure the path(s) through each vertex. As for the scope and formation of the path(s), it is the logical details of these algorithms. Therefore, centrality calculations can be subdivided into many types, such as Degree Centrality, Graph Centrality, Closeness Centrality, Betweenness centrality, Eigenvector Centrality (Prestige Score or Eigencentrality), Cross-clique centrality, etc.

This section mainly introduces three algorithms: graph centrality, proximity centrality, and betweenness centrality.

4.2.1   Graph centrality algorithm

Graph centrality of a node is measured by the *maximum shortest distance* from the node to all other reachable nodes. This measurement, along with other measurements like closeness centrality and graph diameter, can be considered jointly to determine whether a node is literally located at the very center of the graph. Graph centrality takes on values between 0 and 1, nodes with higher scores are closer to the center. Graph centrality score of a node defined by this algorithm is the inverse of the maximum shortest distance from the node to all other reachable nodes. The formula is:

$$C_G(x) = \frac{1}{\max\limits_{y} d(x,y)}$$

The logic of the graph centrality algorithm is very simple. Taking the tiny graph in Fig. 7 as an example, the graph centrality values of each vertex are as follows.

- Vertex 1: 1/4=0.25
- Vertex 2: 1/3=0.3333
- Vertex 3: 1/2=0.5
- Vertex 4: 1/4=0.25
- Vertex 5: 1/3=0.3333

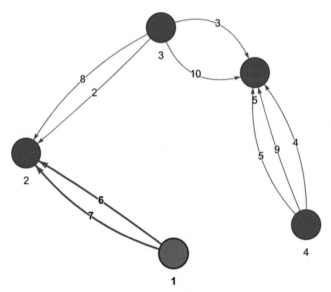

FIG. 7   Centrality sample graph for correctness verification.

The focus of the graph centrality algorithm is to calculate the longest and shortest path from any vertex. If it is in a highly connected graph, or the whole graph has only one connected component, then finding the longest and shortest path from any vertex would involve processing all vertices (and edges), so theoretically the complexity of the algorithm running against a single vertex is $O(|V| + |E|)$, where V is the number of all vertices and E is the number of all edges, and the lowest complexity of running the algorithm against the entire graph would be: $O(|V| * (|V| + |E|))$, which can be formidable in graphs with 10,000 plus entities and edges.

4.2.2 Closeness centrality

Closeness centrality of a node is measured by the *average shortest distance* from the node to all other reachable nodes. The closer a node is to all other nodes, the more central the node is (in the graph). This algorithm is widely used in applications such as discovering key social nodes and finding best locations for functional places. Similar to graph centrality, closeness centrality also takes on values between 0 and 1, and nodes with higher scores have shorter distances to all other nodes, therefore more centrally positioned.

The original closeness centrality score of a node is defined by the inverse of the sum of the shortest distances from the node to all other reachable nodes, with the following being:

$$C(x) = \frac{1}{\Sigma_y d(x, y)}$$

Since the value obtained by such formula is usually relatively small, the actual implementation will improve the readability of the result by calculating the arithmetic mean of all the shortest paths. The revamped formula is as follows:

$$C(x) = \frac{k - 1}{\Sigma_y d(x, y)}$$

With k being the number of all vertices in the current connected component, in a graph that is fully connected or has only one connected component, $k = V$, V is the number of vertices in the graph.

Taking Fig. 7 as an example, the closeness centrality values of each vertex are as follows.

- Vertex 1: $4/(1 + 2 + 3 + 4) = 0.4$
- Vertex 2: $4/(1 + 1 + 2 + 3) = 0.571429$
- Vertex 3: $4/(2 + 1 + 1 + 2) = 0.666667$
- Vertex 4: $4/(4 + 3 + 2 + 1) = 0.4$
- Vertex 5: $4/(1 + 1 + 2 + 3) = 0.571429$

Similar to the degree calculation, multiple limit parameters can also be set in the closeness centrality algorithm (see Table 2).

Since it is necessary to calculate the shortest path distance from the current vertex to all other vertices, the computational complexity is at least $O(|V| * (|V| + |E|))$, which is far greater than $O(|V|^2)$. Closeness centrality tends to consume a considerable amount of computational resources therefore making it impractical on larger graph. To mitigate such challenges, sampling-based approximation method can be used when calculating the closeness centrality of a graph with more than a certain number of vertices (e.g., 1000). For example, if the number of sampled vertices is the logarithm of the total number of vertices, in a graph with 10,000,000 vertices, only seven vertices are sampled—however, this still means that 70,000,000 shortest path queries are to be conducted. If on average each shortest path query takes 1 ms (most graph databases today cannot achieve such performance), 70 million ones would still take 70,000 s, which is about 20 h. These numbers reflect the sheer computing complexity of closeness complexity. In real-world applications, it's very rare for graphs sized over 100,000 vertices to run closeness centrality—at the time the algorithm was invented, back in the 1950s, the graph size was most likely smaller than 100 vertices and no more than a few hundred edges.

TABLE 2 Closeness centrality parameters and specifications.

Name	Type	Spec	Description
ids / uuids	[]_id / []_uuid	/	ID/UUID of the nodes to calculate, calculate for all nodes if not set
direction	string	in, out	Direction of all edges in each shortest path, in for incoming direction, out for outgoing direction
sample_size	int	−1,−2, [1, |V|]	Number of samples to compute centrality scores; -1 samples log(|V|) nodes, -2 performs no sampling; sample_size is only valid when ids (uuids) is ignored or when it specifies all nodes
limit	int	≥−1	Number of results to return, -1 to return all results
order	string	asc, desc	Sort nodes by the centrality score

4.2.3 Betweenness centrality

Betweenness centrality measures the probability that a node lies in the shortest paths between any other two nodes. Proposed by Linton C. Freeman in 1977, this algorithm detects the "bridge" or "medium" nodes between multiple parts of the graph. Betweenness centrality takes on values between 0 and 1, nodes with larger scores have stronger impact on the flow or connectivity of the network. The score of a node is defined by:

$$C_B(x) = \frac{\left(\sum_{i \neq j \neq x \in V} \frac{\sigma_{ij}(x)}{\sigma_{ij}} \right)}{\left(\frac{(k-1)(k-2)}{2} \right)}$$

where x is the target node, i and j are two distinct nodes in the graph (x itself is excluded), σ is the number of shortest paths of pair ij, $\sigma(x)$ is the number of shortest paths of pair ij that pass through x, $\sigma(x)/\sigma$ is the probability that x lies in the shortest paths of pair ij (which is 0 if i and j are not connected), k is the number of nodes in the graph, $(k-1)(k-2)/2$ is the possible number of ij node pairs (in a simple-graph setup).

Still taking the graph in Fig. 7 as an example to calculate the betweenness centrality, we can use vertex No. 3 as an example.

- There are six combinations in total, the number of combinations is 4 * 3/2=6.
- The number of shortest paths between pairwise combinations as follows:
  ○ Vertex 1, 2: # of shortest paths is 2.
  ○ Vertex 1, 5: # of shortest path is 8.
  ○ Vertex 1, 4: # of shortest path is 24.
  ○ Vertex 2, 5: # of shortest path is 4.
  ○ Vertex 2, 4: # of shortest path is 12.
  ○ Vertex 4, 5: # of shortest path is 3.
- The probability that vertex 3 is on the shortest path(s) of any pair of vertices is as follows:
  ○ Vertex1, 2: probability is 0.
  ○ Vertex1, 5: probability is 1.
  ○ Vertex1, 4: probability is 1.
  ○ Vertex2, 5: probability is 1.
  ○ Vertex2, 4: probability is 1.
  ○ Vertex4, 5: probability is 0.
- The betweenness centrality of vertex 3=4/6=0.6666667, which is completely consistent with the running results of the algorithm shown in Fig. 8.

The computational complexity of betweenness centrality is higher than closeness centrality. In large and highly connected graph, the complexity of

betweenness centrality of all nodes	
node	betweenness_centrality
2	0.500000
1	0.000000
3	0.666667
4	0.000000
5	0.500000

FIG. 8 Betweenness centrality algorithm results.

betweenness centrality for a single vertex is $O((|V| * (|V|-1)/2) * (|V| + |E|)/2) \sim = |V|^3$, such complexity is beyond the computing power of any graph databases on today's computing hardware. The option(s) left for the algorithm to be practical is smart approximation via *sampling, coupled with high-concurrency algorithm implementation, and lowering the complexity of the graph* (or reduce to exponentially smaller graphs). It should be noted that using sampling calculations on small graphs will risk lowering the accuracy rate. For example, if sampling is performed on the graph in Fig. 7, only one pair of vertices may be calculated, and the result is shown in Fig. 9, which is significantly off from the centrality results for vertices 2, 3, and 5 in Fig. 8.

Fig. 10 shows how this daunting betweenness centrality algorithm is practical with Ultipa real-time graph database—300 vertices are sampled, each is to be processed against the entire AMZ data set of 400,000 vertices, and still manage to return in seconds. The secret sauces behind this feat mainly are high-density parallel computing, and algorithm optimization geared toward lowered computing complexity, specifically, each sampled vertex can run an exhaustive K-hop to figure out its connectivity (distance from) with all the other vertices, plus the connectivity in between all the other (pairs of) vertices.

betweenness centrality of all nodes	
node	betweenness_centrality
2	1.000000
1	0.000000
3	2.000000
4	0.000000
5	0.000000

FIG. 9 Downside of sampling approximation for betweenness centrality.

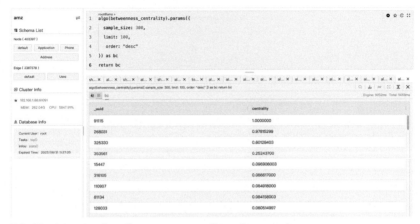

FIG. 10 Approximation of betweenness centrality on larger graph.

The disadvantages of centrality algorithm are as obvious as its simple and easy-to-understand advantages. There are two main disadvantages:

(1) The conclusions drawn by different centrality algorithms may be completely different, that is, the vertices of the maximum centrality may be completely different when operating different centrality algorithms in a graph.

(2) Some centrality algorithms have very high computational complexity and are not suitable for running in large graphs. This also means that the algorithm is limited to academic research and does not have the feasibility of real-world engineering application. To mitigate this, some approximation mechanisms are derived to greatly reduce the complexity of the algorithm. Of course, the approximation algorithm is not limited to the centrality algorithm, but also includes many algorithms such as community identification and graph embedding.

In the 1990s, David Krackhardt, a social network theory researcher at Carnegie Mellon University in the USA, drew a simple graph like a kite, also known as Krackhardt Kite Graph, as shown in Fig. 11, where running different centrality algorithms gives different results.

- Degree centrality: Vertex 3 has the highest degree value 6.
- Betweenness centrality: Vertex 7 has the greatest betweenness centrality.
- Closeness centrality: Vertices 5 and 6 have the greatest closeness centrality.

4.3 Similarity

Similarity algorithms are widely used to quantify and compare the similarity between two entities. In the context of graph theory, the similarity can be

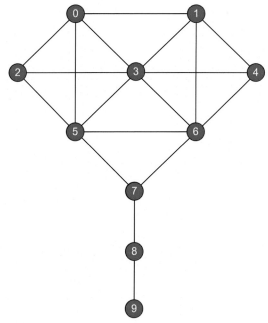

FIG. 11   Krackhardt's kite graph.

regarded as the reciprocal of the distance function of two entities in the topological space, and the value interval can generally be designed as [0,1]. The similarity between two completely similar entity is 1 (100%), while completely
different entities have a similarity of 0 (0%).

There are many types of similarity algorithms, such as Jaccard similarity,
Cosine similarity, Semantic similarity, and others.

4.3.1 Jaccard similarity

Jaccard similarity is often called Jaccard similarity coefficient and used to compare the similarity and diversity of sample data sets. The specific formula is as
follows:

$$J(A,B) = \frac{|A \cap B|}{|A \cup B|} = \frac{|A \cap B|}{|A| + |B| - |A \cap B|}$$

The definition of the Jaccard similarity algorithm is to calculate the result of
dividing the intersection and union between two given sets (Fig. 12). The larger
the result value (≤ 1), the higher the similarity. The smaller the value (≥ 0), the
lower the similarity. In the graph, the 1-hop (direct association) neighbors of the
queried vertex are usually used as a set (but not including the queried vertex) to

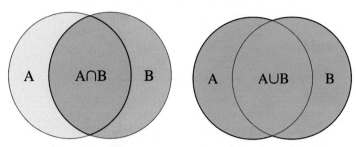

FIG. 12 Jaccard similarity illustrated.

perform operations, that is, the number of common neighbors between the two queried vertices divided by their total neighbors.

We can design a variety of calls for Jaccard similarity, such as calculating the similarity between a pair of nodes; calculating the similarity between multiple pairs of nodes; comparing a node to the whole graph to find the most similar node, sorting and limiting the number of returns, etc.

There are many application scenarios of Jaccard similarity, ranging from recommendation to antifraud. The multipath subgraph query shown in Fig. 28 from Chapter 3 can also be realized by calculating the similarity between two credit application vertices, and the efficiency is very high. If a hash-map table is used, the algorithm complexity is $O(M)$, where M is the set with the larger number of the neighbor set of the two vertices. In fact, a high-performance graph database can complete Jaccard similarity calculations in genuine real-time (milliseconds or even microseconds), and the performance can be independent of the size of the graph, because Jaccard calculations are topologically confined to neighbors, even though it is theoretically to find the most similar vertices in the whole graph, the algorithm complexity is only proportional to the 2-hop neighbors of the subject vertex.

Example 1: Taking Fig. 7 as an example, calculate the similarity between set (ID = 1,2,3) and the set (ID = 4,5). The Jaccard similarity between any two points, sort the results in descending order, and return all the results.

```
algo(similarity).params({
  ids: [1,2,3], ids2: [4, 5],
  type: "jaccard"
})as jacc
return jacc order by jacc.similarity desc
```

According to the definition of the Jaccard similarity algorithm, we know that the common neighbor between vertices 3 and 4 is vertex 5, and all their 1-Hop neighbors are vertices 2 and 5, so the calculation result is $1/2 = 0.5$, similarly,

jaccard similarity result		
node1	node2	similarity
3	4	0.500000
2	5	0.333333
1	4	0.000000
1	5	0.000000
2	4	0.000000
3	5	0.000000

FIG. 13 Jaccard similarity of two sets of entities.

the similarity between vertices 2 and 5 is 1/3 = 0.33333. All calculation results are shown in Fig. 13.

Example 2: For the set of vertices in Fig. 7 (ID = 2,3) for each point, calculate all the vertices most similar to the point and the similarity.

```
algo(similarity).params({
    ids: [2,3],
    type: "jaccard"
})as top
return top
```

Taking vertex 3 as an example, it is directly adjacent to vertices 2 and 5, and there is no common neighbor, so the similarity must be 0; but the similarity between vertices 1 and 3 is 1/2 = 0.5; vertices 4 and 3's similarity is also 0.5 (Fig. 14).

3 top jaccard similarity result	
node_id	similarity
1	0.500000
4	0.500000
2	0.000000
5	0.000000

2 top jaccard similarity result	
node_id	similarity
5	0.333333
1	0.000000
3	0.000000

FIG. 14 Graph-wide Jaccard similarity for vertices.

Jaccard similarity algorithm is very suitable for large-scale concurrency, and the similarity calculation for any pair of nodes or the entire graph of any node can be completed with a very low latency (single digit millisecond level). Performance improvement can be achieved with the utilization of high-concurrency computing and low-latency-n-low-complexity data structures. In other words, if serial execution takes 1 s, 10 concurrent threads can make it in about 100 ms, and 40 concurrent threads can sweeten the pot in 25 ms, and effective data structures can accelerate things for another 10-folds or more.

4.3.2 Cosine similarity

In cosine similarity, data objects in a data set are treated as vectors, and it uses the cosine value of the angle between two vectors to indicate the similarity between them. In the graph, specifying N numeric properties (features) of nodes to form N-dimensional vectors, two nodes are considered similar if their vectors are similar. The prerequisite for understanding cosine similarity is to understand basic concepts such as the dot product (also known as the scalar product and inner product, as shown in Fig. 15) in the Euclidean set.

We usually normalize the similarity result, so that it is between -1 and 1, and the larger the result, the higher the similarity.

The calculation formula of cosine similarity is as follows. On the graph data set, it can be decomposed into the calculation of the inner product between the two entities A and B, and the similarity value can be deduced from this. The specific logic is that the attribute field of A and the attribute field of B participate in the operation, where the numerator part is the dot product—the attribute 1 of A multiplied by the attribute 1 of B, and so on, and sums all the products; the denominator part is the magnitude for A and B.

$$\text{similarity} = \cos(\theta) = \frac{\mathbf{A} \cdot \mathbf{B}}{\|\mathbf{A}\|\|\mathbf{B}\|} = \frac{\sum_{i=1}^{n} A_i B_i}{\sqrt{\sum_{i=1}^{n} A_i^2}\sqrt{\sum_{i=1}^{n} B_i^2}},$$

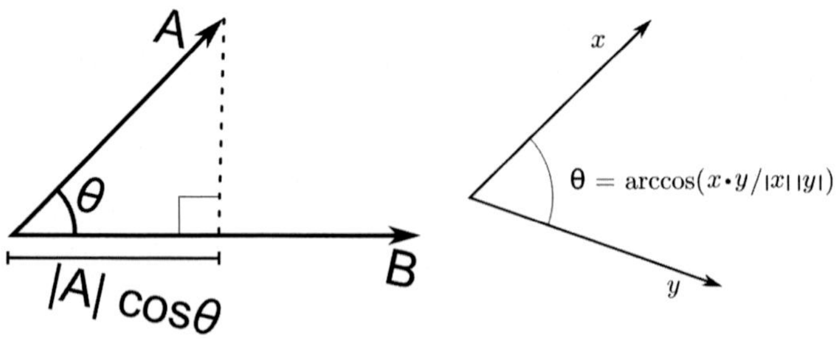

$$\theta = \arccos(x \cdot y / |x||y|)$$

FIG. 15 Inner product (dot product) of vector space.

ID(node-result)	_o	#betweenness_centrality	#degree_all	#out_degree_all	#in_degree_all	操作
1	ULTIPA800000000...	0	2	2	0	✎ 🗑
2	ULTIPA800000000...	1	4	0	4	✎ 🗑
3	ULTIPA800000000...	2	4	4	0	✎ 🗑
4	ULTIPA800000000...	0	3	3	0	✎ 🗑
5	ULTIPA800000000...	3	5	0	5	✎ 🗑

FIG. 16 Vector attributes for cosine similarity calculation.

Still taking Fig. 7 as an example, we can calculate the cosine similarity between vertex 1 and vertex 2, using three attribute fields: #in_degree_all, #degree_all, #out_degree_all as shown in Fig. 16.

According to the above cosine similarity calculation formula, vertex 1 is $A = (0,2,2)$, vertex 2 is $B = (4,0,4)$, then the numerator $A \cdot B = 0+0+8 = 8$; the denominator $||A|| \cdot ||B|| = \text{sqrt}(0+4+4) * \text{sqrt}(16+0+16) = 16$; thus, the cosine similarity value is 0.5.

Let's use a real-world scenario to illustrate the usefulness of Cosine similarity. Fig. 17 has a matrix of products and their properties, to find out which product is most similar to a queried product, we can use Cosine similarity:

```
algo(similarity).params({
    uuids: [1,2,3,4],
    node_schema_property: [price,weight,width,height],
    type: "cosine",
    top_limit: 1
}).stream() as top
return top
```

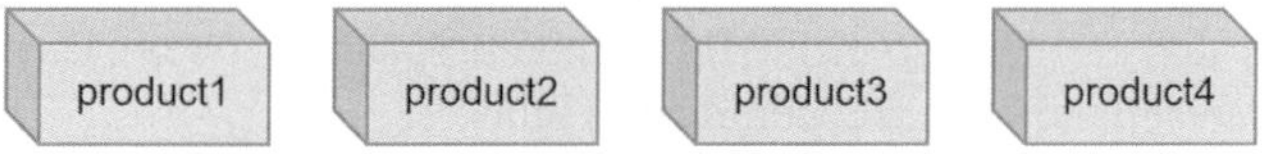

product1 product2 product3 product4

Property	product1	product2	product3	product4
_uuid	1	2	3	4
price	50	42	24	38
weight	160	90	50	20
width	20	30	55	32
height	152	90	70	66

FIG. 17 Three-attribute cosine similarity calculation results.

Note that the calculation of Cosine similarity for any two nodes does not depend on their connectivity (unlike Jaccard similarity).

Cosine similarity has a wide range of applications in the fields of AI and information mining. For example, comparing the similarity of two documents, the simplest calculation method is to use the key words and frequency of occurrence in each document as attributes and attribute values respectively. For the calculation of cosine similarity, it is known from the above detailed algorithm decomposition that its algorithm complexity is equivalent to the Jaccard similarity, and it can be calculated in real time.

4.4 Connectivity calculation

Connectivity is a very important concept in the subject of graph theory, and the connectivity of graph data is involved in operations research and network flow research. The vast majority of academic studies on the connectivity of graphs focus on how to cut the graph. To be precise, when at least a few vertices are removed, two or more subgraphs that cannot communicate with each other will be formed between the remaining vertices (connected components). Therefore, the connectivity reflects the segmentation resistance of the network formed by the graph data. Menger's theorem is an important theorem in the study of graph connectivity. Its core idea (conclusion) is, in a graph, the minimum cut is equal to any maximum number of disjoint paths that can be found between all pairs of vertices.

Fig. 18 illustrates the minimum and maximum cut of a graph data network. Among them, the sum of the edges (weights) of the minimum cut set is equal to the maximum flow between the two subgraphs after cutting—this is also the "Max Flow Minimum Cut Theorem" in network flow optimization theory.

There are many dimensions and algorithms for measuring the connectivity of a graph. Common ones include connected components (divided into weakly connected and strongly connected), minimum spanning tree, triangle count, triple-connected vertices, and common neighbors.

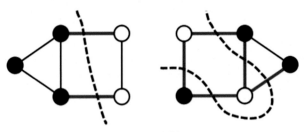

FIG. 18 Minimum cut (left) and maximum cut (right).

4.4.1 Connected component

Connected components are often referred to as components. In an undirected graph, a connected component means that any two vertices in the component have a connected path. If there is only one connected component in a graph, it means that all nodes are connected to each other. The counting logic of connected components is very simple, that is, to count the number of connected Max Subgraphs in the current graph. If the current graph is a fully connected graph, then the connected component is 1. If there are isolated vertices (without any associated edges), each isolated vertex is considered an independent component.

The above definition of a connected component is a weakly connected component (WCC). If it is changed to a directed graph and requires a one-way path from any vertex to another vertex, then the largest subgraph found at this time is a strongly connected component, (SCC). SCC is computationally more expensive than WCC.

Taking the AMZ data set as an example, running connected component algorithm in WCC and SCC mode separately will subjugate the results as shown in Fig. 19: WCC = 7, SCC = 1589, the algorithm can be configured to write back to database (or file system) so that each vertex is attached with a CC attribute indicating which component it belongs to.

On the basis of Fig. 7, we added a vertex 6 and multiple edges, as shown in Fig. 20. At this time, the result of SCC is 3 (Fig. 21)—it's visually clear that vertices 1-4-5-6 are strongly connected, in which there can be two-way

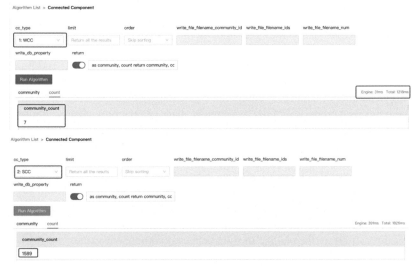

FIG. 19 WCC and SCC.

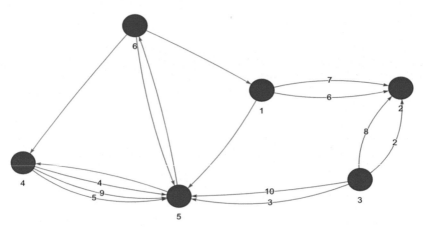

FIG. 20 Schematic diagram of 6-vertex graph.

ID(node-result)	_o	#betweenness_c entrality	#degree_all	#out_degree_all	#in_degree_all	#connected_com ponent
1	ULTIPA800000...	0	2	2	0	6
2	ULTIPA800000...	1	4	0	4	2
3	ULTIPA800000...	2	4	4	0	3
4	ULTIPA800000...	0	3	3	0	6
5	ULTIPA800000...	3	5	0	5	6
6	ULTIPA800000...		0	0	0	6

FIG. 21 SCC results of the full atlas after the connectivity transformation.

connected paths between vertices 1, 4, 5, and 6, whereas vertices 2 and 3 are relatively loosely coupled.

The number of connected components in a statistical graph is an exploration of the intrinsic properties of the topological space of the graph, that is, topological invariant or topological property, or topological characteristic. From a certain point of view, for a graph, if the number of connected components does not change, it can be considered that its topological properties have not changed. Of course, the connected component is a relatively coarse-grained measurement method. It is more about the characteristics of a whole set of graph data from a macro perspective, so it cannot effectively reflect the microscopic changes in the neighbor relationship around each vertex. The change of the neighbor relationship can affect the change of the connected component value.

The calculation of connected components can be achieved using breadth-first or depth-first traversal methods—starting from any vertex, count all the associated vertices in its component. If visiting each vertex's neighbors takes O(1) time, then computing connected components takes O($|V|$) time, where V is the total number of vertices in the graph.

4.4.2 Minimum spanning tree

Minimum spanning tree or minimum weight spanning tree (MST) calculates the minimum connected subgraphs of each connected component of a graph and returns the edges in these minimally connected subgraphs in the form of path. MST is one of the fundamental concepts in graph theory that has important applications in path optimization and identifying lowest cost solutions. The earliest known MST algorithm is Boratvka algorithm, published by Czech mathematician Otakar Boratvka in 1926 when the electrical wiring of the Czech city of Moravia was carried out. The MST algorithm was not known to the world until it was proposed by the Frenchman Georges Sollin in 1965.

For a disconnected graph, traversing the complete MST requires specifying a starting vertex for each connected component in the graph. The algorithm starts from the first starting vertex, and after the MST is fully generated in the current connected components, it goes to the next starting vertex and continues to generate MST until all starting vertices are calculated. If there are multiple starting vertices belonging to the same connected component, only the first calculated starting vertex is valid. Solitary vertex isn't concerned for MST calculation. Different MSTs may exist for one graph.

Note: For a connected graph (WCC = 1), the number of edges contained in its MST is equal to the number of vertices in the graph minus 1. If WCC > 1, all connected components are counted for MST, the result is called minimum spanning forests (MSFs). In addition, MST calculation involves weights, taking the attribute field of edges a weight, if a field is empty, a default weight of 1 is assumed for the edge (Table 3).

Example: Invoke the MST algorithm in a graph with multiple connected components, start from the multivertex set IDs = [1,5,12], use the edge attribute @edgx.rank as the weight to generate the MST, stream the output which is the entire list of edges, find all of these edges and aggregate (sum) by their rank attribute: a possible GQL code is as follows:

```
// MST algorithm call
algo(mst).params({
    uuids: [1,5,12],
    edge_schema_property: @edgx.rank,
    /* limit: 10000 */
}).stream() as mst
with pedges(mst) as mstUUID
find().edges(mstUUID) as edges
return sum(edges.rank)
```

TABLE 3 MST parameters and specification.

Name	Type	Specification	Description
ids / uuids	[]_id / []_uuid	/	IDs or UUIDs of the start nodes of MST; all nodes to be specified if not set
edge_schema _property	[]@<schema>?. <property>	Numeric edge property, LTE needed; mandatory	Edge weight property/properties, schema can be either carried or not; edge with multiple specified properties is calculated by the property with the lowest weight; edge without any specified property does not participate in the calculation
limit	int	>=−1	Number of results to return; return all results if sets to -1 or not set

MST has a wide range of application scenarios in the industries, such as electric power grid cabling, transportation route planning, road network construction, city pipeline laying, communication network laying. For example, in architectural design, a house needs to be prelaid with network cables. How to lay the minimum distance (weight) of network cables from the outdoor hub (or indoor router) to all the rooms that need to be covered without forming a loop? It's a typical MST question.

The MST algorithm has a large span of algorithm complexity in different data structures and implementations (Table 4), and it is generally believed that

TABLE 4 MST complexity under different data structures.

Primary data structures	MST complexity (time)
Adjacency matrix	$O(V * V)$
Adjacency linked list and heap	$O(E * Log\ V)$ or $O(E + V * Log\ V)$ $O(E + V * Log\ V)$
Queue, vector or hash	$\geq O(E+(V*Log\ V/P))$ Where P is the maximum concurrency scale

its reasonable time complexity is O(E * Log V), where E is the number of edges, and V is the number of vertices. For large graphs (10 millions and above), the calculation of this algorithm takes a long time, so research on making the algorithm to run in parallel started early. The earliest parallel MST algorithm is the Prim algorithm, which is based on the Boratvka algorithm and transforms it into parallel style.

It should be noted that, taking Prim's MST algorithm as an example, it cannot be completely parallelized. The following pseudocode describes the flow of Prim's algorithm, and the fourth part can achieve concurrency. In the algorithm time complexity, we ignore the actual intercluster network communication and latency.

```
// Concurrent MST pseudocode
0. Suppose the number of vertices is V, and the number of edges is E;
1. Suppose K processors (threads) for concurrent processing, P(1)to P(k);
2. Assign a subset of vertices (V/P) to each processor;
3.1. Each vertex maintains a min-weight value C[v], and matching edge ID(E[v]),
     Initialize C[v] as infinite, initialize E[v] as -1 meaning no edge is visited yet;
3.2. Initialize two data structures F (empty) and Q (all vertices);
4. While loop until Q is empty: //The concurrency scale of this part≤P
     Remove a vertex v from Q when it is the lowest possible C[in]value;
     /* In concurrent mode, broadcast selected vertices to all processors*/
     Add that v into F, and if E[v]! =-1, add E[v] to F;
     For Vertex w corresponding to each edge e among all associated edges of v:
       If w belongs to Q, and e's weight is less than C[w]:
         C[w] = weight of e;
         E[w] = ID of e // point to edge e
   Return F
```

In Fig. 22, we run the MST algorithm against AMZ data set (4 million vertices and edges) with maximum 64-thread for parallel execution, and we were able to finish the calculation in near real-time (1 s for pure graph computing and 7 s for file-system write back).

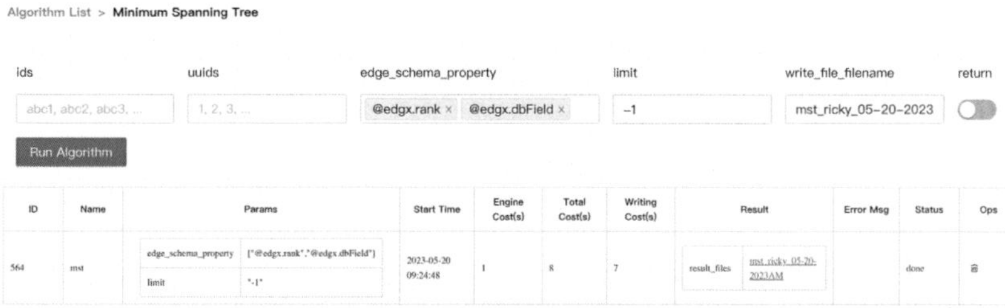

Algorithm List > Minimum Spanning Tree

ids	uuids	edge_schema_property	limit	write_file_filename	return
abc1, abc2, abc3, ...	1, 2, 3, ...	@edgx.rank × @edgx.dbField ×	−1	mst_ricky_05–20–2023	

Run Algorithm

ID	Name	Params	Start Time	Engine Cost(s)	Total Cost(s)	Writing Cost(s)	Result	Error Msg	Status	Ops
564	mst	edge_schema_property ["@edgx.rank","@edgx.dbField"] limit "-1"	2023-05-20 09:24:48	1	8	7	result_files mst_ricky_05-20-2023.AM		done	

FIG. 22   Near-real-time MST on million-scale graph data set.

4.5 Ranking

Graph data ranking (or sorting) algorithms have been widely used in social networks, WWW web page link structure analysis, citation analysis, and other scenarios. Some of the more well-known algorithms include PageRank,[a] SybilRank,[b] Article Rank, Hyperlink-Induced Topic Search (HITS), and Positional Power Function (PPF) .

The above algorithms, except SybilRank which appeared rather late (in 2012), all the other algorithms appeared between 1999 and 2001, and are used to rank pages on the WWW. This section mainly introduces the most representative web page ranking algorithm and Sybil ranking algorithm.

4.5.1 PageRank

PageRank was originally proposed in the context of WWW; it takes advantage of the link structure of WWW to produce a global objective "importance" ranking of web pages that can be used by search engines. This algorithm was first published as a paper in 1998 by Google cofounders Larry Page and Sergey Brin.

PageRank is an iterative algorithm that transfers the nodes' scores along the direction of the directed edges in a directed graph until a convergent score distribution is obtained. In each iteration, the node's current score is evenly distributed to the neighbor node connected to the outgoing edge of the node, the node receives the score from the neighbor node connected to the incoming edge, and the received score is added to the node's score in this round of iteration.

The ranks (scores) of all pages are computed in a recursive way by starting with any set of ranks and iterating the computation until it converges. In each iteration, a page gives out its rank to all its forward links evenly to contribute to the ranks of the pages it points to; meanwhile every page receives ranks from its backlinks, so the rank of page u (where B_u is the backlink set of u) after one iteration is:

$$\text{rank}(u) = \sum_{v \in B_u} \frac{\text{rank}(v)}{\text{outdegree}(v)}$$

To overcome the problems of web pages having no backlinks or forward links, a damping factor, whose value is between 0 and 1, is introduced. It gives each web page a base rank while weakening the ranks passed from backlinks. The rank of page u after one iteration becomes:

$$\text{rank}(u) = d \sum_{v \in B_u} \frac{\text{rank}(v)}{\text{outdegree}(v)} + (1 - d)$$

---

a. The PageRank Citation Ranking http://ilpubs.stanford.edu:8090/422/1/1999-66.pdf.
b. SybilRank https://users.cs.duke.edu/~qiangcao/sybilrank_project/index.html.

TABLE 5 PageRank parameter and specification.

Name	Type	Spec	Default	Optional	Description
init_value	float	>0	0.2	Yes	The same initial rank for all nodes
loop_num	int	>=1	5	Yes	Number of iterations
damping	float	(0,1)	0.8	Yes	Damping factor
weaken	int	1, 2	1	Yes	For PageRank, keep it as 1; 2 means to run ArticleRank
limit	int	≥−1	−1	Yes	Number of results to return, −1 to return all results
order	string	asc, desc	/	Yes	Sort nodes by the rank

Damping factor can be seen as the probability that a web surfer randomly jump to a web page that is not one of the forward links of the current web page.

PageRank is a typical semisupervised graph algorithm, though it can be run completely in unsupervised fashion—meaning the algorithm can run completely on itself without human intervention. Table 5 lists all possible parameters to invoke the algorithm, when no parameter is specified, the algorithm goes into automated (unsupervised) mode. Most of the graph algorithms belong to either unsupervised or supervised category, which make them suitable to work with AI, machine learning and deep-learning frameworks to augment and accelerate things–we'll get to this more in Chapter 6 when we talk about real-world graph applications.

Example: In the small graph of Fig. 23, execute PageRank, set the number of iterations to 5, the damping coefficient to 0.8, and the default score to 1, and write back to database as vertex attributes (new property field "PR" will be automatically generated). The GQL code as below:

```
/* PageRank with customized input parameters */
algo(page_rank).params({
    loop_num: 5,
    damping: 0.8,
    init_value: 1
}).write({
    db:{property: "PR"}
})
```

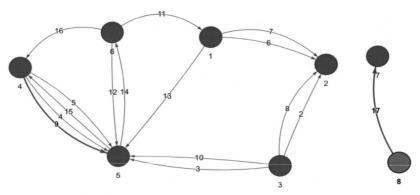

FIG. 23 Small graph for PageRank.

ID(node-result)	#page_rank
1	0.417837037037037
2	0.4866646913580246
3	0.19999999999999996
4	0.961908148148148
5	1.4046775308641974
6	0.744071111111111
7	0.35999999999999993
8	0.19999999999999996

FIG. 24 Write-back result of PageRank.

The calculation results are shown in Fig. 24. Vertices 3 and 8 are the easiest to judge. Since these two vertices have no incoming edges, their PR values are $(1-0.8)=0.2$. The page_rank values of the other vertices need to go through 5 rounds of iterations. For example, Vertex 7's $PR=1-0.8+0.8*0.2=0.36$, which is consistent with the result shown in Fig. 24.

The calculation of PageRank involves the iteration of the whole graph data, and its algorithm complexity is $\mathbf{O(K * |V| * |E| / |V|)}$, where V is the number of vertices, E is the number of edges, E/V can be regarded as the average degree of vertices, and K is the number of iterations. On a large graph (≥ 10 million vertices and edges), it needs to be executed asynchronously in the form of tasks (to avoid cluttering system resources and operational interfaces), and returning values can be written back to the database vertex attribute fields or files. The PageRank algorithm pioneers early implementations of large-scale distributed (graph) computing. In the next chapter, we will introduce in detail how to boost the efficiency of various graph queries and calculations under distributed architectures.

4.5.2 SybilRank

The SybilRank algorithm originally refers to an attack mode called Sybil Attacks initiated by false accounts in online social networks (OSN). Malicious users launch attacks by registering multiple false accounts, causing vicious problems such as the inability to provide services in OSN, and then threatening OSN service providers to pay ransom to avoid total system crash. Due to the rapid development of social networks, sybil attacks are increasing day by day. The SybilRank algorithm can help social platforms to be more efficient in locating false accounts to prevent malicious activities like fake news or contents flooding, harassment, etc.

Fig. 25 illustrates the difference in the "community" formed between normal accounts and fake accounts. There is only a relatively limited relationship between normal accounts and fake accounts, while there may be a relatively closer relationship within the set of fake accounts. The implementation of the SybilRank algorithm is to sort OSN users by performing short-path random walks from credible seed nodes, and screen out the accounts with the lowest scores, which are high-probability potential fake users. Like the PageRank algorithm, the SybilRank algorithm also uses the power iteration method (Power Iteration) to obtain convergence after several rounds of iterative calculations.

The input parameters and specification for SybilRank are shown in Table 6.

Example: Set a maximum of five iterations, credible vertices 3 and 4 in Fig. 26 are used as trustable seeds, intending to find all fake accounts. The GQL code is as below:

```
algo(sybil_rank).params({
  total_trust: 100, /* preferred total ranking "score" */
  trust_seeds: [3,4],
  loop_num: 5
}) as trust
return trust
```

From the above results, it can be known that vertices 2, 7, 8, and 9 are considered Sybil accounts. In small graph, the calculation results may be off hugely

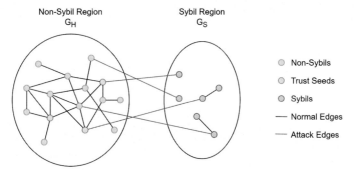

FIG. 25 Schematic diagram of normal account and Sybil accounts.

TABLE 6 SybilRank parameters and specification.

Name	Type	Spec	Default	Optional	Description
total_trust	float	>0	/	No	Total trust of the graph
trust_seeds	[]_uuid	/	/	Yes	UUID of trust seeds, it is suggested to assign trust seeds for every community; all nodes are specified as trust seeds if not set
loop_num	int	>0	5	Yes	Number of iterations, it is suggested to set as log(|V|) (base-2)
limit	int	>=-1	−1	Yes	Number of results to return, -1 to return all results

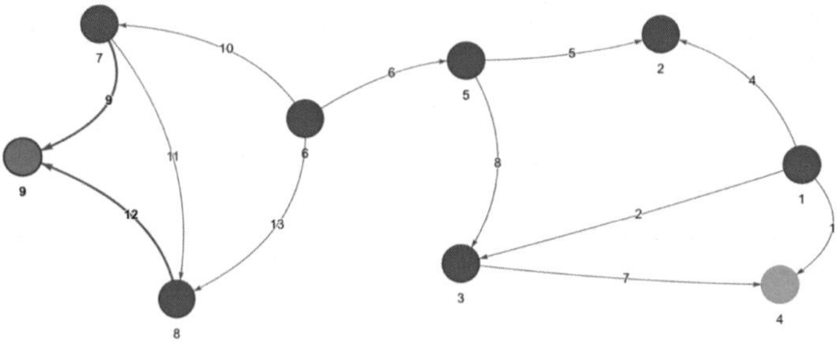

a) Small Graph for SybilRank Test

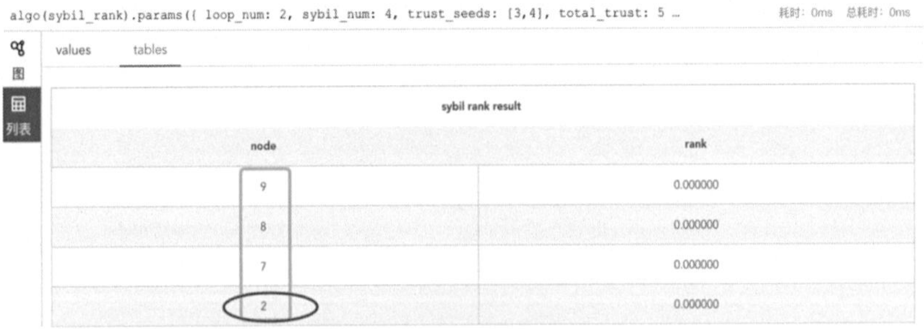

b) SybilRank Results

FIG. 26 Running sybilrank on small graph. (A) Small graph for SybilRank test. (B) SybilRank results.

due to data deviation, for instance, vertex 2 should be regarded as a normal account (with balanced networked behaviors just like vertex 4). On large data sets, the advantage of Sybil accuracy will be very obvious. According to the SybilRank authors in a 2012 report, among the 11 million Tuenti social users, 180,000 of the 200,000 accounts ranked bottom by SybilRank are fake accounts (90%). It can be said that SybilRank's antifraud accuracy stands at 90% (which is impressive).

The complexity of the Sybil algorithm is $O(N * Log\ N)$, N is the number of users, that is, the number of user-type vertices in the graph data set. Generally speaking, the algorithm complexity of SybilRank is lower than that of PageRank, but the scalability is equivalent to that of PageRank.

4.6   Propagation computing

In the complex networks composed of real-world data, the data tends to form a certain community structure, and the algorithm that finds such structure is generally called a community (recognition) algorithm or label propagation algorithm. To be precise, propagation is the process and means, and the community is the result and the goal. Whether an algorithm is classified as a propagation algorithm or a community algorithm depends entirely on whether we focus on the process or the result.

Typical propagation algorithms include Label Propagation Algorithm (LPA), Hop Attenuation and Node Preference (HANP), and other LPA-variant algorithms.

4.6.1   Label propagation algorithm

Label propagation algorithm (LPA) is a community detection algorithm based on label propagation; it is an iterative process of nodes aggregation with the goal of label distribution stabilization—propagates existing labels in the graph (such as the value of a certain attribute of a node) and achieves stability, so as to classify nodes according to labels (communities). In each iteration, a node will get the label value of the vertex with the highest weight among its neighbors. Although the results of the algorithm have a certain degree of randomness, it is very suitable for large complex networks which owes to its low time complexity and no need to specify the number of communities in advance. It is often included in the benchmark testing in both academia and industry.

Label is the value of a specified property of node. Label can be propagated to the 1-step neighbors of the node, nodes with label can adopt label from its 1-step neighbors. When detecting the communities, label represents the community of the node, and the propagation of label is the expansion of the community.

Consider the propagation of labels a and b of the red and yellow nodes in Fig. 27:

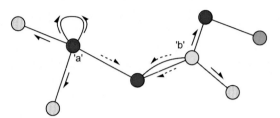

FIG. 27 Label propagation illustrated.

- Blue node is the common neighbor of the two nodes; it will decide which label to adopt based on the weight of label a and b, or to adopt both labels if it is allowed. If a node can only adopt only one label from its neighbors, it is called *single-label propagation*, while if multiple labels are allowed to adopt, it is called *multilabel propagation*.
- Red node has self-loop edge, so label a also propagates to itself.
- Label b cannot propagate to the green node during one round of propagation; if the purple node (green node's only neighbor) adopts label b, the green node will adopt label b during the next round of propagation.

It is worth noting that the neighbors in the LPA algorithm are neighbors found based on each edge directly associated with the current node, that is, the same neighbor found through multiple edges does not need to be deduplicated; for nodes with self-loops (pointing back to the node itself), the node itself also participates in the weight calculation as a neighbor, and each self-loop corresponds to two nodes (the self-loop is regarded as two edges, one outbound edge and one inbound edge). The advantage of this processing logic is that it can handle multigraph, which is different from the OSN's simple-graph that do not address multiedge or self-loop challenges.

Example: Run the label propagation algorithm on the AMZ data set, set the number of iterations to 5, select the node schema @nodx's ccid field as propagating label, allowing for up to four labels (k value), return the results and count of labels propagated, the GQL is as below (Table 7):

```
/* LPA with results and count */
algo(lpa).params({
  node_label_property: @nodx.ccid,
  k: 4,
  loop_num: 5
}) as results, count
return results, count /* limit 100 */
```

The sample code above runs a multilabel propagation, and the results in Fig. 28 reflect that each vertex is assigned with 4 labels, each with an associated probability between 0 and 1, and higher probability means the matching label is more likely to stay.

TABLE 7 LPA parameters and specification.

Name	Type	Specification	Description
node_label_property	@<schema>.<property>	Numeric or string class node property, LTE needed	Name of the node schema and property as label; node without this property will not participate in the calculation; random number is used as label if not set
node_weight_property	@<schema>.<property>	Numeric node property, LTE needed	Name of the node schema and property as node weight; node without this property will not participate in the calculation; weight is 1 if not set
edge_weight_property	@<schema>.<property>	Numeric edge property, LTE needed	Name of the edge schema and property as edge weight; edge without this property will not participate in the calculation; weight is 1 if not set
k	int	>=1	The maximum number of labels each node can keep; labels are ordered by probability
loop_num	int	>=1	Number of iterations

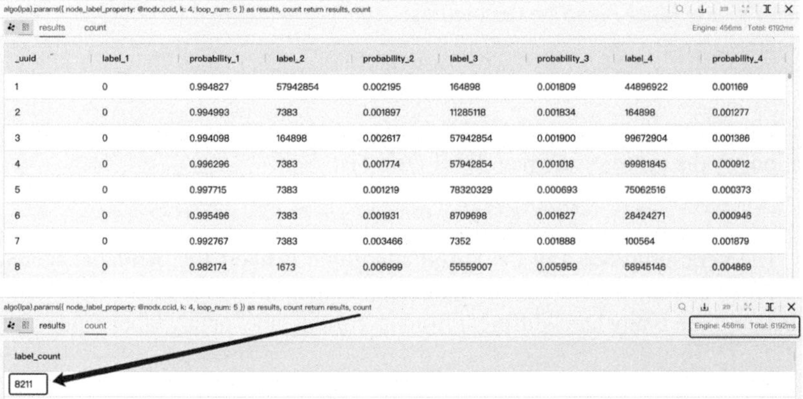

algo(lpa).params({ node_label_property: @nodx.ccid, k: 4, loop_num: 5 }) as results, count return results, count

results   count

Engine: 456ms Total: 6192ms

_uuid	label_1	probability_1	label_2	probability_2	label_3	probability_3	label_4	probability_4
1	0	0.994827	57942854	0.002195	164898	0.001809	44896922	0.001169
2	0	0.994993	7383	0.001897	11285118	0.001834	164898	0.001277
3	0	0.994098	164898	0.002617	57942854	0.001900	99672904	0.001386
4	0	0.996296	7383	0.001774	57942854	0.001018	99981845	0.000912
5	0	0.997715	7383	0.001219	78320329	0.000693	75062516	0.000373
6	0	0.995496	7383	0.001931	8709698	0.001627	28424271	0.000945
7	0	0.992767	7383	0.003466	7352	0.001888	100564	0.001879
8	0	0.982174	1673	0.006999	55559007	0.005959	58945146	0.004869

algo(lpa).params({ node_label_property: @nodx.ccid, k: 4, loop_num: 5 }) as results, count return results, count

results   count

Engine: 456ms Total: 6192ms

label_count

8211

FIG. 28 Multilabel LPA results and count.

In Fig. 28, we also noticed a feature of heterogeneous data processing, that is, after running the LPA algorithm, at least two data structures are returned: results and count, where results contain the actual table of vertices and their labels and labels' probabilities, and count is the total number of labels propagated. Note that even though the LPA is required to run against the entire data set, the results sent back the client side can be trimmed by adding a limit (see the above GQL code snippet), so that we don't have to wait long time for large amount of data to be transferred over the network back to the client terminal. As highlighted in the bottom right portion of Fig. 28, the graph computing engine takes <0.5 s to finish the calculation, but the writing back and steaming may take 10× or more time.

The LPA algorithm has a wide range of application scenarios, such as virtual community mining, multimedia information classification. Its biggest advantage is its low algorithmic complexity ($O(K * |V| * |E| / |V|)$), assuming K is the number of iterations, V is the number of vertices, and E is the number of edges. As K is generally a small integer, and E/V in most cases is comparable to K, so the time complexity of the algorithm is approximately equal to $O(|V| * Log|V|)$, which is suitable for running on large graphs in real-time or near real-time with appropriate engineering optimization (such as high-concurrency).

4.6.2 HANP algorithm

Hot Attenuation and Node Preference (HANP) algorithm is an extension of the LPA algorithm, which considers the activity of the label and the degree of the neighbor node when calculating the label weight. The label value calculated by the HANP algorithm for a node in each iteration is as follows:

$$w_l = \sum_{j \in N_l} S_j \cdot (d_j)^m \cdot A_{ij}$$

where $d(j)$ is the degree of j, m is the power exponent, $A(ij)$ is the total weight of all edges between i and j. For any node i in the graph, label l is on i's neighbor, then i's new label l' can be denoted as:

$$l'_i = \underset{l}{\mathrm{argmax}}(w_l)$$

Let $s(l')$ be the score of the label on the neighbor nodes, δ is the hop attenuation, so the score of the new label l' on i is:

$$s'(l') = \max s(l') - \delta$$

The iteration process of the HANP algorithm is like LPA. All label scores are set to 1 at the beginning; during each iteration, it calculates for each node whether there is any label different from its current label or any label same with

TABLE 8 HANP input parameters and specification.

Name	Type	Default	Specification	Description
loop_num	int	5	>=1	Number of iterations
node_label_property	@<schema>.<property>	/	Numeric or string class node property, LTE needed	Name of the node schema and property as label; node without this property will not participate in the calculation; random number is used as label if not set
edge_weight_property	@<schema>.<property>	/	Numeric edge property, LTE needed	Name of the edge schema and property as edge weight; edge without this property will not participate in the calculation; weight is 1 if not set
m	float	/	Mandatory	The power exponent of the degree of neighbor node, which indicates the preference of node degree; m = 0 means to ignore node degree, m > 0 means to prioritize neighbors with high degree, m < 0 means to prioritize neighbors with low degree
delta	float	/	(0,1); mandatory	Hop attenuation of label score during propagation
limit	int	−1	>= −1	Number of results to return; return all results if sets to -1 or not set

its current label but has higher score that can be adopted from its neighbors; and after all nodes are calculated, updates the nodes which can be updated. Iterating and looping by this rule until no node could adopt new label or update label score, or the number of iterations reaches the limit (Table 8).

Example: Run HANP on the modified AMZ data set, set the maximum number of iterations to 5, the edge weight attribute field is rank, the vertex label attribute field ccid, set $m = 500$, delta $= 0.2$, have it write back to the database. The result of the algorithm run as an asynchronous task is shown in Fig. 29.

FIG. 29 HANP algorithm execution results.

The HANP algorithm from Fig. 29 was executed in a no-code fashion, which can be quite convenient, but for those of you who are adhesive to the underpinning graph query code, here it is:

```
/* Launching HANP */
algo(hanp).params({
    loop_num: 5,
    node_label_property: "@nodx.ccid",
    edge_weight_property: "@edgx.rank",
    m: 500,
    delta: 0.2
}).write({
    db:{
        property: "HANP_ricky"
    }
})
```

4.7   Community computing

In the analysis and research of complex networks, if there are different communities in a network (graph), that is, vertices with different characteristics gather to form different tightly combined sets (communities), then this graph has a Community Structure. Community computing includes a variety of algorithms, such as the Minimum Cut, K-Core, K-means, Louvain, Triangle Counting, Clique), and others. This section mainly introduces the Triangle Counting and Louvain community identification algorithms.

4.7.1   Triangle counting

Triangle counting calculates the number of triangles in the entire graph and returns the vertices or edges that make up each triangle. In graph theory, triangle computation is widely used, such as social network discovery, community iden-tification, closeness analysis, stability analysis. In a financial transaction sce-nario, the number of triangles formed by account (card) transactions identifies the degree of connectivity and connection tightness among the accounts of the financial institution. In biological research, the biological path-ways of the interaction between cell molecules are also often expressed in the topology of multiple continuous triangles.

Triangle counting can be seen as a general-purpose graph algorithm, and in different contexts, it can also be classified as connectivity, community recog-nition, link prediction, and garbage filtering algorithms.

Triangle calculation is divided into two modes: one is to assemble triangles according to edges; the other is to assemble triangles according to vertices. Among them, the number of triangles assembled by edges may be many times more than the number of triangles assembled by vertices, and the computational complexity is correspondingly higher. For example, running the triangle count-ing on the AMZ data set, there are close to 4 million triangles by vertices and over 14 million triangles by edges (Fig. 30).

Interested readers can first use a small graph to validate the difference between the two modes of execution. Using Fig. 23 as an example, there are 10 triangles if counted by edge, but only 2 triangles by node (the result is shown in Fig. 31, the upper part is by node and the lower part is by edge).

Graph theory research dominated by social network analysis generally adopts the principle of counting triangles by nodes, while completely ignoring the possibility of multiple edges between any two nodes. We have illustrated the huge difference between the two modes in Chapter 1 of this book, and the igno-rance has a great impact on the design and implementation of the algorithm, as well as the algorithm complexity and optimization process. Essentially the dif-ference is rooted in simple graph vs. multigraph, or the difference between a social network and a financial and industrial knowledge graph network.

From the perspective of algorithm design, the write-back mechanism of the triangle calculation is not suitable for writing back to the database attribute

FIG. 30 Two modes of triangle counting (by node or edge).

triangle list		
nodeA	nodeB	nodeC
4	5	6
1	5	6

a) Number of triangles counted by vertices (result list)

triangle list		
edgeA	edgeB	edgeC
15	14	16
15	12	16
9	14	16
9	12	16
5	14	16
5	12	16
4	14	16
4	12	16
13	14	11
13	12	11

b) Number of triangles counted by edges (result list)

FIG. 31 Triangle counting results per node vs edge. (A) Number of triangles counted by vertices (result list). (B) Number of triangles counted by edges (result list).

field, but it can be written back to the disk file in batches, or return the triangles corresponding to the vertex or edge mode when being invoked through API/SDK.

Triangular topology (Triangles) is composed of more basic topological structure triples (vertex-edge-vertex-edge-vertex) connected end to end, so triangles are also called closed triples. If each triple is uniquely identified by the middle vertex, then there must be three triples to form a triangle. The transitivity of the graph is defined as the number of closed triples (numerator) divided by the number of all triples (denominator, as follows):

$$\gamma(G) = \frac{|\Pi|}{|\Pi|} = \frac{|\Pi|}{|\Pi^{\angle}| + |\Pi|}$$

Among them, if only simple graph is considered, the number of closed-loop triples is three times the number of triangles. The complexity and timeliness of the triangle counting algorithm depend on the following factors:

- Data structure.
- Algorithm.
- System architecture (including programming language, engineering implementation, network topology, etc.).

The complexity of triangular calculations achieved by traversing (closed loop) combinations of triples through loops is $O(|V|^3)$; obviously this complexity is too high even for a graph with only 1000 vertices. There are many optimization techniques to accelerate, one such technique is to traverse (once) and sort by nodes and/or edges then traverse again, which can effectively lower the complexity by $|V|$ or $|V|^2$.

4.7.2   Louvain community recognition

The Louvain algorithm is a community identification algorithm based on modularity, which is an iterative process of clustering vertices with the goal of maximizing modularity. The algorithm was developed by Vincent D. Blondel of the University of Louvain in Belgium.[c] It was first published in a paper in 2008, because it can calculate satisfactory community detection results with high efficiency and has become the most mentioned and used community detection algorithm in recent years.

The overall logic of the Louvain algorithm is relatively complex. We can split it into four parts: weighted degree, community compression, modularity, and modularity gain, and divide and conquer each one of them one after another.

The ***weighted degree of node*** is a degree calculation that takes into account the weights on the edges. The Louvain algorithm uses two weights: node weight and community weight when calculating modularity. Node weight refers to the weight sum of all edges related to a vertex (with the vertex as the endpoint), including the adjacent edges of the vertex (connected to neighbors) and the self-loop edges of the vertex (connecting to the vertex itself).

As shown in Fig. 32, the red node has three adjacent edges and one self-loop edge, so the weight of this vertex is $1+0.5+3+1.5=6$ (note: the weight of the self-loop is only calculated once). Community weight refers to the sum of weights of all nodes in a community.

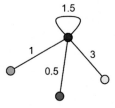

FIG. 32   Louvain node weight.

---

c. Fast unfolding of communities in large networks http://arxiv.org/pdf/0803.0476.pdf.

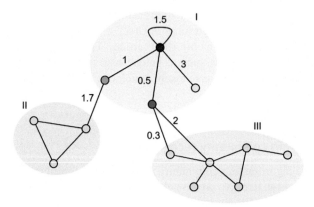

FIG. 33 Louvain community weight.

As shown in Fig. 33, the red node's weight is $1+0.5+3+1.5=6$, green node's weight is $1+1.7=2.7$, blue node's weight is $0.5+0.3+2=2.8$, yellow node's weight is 3, so the weight of community is $6+2.7+2.8+3=14.5$.

The internal weight of the community (or **weighted-degree inside community**) means that when calculating the weight of a community, only consider the edges whose two endpoints are in the community, or remove edges that connecting the current community with other communities. As shown in Fig. 33, community I has three edges connecting with community II and III, weights 1.7, 0.3, and 2, so the internal weight of the community I is $14.5-1.7-0.3-2=10.5$.

Note that the weight within the community is not the sum of the weights of the edges whose two endpoints are in the community, but twice the sum of the weights of the nonself-loop edges in these edges plus the weight of the self-loop edges. The reason is that the two endpoints of a nonself-loop edge cause the edge to be counted twice. In other words, the weights of the edges in the community except the self-loop type are only calculated once, and the other edges' weights will be counted twice—because each edge will connect two endpoints. Use Fig. 33 to verify, the internal weight of community I is: $(1+0.5+3)\times 2+1.5=10.5$. .

The weight of the whole graph (**Weighted-degree of the Whole Graph**) refers to the sum of the weights of all nodes in the graph. If the whole graph is divided into multiple communities, since each node in the graph belongs to only one community, the weight of the whole graph is also equal to the sum of the weights of these communities.

As shown in Fig. 34, the weight of community I is 14.5, the weight of community II is $0.7\times 3\times 2+1.7=5.9$, the weight of community III is $1\times 6\times 2+0.3+2=14.3$, so the weight of the whole graph is $14.5+5.9+14.3=34.7$. If the whole graph is regarded as a community, then the weight of the whole graph can also be understood as the internal weight of the

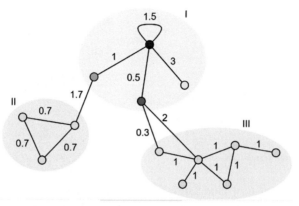

FIG. 34 Weighted-degree of the whole-graph in Louvain.

community. Also use Fig. 34 for verification, the weight of the whole graph is $(1+0.5+3+1.7+0.7 \times 3+0.3+2+1 \times 6) \times 2+1.5=34.7$.

The results of the above two calculation methods can be mutually verified and should be consistent.

A large amount of community compression is used in the Louvain, and the subsequent (iterative) calculation speed is improved by minimizing the number of nodes and edges of the graph through compression without changing the local weight degree and the weight degree of the full graph. After compression, the nodes in the community will be calculated as a whole for modular optimization, and will not be split, so as to achieve the hierarchical (iterative) community division effect.

Community compression is to represent all nodes in each community with an aggregation node, the internal weight of the community is the weight of the self-loop edge of this aggregation node. The weights of the edge(s) between each two communities will be weight of the edge between the two corresponding aggregation nodes.

As shown in Fig. 35, compressing the three communities on the left will subjugate 3 aggregation nodes on the right. The whole-graph weight after compression still is 34.7.

Geometrically speaking, modularity attempts to compare the closeness of node connections within and between communities by calculating weights. Make $2m$ as the weight of the whole graph, if C is any community in the graph, Σ_{tot} is the weight of C, Σ_{in} is internal weight of C, then the modularity Q can be expressed as:

$$Q = \sum_c \left[\frac{\sum_{in}^c}{2m} - \left(\frac{\sum_{tot}^c}{2m} \right)^2 \right]$$

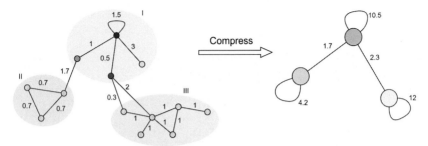

FIG. 35 Louvain community compression.

The value range of modularity is $[-1,1]$. For connected graphs (or connected subgraphs), the range of modularity is $[-1/2,1]$.

The significance of modularity is to reflect the quality of community division. The higher the value of modularity, the more reasonable the community division is. Comparing the two community division methods in Fig. 36, the difference is which community the red node belongs to; our intuition tells us that the division on the left side is more reasonable.

Assuming that the weight of all edges in Fig. 36 is 1, the weight of the whole graph is $1+8 \times 2 = 17$, and let us examine the left division vs. the right division:

- Left: The weight of community I on the left is 10, and the internal weight is 9; the weight of community II is 7, and the internal weight is 6; the modularity after division is 0.3668;
- Right: The weight of community I on the right is 13, and the internal weight is 11; the weight of community II is 4, and the internal weight is 2; the modularity after division on the right is 0.1246.

The calculated results are as expected (division on the left is more reasonable).

The modularity gain refers to how much the modularity increases after the community division is changed. When the Louvain algorithm adjusts the community belonging of a certain node, it decides whether to adjust the node by examining the modularity gain.

Making i a node in the graph, c as a community that does not contain i, $k(i)$ for the weight of i (or the contribution to the community's weight when i joins a community), $k_{i,\text{in}}$ for $t$wice the sum of the weights of the edges between i and a

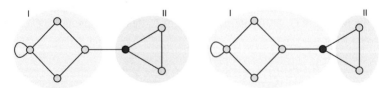

FIG. 36 Louvain modularity (community division).

certain community. If i does not belong to any community, when it joins a community c, the resulting modularity gain ΔQ can be expressed as:

$$\Delta Q = \frac{k_{i,in}^{c}}{2m} - \frac{2k_{i}\sum_{tot}^{c}}{(2m)^{2}}$$

Assuming a and b are two communities that do not contain i, when i is transferred from a to b, the resulting modularity gain is:

$$\Delta Q = \frac{k_{i,in}^{b} - k_{i,in}^{a}}{2m} - \frac{2k_{i}\left(\sum_{tot}^{b} - \sum_{tot}^{a}\right)}{(2m)^{2}}$$

Now, in Fig. 37, if we are transferring node A (red one) from community a to community b, the modularity gain is calculated in the following steps:

- Full graph weight $2m$ is 17, the red node A's k_i is 3.
- Community a's Σ_{tot} is 10, $k_{i,in}$ is 2.
- Community b's Σ_{tot} is 4, $k_{i,in}$ is 4.
- Putting it into the formula, the modularity gain is $(4-2)/17 - 2*3*(4-10)/(17*17)=0.2422$, which is consistent with the previous calculation result of $0.3668 - 0.1246 = 0.2422$.

When programming an algorithm, the modularity can be calculated using the following formula:

$$Q = \frac{1}{2m} \sum_{i,j \in V} \left[A_{ij} - \frac{k_i k_j}{2m} \right] \delta(C_i, C_j)$$

In the formula, $2m$ is the global weight, i and j are any two nodes in the graph, when i and j belong to the same community, δ is 1, otherwise δ is 0:

- When i and j are different nodes, A_{ij} is the weighted sum of the edges between i and j.
- When i and j are the same node, A_{ij} is the weighted sum of the self-loop edges of the node.

Combining the above two situations for A_{ij}, the sum of it is the internal weight Σ_{in} of the community; in the above formula k is the weight of a node, $k_i \cdot k_j$ can be

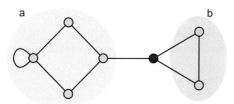

FIG. 37 Louvain modularity gain.

decomposed into the sum of k_i multiplied by the sum of k_j, that is, the square of the weight of the community $(\Sigma_{tot})^2$. Thus, the above formula is completely equivalent to the previously defined modularity.

The optimization of modularity is an NP-hard problem, and the Louvain algorithm uses a heuristic algorithm to optimize modularity in multiple rounds of compound iterations. Each iteration is divided into two phases:

Phase 1: Iteration. At the beginning of this phase, each node is regarded as a separate community; in each round of iteration, it is calculated for each node whether it can find a community where its neighbors are located, and if the placement of the node can generate the largest and positive ΔQ, which maximally increases the modularity, transfer the node to the new community; use the same method to calculate and adjust the next node; after all nodes are calculated and adjusted, enter the next round of iteration. The first phase loops and iterates according to this rule until no node can be reassigned, or the number of iterations reaches the limit.

Phase 2: Community compression. Compress each community divided in the first phase to get a new and smaller graph. If the compressed new graph is consistent with the graph structure at the beginning of the current round of the iteration cycle, that is, the modularity has not been improved, then the algorithm ends, otherwise the new graph is used as the initial graph of the next round of iteration (the outer cycle).

Considering the calculation convergence of large graphs, the program introduces a modularity gain threshold when judging whether the modularity is improved. This value is a floating-point data greater than 0. When ΔQ does not exceed this value, the modularity gain is not improved so that the iteration can end.

Let's manually carry out Louvain community detection using Fig. 37, assuming the weight of each edge in the graph is 1, and the modularity gain threshold is 0.

- First round of calculation
  - First phase, and the first iteration: ab, cd, e, fg.
  - First phase, and the second iteration: ab, cd, ef.;
  - First phase, and the third iteration: no node can be moved.
  - Second phase: Compression into ABC, each with weighted self-loops.
- Second round of calculation
  - First phase, and the first iteration: AB, C.
  - First phase, and the second iteration: no node can be moved.
  - Second phase: Compression into two communities alpha and beta, as shown in Fig. 38.
- Third round of calculation
  - First phase, and the first iteration: No node can be moved.
  - Second phase: no compression is needed, algorithms ends.

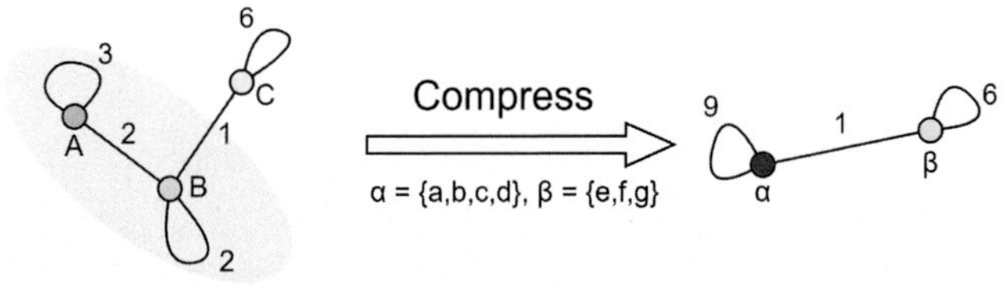

FIG. 38   Louvain community state at the end of second round.

Conclusion: The original graph from Fig. 37 is divided into two communities, namely {a,b,c,d} and {e,f,g}.

If there are isolated nodes in the graph, then the solitary node will form its own community, no matter how many rounds of iterations are done, the node will not merge with anyone else. The reason is that the isolated node has no adjacent edge, and its $k_{i,in}$ is 0, moving it into another community will only create a negative ΔQ.

For a disconnected graph, there are no adjacent edges between disconnected communities, and nodes from different communities cannot be merged, so each disconnected community is independent, and the community division of the Louvain algorithm is only meaningful within the connected component.

The Louvain algorithm treats the self-loop edge differently from the degree algorithm when calculating the weighted degree. In the degree algorithm, each self-loop edge is counted twice, and in the Louvain algorithm, each self-loop edge is counted once. As shown in Fig. 32, red node's degree is $1+0.5+3+1.5\times2=7.5$, and the weight of this node in the Louvain algorithm is $1+0.5+3+1.5=6$.

The implementation process of the Louvain algorithm mandates that the graph can be regarded as an undirected graph, and the community id of each node and the value of the final modularity are calculated according to the above process and written back to the node attribute, and the final modularity is stored as task information. In addition, a result file is also generated to record the number of nodes in each community for subsequent feature engineering processing.

The computational complexity of the Louvain algorithm is high. Usually, a program implemented in an interpreted language such as Python takes several hours to run on a graph with only millions of vertices and edges. Running against larger graphs may take days, weeks, or fails due to insufficient memory. Therefore, the gists to speed up the Louvain algorithm can be summarized as:

- Implement in a compiled language, which is close to the hardware and has high execution efficiency.
- Go parallel processing as much as possible, low-latency data structure support is required.
- Use in-memory computing when possible.
- Watch out for the possible down-side and cost of data migrations on horizontally scalable architectures.

TABLE 9 Louvain input parameters and specification.

Name	Type	Default	Specification	Description
phase1_loop_num	int	5	>=1	The maximum number of iterations in the first phase of the algorithm
min_modularity_increase	float	0.01	0~1	The threshold of the gain of modularity
edge_schema_property	[]@<schema>?.<property>	/	Numeric edge property, LTE needed	Edge weight property/properties, schema can be either carried or not; edge without any specified property will not participate in the calculation; all weights are 1 if not set
limit	int	−1	>=−1	Number of results to return; return all results if sets to −1 or not set
order	string	/	ASC or DESC, case insensitive	The sorting rule, only effective with the *Streaming Return* execution method when the mode set to 2 (community:id/count)

According to the author's experience, making full use of the advantages of the architecture and data structure optimizations can achieve exponential performance improvement (hundreds of times)—which means Louvain can be run on large-scale graphs with near-real-time performance (single-digit minutes on billion-scale graphs).

The input parameters of the Louvain algorithm are shown in Table 9.

Example: On the AMZ data set, launch Louvain, set stop modularity threshold to 0.000001, loop count to 100, return with statistical information:

```
/* Louvain Algorithm in GQL */
algo(louvain).params({
  phase1_loop_num: 100,
  min_modularity_increase: 0.000001,
  order: "desc"
}).stats() as stats
return stats
```

FIG. 39 Run Louvain in web console with stats.

The results of the above GQL code snippet is shown in Fig. 39, the total number of communities is 171. Note that even though the iterations is 100, which is unnecessarily large, because most of the time, convergence happens within ~5 iterations. The same is true for modularity threshold, 0.000001 may not be much different from 0.0001 or 0.01 in terms of computation power needed—this is a piece of good news for Louvain developers and users—it does create an illusion that you graph algorithm can return sooner than expected when the two parameters are aggressively set.

Fig. 40 shows Louvain runs on Twitter data set (1.5 billion nodes and edges), and execution time is around 3 min (187 s), as the data set is about 400 times larger than the AMZ data set, the execution time is about 300 times longer. The cluster info portion captures that the CPU concurrency has reached 5863% meaning at least 59 threads (or vCPUs) were involved in parallel processing; this is at least 50–60 times faster the original Louvain C++ implementation (see Fig. 41).

Another useful feature with Louvain is visualization which helps users to understand how the entities in the graph are forming communities. The challenge with Louvain visualization is that there may be too many nodes and edges to display, even transferring 100,000 plus nodes and edges over the network takes a long time, and rendering that much data would be very challenging.

FIG. 40 Running Louvain on billion-scale large graph.

	Karate	Arxiv	Internet	Web nd.edu	Phone	Web uk-2005	Web WebBase 2001
Nodes/links	34/77	9k/24k	70k/351k	325k/1M	2.6M/6.3M	39M/783M	118M/1B Smaller than Twitter
Clauset, Newman, and Moore	.38/0s	.772/3.6s	.692/799s	.927/5034s	-/-	-/-	-/- No return in 24-hr
Pons and Latapy	.42/0s	.757/3.3s	.729/575s	.895/6666s	-/-	-/-	-/- No return in 24-hr
Wakita and Tsurumi	.42/0s	.761/0.7s	.667/62s	.898/248s	.56/464s	-/-	-/-
Louvain Method	.42/0s	.813/0s	.781/1s	.935/3s	.769/134s	.979/738s	.984/152mn

FIG. 41 Performances of different community detection algorithms.

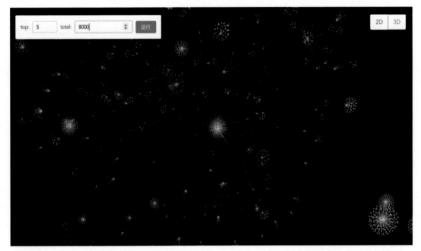

FIG. 42 Real-time visualization of Louvain.

The solution to this is to do data sampling and data-rendering acceleration. The former is to sample a handful of communities and with <10,000 entities to display, the latter is to leverage, for instance, WebGL for GPU rendering acceleration; Fig. 42 illustrates real-time Louvain visualization with top-5 communities and 8000 vertices (and associated edges).

4.8 Graph embedding computing

In mathematics, an embedding (or imbedding) is one instance of mathematical structure contained within another instance. In graph theory, graph embedding is to transform a graph into vector space. The benefit of such transformation is that vector operations are much simpler and faster. This transformation essentially is to compress high-dimensional data to be represented in lower dimensions while still preserving certain properties, such as graph topology and node/edge relationships. Graph embedding algorithms often serve for downstream AI/ML tasks, such as classification and link prediction.

Graph embedding is very popular in the field of graph computing in recent years. A lot of AI researchers have changed their research direction from deep learning and neural network to graph embedding and graph neural network. It is also increasingly found that graph embedding leads to better antifraud or smart marketing effects.

Graph embedding is an ever-expanding category of algorithms.

One of the main methods to do graph embedding is random walk. Random walk refers to the process that a node moves along the edges for several steps, each step it picks one of its adjacent edges randomly; the process is repeated multiple times to generate a sequence of paths; then the sequence can be transported to a model as samples for training, resulting in node embedding vectors of the graph.

This section introduces and analyzes the complete random walk algorithm and Struc2Vec algorithm.

4.8.1 Complete random walk algorithm

A completely random walk (Fully Random Walking, referred to as RW or FRW), as the name implies, means starting from a certain vertex (any or all vertices) in the graph, traversing its connecting edges randomly to reach other vertices, and repeating this process until either all reachable vertices in the graph have been traversed or the walk depth or time has been satisfied. The concept of "random walk" was formally proposed by British mathematician and biostatistician Karl Pearson in 1905.

Fig. 43 vividly expresses the effect of random walk. (A) is a one-dimensional random walk, where the X-axis is time (steps), and the Y-axis (one-dimensional) expresses the deviation between the current value and the original value; (B) is a representation of two-dimensional (plane) random walk; (C) is the superposition effect of three random walks starting from a certain vertex in three-dimensional space.

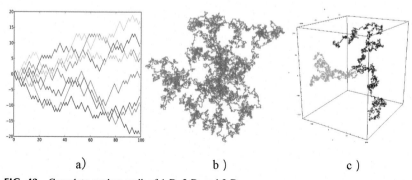

a) b) c)

FIG. 43 Complete random walk of 1-D, 2-D, and 3-D.

Random walks have a wide range of applications in the field of scientific research, such as being used to approximate the trend of stock prices in financial markets, the financial status of investors, the foraging routes of animals in zoology, and the movement routes of molecules in a medium in the field of chemistry. In fact, the randomness of molecular trajectories is Brownian motion in physics—in stochastic analysis, Brownian motion is a continuous random process that is normally distributed.

From 1827, when Scottish botanist Robert Brown first proposed the concept of plant pollen molecules doing Brownian motion in water, to 1905, when theoretical physicist Einstein (who won the Nobel Prize in Physics in 1921) demonstrated through models that pollen molecules are pushed by water molecules for "random" motion. In 1908, when French physicist Jean Perrin experimentally confirmed Einstein's model and proved the existence of molecules and atoms, and in 1926 he won the Nobel Prize in Physics. These pioneers' study on "random walk" spanned a full century.

In the graph data set, the random walk of the whole graph starts from each vertex, according to the limited walk depth (that is, the number of vertices passed, or the number of edges +1), walk filter conditions—generally can be set to a certain attributes of the edges, and the number of walks, finally return random walk paths starting from each vertex (Fig. 44).

The typical configuration of the complete random walk algorithm is shown in Table 10.

Example: On the AMZ data set, launch the random walk algorithm to walk against the entire graph for 3 iterations, and walk depth as 8, return 10 walks. The GQL code is as below:

```
/* Call Random Walk Algo */
algo(random_walk).params({
  walk_length: 8,
  walk_num: 3,
  limit: -1
}) as walks
return walks limit 10
```

In the above example, the full graph has 400,000 vertices, walking length of 8 and walking number 3 will generate 1,200,000 paths, each path has a length of 8 edges, so there are 9,600,000 total edges in results.

The complexity of the random walk algorithm is $O(K * D * |E|)$ when the optimal complexity of visiting the neighbors of each vertex is $O(1)$ (in fact, the average time complexity of locating all connecting edges attributes from a vertex is $O(|E|/|V|)$), where V is the number of vertices, E is the number of edges, K is the number of walks, and D is the depth of the walk. Obviously, the random walk algorithm can achieve a significant boost in performance and reduction in time consumption by starting from multiple vertices in a high-concurrency way. However, the horizontally distributed system can hardly

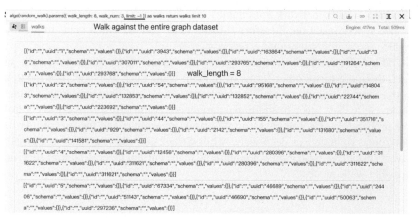

FIG. 44 Random walk results.

TABLE 10 Random walk input parameters and specification.

Name	Type	Default	Specification	Description
ids / uuids	[]_id / []_uuid	/	/	IDs or UUIDs of nodes to start the walk; all nodes to be selected if not set
walk_length	int	1	>=1	Depth of each walk, i.e., the number of nodes to visit
walk_num	int	1	>=1	Number of walks
edge_schema _property	[]@<schema> ?.<property>	/	Numeric edge property, LTE needed	Edge weight properties; nodes only walk along edges with the specified properties and the probability of passing through these edges is proportional to the edge weight; if edge has multiple specified properties, the edge weight = sum of these property values; the weight of all edges is 1 if not set
limit	int	−1	>=−1	Number of results to return; return all results if sets to −1 or not set

generate a lot of network communication due to data migration and interinstance synchronization, therefore causing huge delays, which tends to greatly reduce the overall algorithm efficiency.

The above discussions equate complete random walk with random walk. In fact, there are many other forms of random walk, such as Node2Vec random walk, Struc2Vec random walk, GraphSage Train, Graphlets sampling. Among them, Node2Vec sets two additional parameters p and q, respectively to control the probability of going back (turning back, returning) and the probability of horizontal right–left or deep walk; while Struc2Vec uses the number of walk layers (k) and the probability of remaining in the current layer to control specific possible walk modes.

4.8.2   Struc2Vec algorithm

Struc2Vec[d] algorithm appeared very late and was first published at the SIGMOD 2017 conference. As its name suggested, literally, it is a "(topological) structure to vector" algorithm. This algorithm describes a framework for generating vertex vectors on a graph while preserving graph structure. Its main difference with the Node2Vec algorithm is that Node2Vec optimizes the vertex embedding so that adjacent vertices in the graph have similar embedding expressions (values), while Struc2Vec collects the roles of vertices (even if two vertices are very far away from each other in the graph, they could be having similar "roles" or structural or topological similarity), and random walk to construct a multidimensional network graph (Multilayer Graph or Multidimensional Network) corresponding to each vertex to form a vector value that can express the structural characteristics of each vertex.

To be precise, Struc2Vec judges the role a vertex plays by the high-dimensional spatial topology of the graph in which each vertex resides. This differs from the traditional role differentiation logic per vertex attributes or neighbors. Structural similarity means that the neighborhood topology of two nodes is homogeneous or symmetrical. As shown in Fig. 45, nodes v and u are structural similar (degrees 5 and 4, connected to 3 and 2 triangles, connected

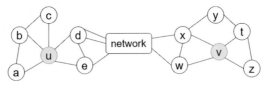

FIG. 45   Structural similarity in Struc2Vec.

---

d. Representation learning on graphs https://arxiv.org/abs/1709.05584.

to the rest of the network by two nodes), but very far apart in the network, they neither have direct connection nor any common neighbor.

The Struc2Vec algorithm framework is very complicated. In the last section of this chapter, we will introduce the concept of white-box interpretable graphs by analyzing the Node2Vec and Word2Vec algorithms. However, graph embedding does make graphs gray-boxed due to their inherently complex logic. Interpretability is a little more complicated, and Struc2Vec (and GraphSage) definitely makes it worse.

The complexity of Struc2Vec can be seen from its invocation configurations as listed in Table 11.

The complexity of the Struc2Vec algorithm is high among all embedding algorithms, about $O(K * |V| * (|E|/|V|) * D)$, where K is the number of walks, D is the walking depth, V is the number of vertices, *and* E is the number of edges. In a graph with high connectivity, the time complexity of Struc2Vec is about $O(|E| * |N|) \gg O(N^2)$. Taking the AMZ graph as an example, there are about 400,000 vertices and 3.4 million edges (two-way), and the time complexity of doing multiround deep walks easily in the range of trillions.

Example: Still on the AMZ data set, launch Struc2Vec with walk length = 80, walk_number = 10, $k = 3$ layers, with stability = 0.5 (50%), window size of 10, dimension of 128, walk against the entire graph, return 10 embedding results (to save some network bandwidth), the code snippet is as follows with results captured in Fig. 46:

```
algo(struc2vec).params({
    walk_length: 80,
    walk_num: 10,
    k: 3,
    stay_probability: 0.5,
    window_size: 10,
    dimension: 128,
    learning_rate: 0.025,
    min_learning_rate: 0.00025,
    min_frequency: -1,
    sub_sample_alpha: -1,
    resolution: 2,
    neg_num: 0,
    loop_num: 2,
    limit: -1
}) as results
return results limit 10
```

Note that for Struc2Vec execution, most majority time is in the random walk phase which accounts for 90%–95% of all algorithm running time. It is consistent with the conclusion in the original paper of Struc2Vec. The caveat in the algorithm is that the original paper had it executed in a serial manner and takes ~250,000 s on a graph of 1,000,000 vertices. By appropriately optimize and

TABLE 11 Struc2Vec input parameters and specification.

Name	Type	Default	Specification	Description
ids / uuids	[]_id / []_uuid	/	/	IDs or UUIDs of nodes to start the walk; all nodes to be selected if not set
walk_length	int	1	>=1	Depth of each walk, i.e. the number of nodes walking through
walk_num	int	1	>=1	Number of walks
k	int	/	[1,10]	Number of layers of the multilayer graph, it should be equal or lower than the diameter of the graph
stay_probability	float	/	[0,1]	The probability of staying at the current level
window_size	int	/	>=1	Size of the sliding window, to sample window_size nodes on left and window_size nodes on right
dimension	int	/	>=1	Dimension of node embedding vector
learning_rate	float	/	(0,1)	Initial learning rate, it decreases until reaches min_learning_rate with increase in training iterations
min_learning_rate	float	/	(0,learning_rate)	Minimum learning rate
min_frequency	int	/	>=0	Minimum number of occurrences of a node in the walk sequences to be included in the model, <=1 means to keep all nodes
sub_sample_alpha	float	/	/	Threshold for sub sampling, <=0 means no sub sampling
resolution	int	/	>=1	Such as 10, 100
neg_num	int	/	>=0	Number of negative samples, it is suggested 0~10
loop_num	int	/	>=1	Number of training iterations (epochs)
limit	int	−1	>=−1	Number of results to return; return all results if sets to −1 or not set

FIG. 46   Struc2Vec results on AMZ Data set (4,000,000 Nodes+Edges).

accelerate the algorithm, 100 times or better performance enhancement can be achieved. Fig. 46 shows a performance boost of ∼500× achieved by Ultipa graph on its v4.3 HTAP architecture platform.

4.9   Graph algorithms and interpretability

In the era of big data, more and more merchants and enterprises like to use machine learning to enhance their predictability of business prospects, and some use deep learning and neural network technologies to obtain greater predictive capabilities. Around these technical means, three questions have been lingering:

- Siloed systems within AI ecosystem.
- Low-performance AI.
- Black-box AI.

Fig. 47 perfectly echoes the three major issues mentioned above. First, there are quite a few AI learning techniques, but they are NOT offered in a one-stop-shop fashion, most AI users and programmers are dealing with multiple siloed systems, software and hardware wise, which means a lot of data ETL are needed all the time (we'll give detailed examples to illustrate how serious the problem is next).

Secondly, AI (deep/machine) learning seems to be facing the dilemma of performance and explainability contradictory to each other, you can NOT have the best of both, and needless to say that as we go deeper into the deep-learning and ANN/CNN space, things are so black-box that human decision makers are NOT okay with this status-quo.

Lastly but certainly not the least important problem is most of these learning processes are NOT high performance; the data preparation, ETL, training, and application against production data can be a very complex and lengthy process.

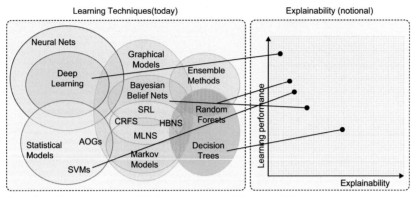

FIG. 47 AI/ML techniques and their performance and interpretability.

Some large retail banks tried to use machine learning and AI to predict monthly credit card spending (turnover) for their hundreds of millions of card holders, but the prediction process takes weeks (as you would imagine, if they start the process mid-month every month, they won't get a result until a few days passing the deadline, which renders the whole effort pointless). Besides, such ML/AI-based prediction method is problematic in terms of accuracy which implies higher-cost associated with the bank's liquidity management.

All in all, these problems deserve elevated attention, and ideally, infrastructure level innovation should be attained to address and solve these challenges.

To be a bit analytical on the needed innovation to address the problems we are facing—the trend we are seeing, and what we need, is the convergence of AI and Big-Data. We desire AI to be smarter, faster, and explainable on the one hand, and we request Big-Data to be fast, deep, and flexible in terms of data processing capabilities on the other hand. The greatest common factor of this AI+Big-Data convergence leads to graph augmented intelligence and XAI. Graph database is the infrastructure and data-intelligence innovation that we have been looking for. We'll show you why and how in this section (Fig. 48).

Traditionally, mathematical and statistical operations are rather limited on graphs in either adjacency matrix or adjacency list centric data structures, very few math or statistical operations can be done. Due to these limitations, developers have invented the concept of graph embedding, which basically transforms a property graph into vector space, so that a lot more mathematical and statistical operations can be performed with higher efficiency. This transformation essentially is to transform high-dimensional data to be represented in lower dimension while still preserving certain properties (such as graph topology and node/edge relationships). The other benefit of such transformation is that vector operations are much simpler and faster.

Slightly expanded discussions around vector computing are necessary. It's an effective way to accelerate graph computing. A major advantage that's

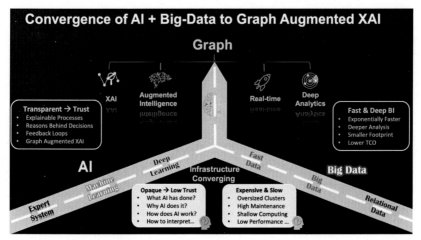

FIG. 48   Convergence of AI and Big Data to graph augmented XAI.

highly relevant to graph embedding is that the underlying data structure is vector-space comparable, which enables convenient mathematical and statistical computing operations. Moreover, from a technological stack perspective, it's possible to parallel everything, from infrastructure, data structure, algorithm to engineering implementation. The pursuit for high concurrency has never ceased to thrust forward, and this will ensure the otherwise time-consuming operations like graph embedding and learning can be completed as fast as timely as possible!

We'll use the Word2Vec example to explain why vector-computing empowered graph embedding can be orders of magnitudes faster, and simpler. Word2Vec method can be seen as the basis, and prototypical, for all node, edge or struct centric graph embedding methods. It basically transforms words into embedding vector space, so that mathematically in the lower dimensional vector space, similar words have similar embeddings—and "You shall know a word by the company it keeps" as coined by the famous English linguist J.R. Firth in 1957—this is to say that similar words with similar meanings tend to have similar neighboring words!

In Figs. 49 and 50, a Skip-gram algorithm and SoftMax training method implementing Word2Vec are illustrated; there are other algorithms (i.e., Continuous Bag of Words) and methods (i.e., Negative Sampling) which are abbreviated for discussion simplicity. Fig. 49 shows some of the training samples extracted from a sentence with a window size of 2 (meaning 2 words pre- and 2 words post-the focused word). Note that in improved Word2Vec method, techniques like subsamplings are used to remove samples that contain words like "the", so that to improve overall performance.

To represent a word and feed it to a neural network, the one-hot vector method is used, as illustrated in Fig. 51, the input one-hot vector for the word

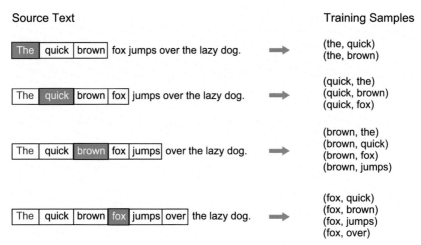

FIG. 49 Word2Vec from source text to training samples.

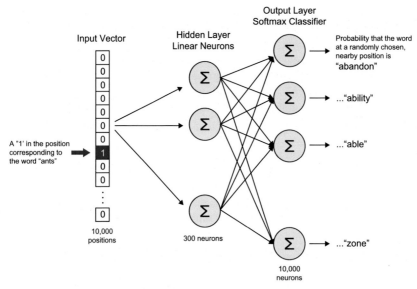

FIG. 50 Architecture of Word2Vec neural network.

"ants" is a vector of 10,000 elements, but with only 1 and all the other positions are 0s. Similarly, the output vector also contains 10,000 elements, but filled with probability (floating points) of each nearby word.

Assuming we have only 10,000 words in our vocabulary, and 300 features, or parameters that you would use to fine tune this Word2Vec model. This translates into 300 neurons in the hidden layer and 300 features in the word vector (see Fig. 51).

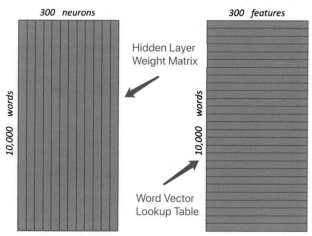

FIG. 51   Huge word vectors (3,000,000 weights).

As you can see from Fig. 51, the hidden layer and output layer each carries 3,000,000 weights, this is NOT feasible for real-world applications. In real-world circumstances, the vocabulary size can be as large as in the range of millions, multiplied by hundreds of features, which will make the size of one-hot vector reaching billions, even with the help of subsampling, negative sampling, and word-pair-phrase techniques, this sheer computational challenge is NOT acceptable. The one-hot vectors in both the input and output layers are clearly way too sparse a data structure, it's like the adjacency matrix, it may be partitioned for parallel computing on GPUs when it's very small in terms of the number of elements (<32,768 for instance) the matrix can hold for each dimension, but for much larger real-world applications, this will be overwhelming in terms of resource and time consumption.

Now, recall what we wanted to achieve with Word2Vec, essentially, we are trying to predict, in a search and recommend engine setup, when a word (or more) is entered, what other words we would like to recommend. This is a statistical game, and underlying, it's a mathematical problem, for one word asked, a handful of other words with the highest probability of appearing as close neighbors for this asked word. This is naturally a graph challenge, every single word is considered a node (or vertex) in the graph, and its neighbors are forming edges connecting with it directly, this goes on recursively. And each node and edge can have its own weight, type, label, and attributes (as many as needed).

What we just described in the above paragraph can be represented in Fig. 52, with two data structure variations. On the surface, they look like big-table data structures, but under the hood, it's the high-concurrency multidimensional vector-space implementations that make things fly.

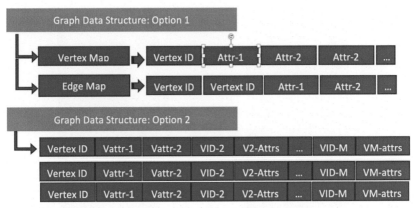

FIG. 52 Graph data structures with natural graph embeddings.

Given the vector-space data structure, Word2Vec problem really boils down to two steps:

- Assigning a probability weight for each directed edge connecting two words, and this can be done statistically.
- Search and recommendation are simple and short graph traversal tasks which simply pick the top-X number of vertices connecting with the starting vertex, and this operation can be done recursively to traverse deeper in the graph—this is like to search and recommend in real-time (similar to real-time suggestion, but potentially more accurate and intelligent, if you recall this exactly resembles how human brain functions with NLP).

What we just described is a complete white-box process, every single step is specific, explainable, and crystal clear. It's very much unlike the black-box and hidden layer(s) that you have to deal with cluelessly as in Figs. 50 and 51. This is why graph will drive XAI (short for eXplainable Artificial Intelligence) forward! More specifically, vector-space computing is naturally graphically embedded—there is no further computation or operations needed, when the data are presented in the graph in the format of vectors (a form of index-free adjacency), the embeddings are done.

DeepWalk is a second example to investigate. It is a typical vertex-centric embedding approach and uses random walks to produce embeddings. With the help of vector-space computing, random-walk is as easy as selecting a specific node and move on to a random neighboring node and deeply traverse the graph for a random number of steps (i.e., ~ 40 steps). Pay attention to the depth/length of the walk, we understand that every hop deeper into the graph, the computational challenge is exponential, the simple math here is that, if an average node has 20 neighbors, going 40 steps deep is to look through neighbors as many as 20^{40}, no production system today can handle this in a brutal-force way.

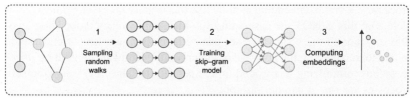

Phases of DeepWalk approach

FIG. 53 Breakdown of depth walk.

However, with the help of highly concurrent vector computing, each step of traversing (sampling) in the graph takes O(1), going 40 step further is basically O(40 * 20) in terms of overall latency (though compute-wise, all the concurrency tasks still parallelly deal with all neighbors encountered and filter out any duplicated or revisited nodes effectively).

The step-by-step breakdown of DeepWalk is illustrated in Fig. 53. With highly concurrent vector computing, the random walking, training, and embedding processes are reduced into native graph operations like K-Hop-ing and Path-Finding, which are straightforward, explainable therefore XAI friendly!

DeepWalk has been criticized with (1) the inability of generalization, not capable of handling highly dynamic graphs (each new coming node must be retrained) well and (2) not possible to preserve the local neighborhood (because deep random walks do NOT preserve neighborhood information).

Node2Vec method has been invented to address at least the second shortcoming of DeepWalk. Fig. 54 illustrates the many steps taken to make the whole learning process complete. Please note that there are 10 steps (subprocesses) involved, each step by itself can further breakdown into a series of computing

```
6
7  void node2vec::runAlgorithm(adjacency_hashstar& embedded_vector){
8      // node2vec random walk
9      walker w(convertedRelations,edges);
10     for(size_t i=0; i<walk_num; ++i)
11         w.node2vec_walk(walks,walk_length,p,q);
12     // Word2Vec
13     word2vec w2v(walks,dimension_word_vector,learning_rate,min_learning_rate);
14     w2v.calExpTable();
15     w2v.learnVocab();
16     w2v.removeLowfrequencyVocab(min_frequency);
17     w2v.subSample(sample);
18     w2v.initNet();
19     w2v.initUnigramTable(resolution);
20     w2v.train_skip_gram(window_size,neg_number,iter_num);
21     w2v.get_word_vectors(embedded_vector);
22 }
23
```

FIG. 54 Outer steps of Node2Vec relying on Word2Vec.

processes. A graph embedding method like this usually takes long time to complete. But this is NOT the case with truly high-concurrency embedding algorithms, when all these subprocesses can be converted into concurrent fashion so that to maximizing the leverage of underpinning hardware computing juices as well as minimizing turnaround time. Specifically, all of the following subprocesses are run in parallel fashion:

- Node2Vec's random walks.
- Preparation of precomputed Exp. Table.
- Learning of Vocabulary (from Training File).
- Removal of Low-Frequency Vocabularies.
- Subsampling of Frequent Words (Vectors).
- Initialization of Network (Hidden Layer, Input and Output Layers).
- Initialization of Unigram Table (Negative Sampling).
- Training of Skip-Gram.

The parallelization of these steps helps achieve much better latency and performance improvement, always remember that every single second wasted on not achieving high concurrency is a waste of resources, therefore not environment friendly, and we are not kidding.

There are other graph embedding techniques that bear similar concepts and steps:

- Sampling and labeling.
- Training of certain model (i.e., Skip-Gram, n-Gram, etc.).
- Computing embeddings.

Some are done locally at vertex/subgraph level, others are done against the entire graph therefore mathematically more time consuming, whichever approach we take, being able to achieve the following goals would be meaningfully fantastic:

1. Attain to high performance via high concurrency on data structure, architecture and algorithm level.
2. Explain every step of operations in a definitive, specific, and white-box way.
3. Ideally, complete the operations in one-stop-shop fashion.

If we recall Fig. 47 from the beginning of this section, the Explainability-vs-Performance chart is intriguing. Decision trees have better explainability, because we understand what they do step-by-step, trees are essentially simple graphs, they don't contain circles which are considered perplexing mathematically, therefore harder to explain, and most so-called graph algorithms or embedding implementations today are based on low-performance adjacency list or academic-oriented adjacency matrix. On the other hand, overly sophisticated neural networks and deep learning techniques are going into a direction that's black-box centric and difficult to explain. In real-world applications, AI can

NOT go away with the human beings not being able to understand what exactly is going on under the hood—it's against the XAI rule of thumb—Explainability!

Graphs are used to express the high-dimensional relationship between information and data. Graph databases are the ultimate solutions to faithfully restore high-dimensional spatial expression. If the human brains are the ultimate database, the graph databases are on the shortest paths leading to the ultimate database.

Chapter 5

Scalable graphs

Like all other types of databases, the Scalable Graph Database is an inevitable stage in the development of graph databases. The maximum amount of data, throughput rate, and system availability that a single machine (single instance) can carry are obviously limited, and it is precisely this limitation that almost all new database systems will take the expansion capacity, especially the horizontal cluster expansion ability formed by multiple examples, as an important capability measure. This is the core reason why distributed databases are in the ascendant. This chapter will analyze in detail the underlying architecture logic of building a distributed graph database oriented to business needs and pain points, as well as the pros and cons of different architecture choices.

5.1 Scalable graph database design

When discussing the design of scalable graph databases, it must be made clear that the construction and iteration of horizontally scalable systems should only be pursued when vertical scaling is not possible. In other words, for many business problems, whether considered in terms of data volume, throughput, performance, or other factors, solutions in vertical scaling usually come with the least complexity and the least cost. In fact, the design, implementation, and operational complexity of horizontally scalable systems far exceeds the expectations of most developers, and in many cases, the stability and timeliness of the system become unpredictable due to real-world business complexity, which can form an ironic effect: many so-called large-scale horizontal distributed systems are often less effective than expected, and even weaker than the monolithic (so-called standalone or centralized) systems they replace—Do the performance, stability, total cost of ownership, and efficiency of distributed relational databases built out of the short-chain transaction scenario really surpass the Oracle databases they "replace" in broad-spectrum scenarios?

Perhaps, when we pay attention to the design of the distributed graph database architecture, we can start from the real business needs and avoid making the architecture design an infinite loop for the pursuit of distribution. In this section, we first explore the possibility of vertical scalability and then discuss the significance and advantages and disadvantages of horizontal scalability.

The Essential Criteria of Graph Databases. https://doi.org/10.1016/B978-0-443-14162-1.00006-4

5.1.1 Vertical scalability

Vertical scalability in the traditional sense refers to upgrading the three elements—storage, computing power, and network on a single node (single instance)—to obtain higher system performance and throughput rate to achieve the business goal of serving more customer requests. Until the concept of horizontal scalability is widely accepted by developers and customers, vertical scalability is almost the only way to upgrade our systems. It is generally believed that horizontal scalability began in 2006, when AWS began to provide large-scale cloud computing services, and its technology stack, was inspired by Yahoo! MapReduce+HDFS, the core component of the Apache Hadoop system contributed to the open-source community. Hadoop (or MR) allows low-cost cheap PCs to batch process massive amounts of data through large-scale networking, and the core concept of Hadoop is inspired by Google's two papers on Map Reduce and Google File System developed out of its internal closed-source system in 2003–2004. With the widespread penetration of horizontally scalable architectures and concepts, many people think that vertically scalable systems are useless. However, in the continuous development of computer architecture, the vertically scalable approach still has its tenacious vitality, as shown in Fig. 1.

Many people simply equate the vertically scalable system to the "hardware upgrade" of the computer system and its peripheral equipment, that is, to expand the following *three elements of cloud computing*:

- Storage: hard disk, network disk, memory, etc.
- Computing: CPU, GPU, etc.
- Network: network card, gateway, router, etc.

By upgrading storage, computing, and network resources at the hardware level, as well as upgrading motherboards and bus structures, higher "computing power" can indeed be obtained. However, this increase in computing power at the hardware level is only at the book level. It is not proportional and self-explanatory, and will not allow business systems and applications to automagically achieve higher throughput and lower latency without any "software

Concurrency: 10   Concurrency: 100   Concurrency: 1,000

FIG. 1 Schematic diagram of vertically scalable system.

upgrade." This is a problem with many people who criticize vertically scalable systems often ignore intentionally or unintentionally.

It is undeniable that upgrading the 5400-RPM disk to 15,000-RPM ones will indeed increase the data throughput by nearly 300%. Upgrading the main frequency of the CPU from 2 to 3.2 GHz can also increase the computing efficiency by nearly 60%. Similarly, upgrading 100 Mbps of the network card to 1 Gbps one will also have a 10 times network throughput increase. However, there are still many hardware capability indicators that need software upgrades to unleash. We give three scenarios to illustrate this.

1. Upgrade from traditional enterprise-class hard disk (HDD) to enterprise-class solid-state drive (SSD)
2. Upgrade from 8-core CPU to 40-core*2 CPUs
3. Upgrade from 1000 Mbps network card to 10 Gbps network card with Zero-Copy RDMA function

In the first scenario, when upgrading from HDD to SSD, for database-level server-side software, the biggest difference lies in whether the capability advantage of SSD can be fully leveraged. As we introduced in Chapter 3, the performance of SSD in random read and write is two orders of magnitude ($\sim$100 times) better than that of HDD, but the advantage in sequential read and write is only several times (e.g., three to five times), that is to say, under different read and write modes, the performance difference between SSD and HDD will vary by 1–1.5 orders of magnitude (e.g., 100 times difference vs 5 times difference). For the database, the most direct difference will be reflected in the following two scenarios:

- The first type of scenario: continuous data writing or reading
- The second type of scenario: random read or update of data (delete, insert)

The first type of scenario is most typical of database batch loading data or continuous update log file operations; the second type of scenario involves almost all other operations, especially for access to index files or main storage files—index-file operations are mainly read-based, and the latter, at least for OLTP-type graph databases consider the efficiency of read and write operations.

Now, back to our original question, does simply upgrading from HDD to SSD automatically get all the performance benefits that an SSD claims to provide? The answer is, not necessarily. Software is often a restrictive factor and an optimized software architecture can more fully unleash the capabilities of the underlying hardware.

In the storage engine part of Chapter 3, we introduced WiscKey. By transforming the algorithm architecture of LSM-Tree and targeting the hardware characteristics of SSD, the performance of WiscKey has been greatly improved over LSMT. This transformation can be regarded as a typical business data-oriented performance optimization—to be precise, the system throughput rate is greatly improved after the size of the data field (the value in the KV store) exceeds 4 KB—the optimization at the software level can effectively expose the power of the underlying hardware.

The database query optimizer usually optimizes the internal logic of the query statement and completes the query through the execution path it deems optimal. The query optimizer designed based on the hardware characteristics of HDD can run on top of SSD without any modification. However, for data files, index files, cache data, and database log access and modification scenarios, through targeted optimization of SSD-oriented access paths—for example, reducing I/O requests, merging multiple operations, optimizing random reads and writes into sequential reads and writes, etc., these optimization operations can allow the system to achieve higher data throughput and lower query time consumption. Similarly, with multilevel storage and cache logic, lower-cost hardware can be used to achieve performance yields comparable to higher-cost systems.

In the second scenario, the upgrade from an 8-core CPU to an 80-core CPU appears to be a 10-fold increase in computing resources. However, without the full utilization of this computing power by database software, this upgrade will not bring about any change. This is also why many people selectively ignore the true meaning of vertical scalability—by writing high-concurrency software to make full use of multicore CPUs and multithread concurrency to achieve higher system throughput—and the effect of high concurrency is often times better than the performance of those so-called horizontally distributed systems because horizontally distributed systems usually have a large amount of network communication, which increases the system delay and decreases system throughput, and offsets the other benefits that a horizontally distributed system brings forward.

However, the difficulty of designing and implementing high-concurrency software is no less difficult than the construction of horizontal distributed systems, and although the differences in the focus of the two at the architectural level determine the difficulties in the respective implementation processes, their ultimate goals are to achieve higher system throughput through "concurrent and parallel execution"—be it over a centralized model or distributed model—the goal is to offer lower TCO (total cost of ownership), better overall performance, and ideally shorter time-to-value and time-to-market. The author suggests that from this moment on, abandon the debate between horizontal and vertical scaling and focus on how to implement concurrent software. There are three "pitfalls" in the design and development of high-concurrency software:

- Inability to Think Concurrently
- Algorithm (and Data Structure) Design
- Locks and Shared Mutable State

For most of our programmers (business personnel as well), the lack of concurrent thinking (as in parallel programming) makes us underestimate the benefits of vertical scaling and overestimate the returns of horizontal expansion. It may also make us short-sighted when designing concurrent system architectures that prevent us from thinking about longer-term goals, which in turn leads to system architecture that needs to be repeatedly overturned, and ultimately fails to achieve the preset design goals.

Similarly, what data structures can be selected to better adapt to and meet the customer's business needs (such as read-write allocation, access behavior patterns, maximum load conditions and common load patterns, etc.)? We need to admit that efficient algorithms can bring about exponential efficiency improvements, just as we introduced in earlier chapters that most graph queries and algorithms (such as K-Hop, Path queries with supernodes, Louvain, triangle computing, graph embedding, etc.) can achieve exponential performance boost via full concurrency with multicore CPU, and low-latency data structure utilization. For example, a K-Hop query starting from a node, assuming $K = 5$, in a horizontally distributed graph database system, whether starting the query from one instance or from multiple instances, as long as the multilayer neighbors of the initial node are distributed on multiple instances in the entire cluster, every hop deeper traversing the distributed dataset will bring about an exponential increase in latency—the worst scenario for a distributed system is when the data are *evenly distributed*, which means the interinstance communication cost and data migration cost will be at the climax—in such case, the cost of communications and data migrations within the distributed cluster fully offset the advantages of distributed storage, and eventually cause the query operation to either fail or take extremely long time to complete.

The existence of locks (and shared mutable state) is to ensure that multiple concurrent tasks (threads or processes) will not cause inconsistent or unpredictable data in the database due to out-of-control access. The existence of a lock means that the data resource it protects can only be accessed in a serial manner, which also implicitly shows that concurrent systems are not completely parallel. In fact, in any distributed system, there must be some links, components, and processes that are carried out serially. This may be the underlying truth of all distributed systems—it is just that in the overwhelming propaganda waves of distributed systems, we have turned a blind eye to these truths.

Fig. 2 shows that the running time of a completely serial program can be reduced from 50 to 40s after partial parallel transformation because 2/5 of the program is executed concurrently at two times the speed, which is an overall 1.25× acceleration. If all components can be modified, the efficiency of the entire program can be increased by 2×. If all components can be executed at 10× speed, the overall program will also gain 10× efficiency.

$$S_{\text{latency}}(s) = \frac{1}{(1-p) + \frac{p}{s}}$$

In computing architecture, Amdahl's Law describes the mathematical model of the efficiency improvement that can be obtained in the process of transforming the program from serial to parallel. Assuming that a program takes

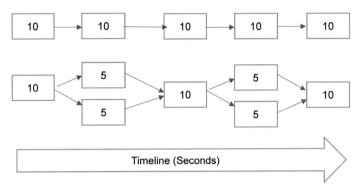

FIG. 2 Serial program and partially parallel program.

50 s to complete serially, but most of it (40 s) can be modified to make full use of the concurrent processing capabilities of the underlying hardware, but the 10-s portion cannot be modified, no matter how fully concurrent, the shortest time for the entire program cannot be lower than 10 s, then $P = 40/50 = 0.8$ in the Amdahl formula as shown previously. If concurrency can be transformed from serial $1\times$ to $80\times$, then $S = 80$, $S_{\text{latency}} = 1/(0.2+0.8/80) = 1/0.21 = 4.76$, which is very close to a theoretical $5\times$ speedup.

The challenge of transforming from serial to parallel is that when data is changed, errors will occur if concurrent access is not controlled. The following table (Table 1) demonstrates the complexity of the problem under multithreaded read and write conditions.

In the last (third) scenairo, when the network is upgraded from 1 to 10 Gbps, on the surface, a 10-fold increase in bandwidth is obtained, and the 10 Gbps network card and this theoretical network speed make it easy for us to have an "illusion" that the network speed has surpassed HDD and even SSD (Note: The performance of sequential read and write of general enterprise-level SSD is 500 MB/s, and the performance of random read and write will be lower)—this illusion will make us feel that the bottleneck of the distributed system is no longer existent because even if multiple node instances communicate with each other, with such a fast network, there is no need to worry about network delays at all. However, the reality is that, especially with database systems, the network is almost never the bottleneck of the system. To better illustrate this problem, let us do a thought-provoking experiment: suppose 10 Gbps network bandwidth is used to transfer 1 million small files, each $\sim$4 KB in size, and to transfer a 4 GB file (at the file size level 4 GB = 4 KB * 100 10,000), what is the difference in transmission time between the two?

The difference between the two transfer scenarios may be at least 100 times, or even as much as 300 times—1 million 4 KB files is much slower. Why is this so? Because most of us will habitually ignore the cost of I/O—when transferring a large file, assuming that the maximum and effective utilization rate of 10 Gbps

TABLE 1 Multithread contention problem.

Contention: Thread A goes first		
Thread A	*Thread B*	*V =*
READ V(11)		11
ADD 100	READ V(11)	11
WRITE V(111)	SUB 110	111
	WRITE V(-99)	-99
Contention: Thread B goes first		
Thread A	*Thread B*	*X =*
	READ V(11)	11
READ V(11)	SUB 100	11
ADD 100	WRITE V	-99
WRITE V(111)		
Ideally: A-B sequential execution		
Thread A	*Thread B*	*X =*
READ V(11)		11
ADD 100		11
WRITE V(111)		111
	READ V(111)	111
	SUB 100	111
	WRITE V(11)	11

bandwidth is $80\% = 8$ Gbps (1 GB/s), then the transfer can be completed in 4 s, but when transferring 1,000,000 small files, there will be millions of I/Os. Assuming it only takes 400 μs (0.4 ms) to read 1 file (block), it takes 400 s to read 1 million times. That is to say, 8 or 10 Gbps bandwidth is not a bottleneck at all, even 1 Gbps is more than enough for transferring these small files.

In fact, in the database system, the challenge of information synchronization among multiple instances is quite similar to the millions of small files scenario, and because these small files and small data packets are often only a few bytes to a few hundred bytes, the synchronization between multiple instances will generate a large number of frequent I/O operations. We can even say that in the design of distributed systems, the minimum use of network synchronization is a magic weapon to improve system performance (reduce latency). However, *avoiding network synchronization within a distributed system seems to be the same as pursuing vertical expansion instead of horizontal expansion*; and both the vertically and horizontally scalable systems seek to increase efficiency and speed through concurrency or parallelism—it is just that vertical system wastes less time and resources on synchronization. So, whether it is vertical scalability or horizontal scalability, the two seem to be more like Schrödinger's cat. Pursuing the ultimate concurrency is a lower-level challenge—it is the hard-core issue we need to pay attention to.

In the third scenario, there is another interesting technology called Zero-copy RDMA (Remote Direct Memory Access). Fig. 3 shows the advantages of the RDMA technology, which allows the application program of the request initiator to directly write data to the application program of the receiver instance through the network card supporting RDMA bypassing the multilevel buffering. In this process, compared with the traditional data exchange mode between network instances, RDMA has the following advantages:

- Zero-copy.
- Low CPU intervention.
- No Operating System kernel intervention required.
- Almost no network performance loss.
- Convergence of storage and computing.

However, the prerequisite for RDMA to be most effective is that the application needs to be greatly transformed to get rid of the dependence on disk I/O. That is, it is possible for in-memory databases to more fully utilize the hardware acceleration capabilities of RDMA. In fact, any database can achieve better speed by expanding on and fully utilizing memory. For graph databases, there are many channels for speeding up through memory, listed as follows:

- In-memory Indexing.
- Partial Data-persistency in memory: such as metadata.
- Intermediate Data Memory Storage.
- Cache Data resides in memory.
- Partial Log data in memory.
- All data is persistent in memory.

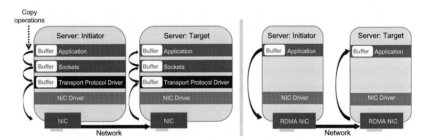

FIG. 3 Regular network communication vs RDMA.

Regarding memory, there are a few important things to know:

1. Memory will be the new disk. Large memory technology will undoubtedly be one of the most important computer architecture upgrades in the next 10 years.
2. Upgrading memory to obtain higher system performance is often the easiest and lowest-cost solution.
3. Memory, external memory, and the network can construct a whole set of multilevel storage acceleration architecture, and most of the existing database system transformation and acceleration have not fully considered such a strategy.

In computer architecture, there are HDD, HDD, Memory, and CPU cache components, if network storage is also included, there are at least five layers of storage, and there is an exponential performance gap between any two adjacent layers, as shown in Fig. 4. The physical existence of this performance gap exists objectively whether it is vertical scalability or horizontal scalability. What really allows us to improve performance is the realization of high concurrency and low latency. Perhaps this is the essence of any high-performance system.

The performance improvement of the graph database system is achieved through the following methods:

- Better hardware.
- System architecture that matches with the hardware, with maximum concurrency and low-latency design, including the choice of architecture, data structure, programming language, etc.
- Specifically, all possible operations in the graph database are optimized for concurrency to achieve the lowest latency and maximum throughput.

In the field of high concurrency (or high parallelism), there has always been an interesting and unintentional cognitive misunderstanding. High concurrency

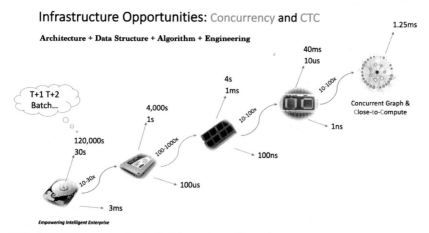

FIG. 4 Multitier acceleration in database system architecture.

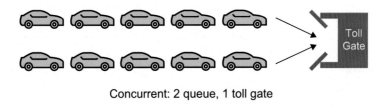

Concurrent: 2 queue, 1 toll gate

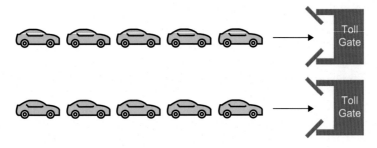

Parallel: 2 queues, 2 toll gates

FIG. 5 The original meaning of concurrency and parallelism.

and high parallelism are often used interchangeably, but they actually express different meanings, as shown in Fig. 5.

As shown in Fig. 5, high concurrency has an obvious single-node resource barrier, that is, the shared mutable state area, while a high-parallelization system does not have such a single resource bottleneck. However, in most cases, everyone has already equated high concurrency with high parallelism, and it does not explicitly reflect the existence of restrictive bottlenecks anymore. This book also no longer distinguishes between the two in accordance with the convention.

In the previous chapters, we have explained many times that the high concurrency of some operations in the graph database can achieve exponential performance improvement. The upper right part of Fig. 4 shows that in theory, if a serial execution at the CPU level takes 40 ms, making full use of 32× (3200%) concurrency will complete the work in 1.25 ms, but if it is at the memory level, it will take 4 s, at the SSD level, it will take more than 1 h, and at the HDD level, it will take more than a full day. The takeaway should be: Never underestimate the effect of concurrency acceleration, but this effect is relatively easier to obtain on a vertically and centrally scalable system because it is related to the characteristics of complex queries in graph databases—deep queries and complex graph algorithms all have a common feature that nodes tend to connect with large numbers of edges if these nodes and edges cannot be stored in the same data structure as a whole, there will inevitably be a large number of network requests or a lot of I/Os due to file system access, these additional requests and I/O are the worst enemies of any distributed system.

5.1.2   Horizontal scalability

Horizontally scalable systems thrive with the emergence of concepts such as cloud computing and big data. It must be said that scalability is a job that almost all contemporary IT practitioners must consider and face.

Before we start discussing the construction of a horizontally scalable graph database system, we need to have a clear understanding of the advantages and disadvantages of horizontally and vertically scalable systems. Table 2 lists the pros and cons between the two schools.

TABLE 2 Pros and cons of horizontal and vertical scalable systems.

Scalable system	Pros	Cons
Horizontal scalability	System performance improvements can be achieved in small steps (instance superposition mode)	Software has to deal with complex issues such as data (re)distribution, synchronization, and parallel processing
	Brings infinitely (substantially) scalable possibilities	There are very few systems that truly realize horizontal distribution
	Compared with vertical systems, low-configuration hardware preferred over expensive ones	The cost of hardware, software, and maintenance may not be as low as imagined
	High availability is the most important appeal and promise of a horizontally scalable system	**Data consistency** is very complex and hard to realize. Tradeoffs are often made
Vertical scalability	Most software does not need to be rewritten to directly obtain the system performance improvement brought about by hardware upgrades	The cost may increase significantly due to the purchase of high-end hardware (such as CPU)
	Relatively simple system installation, management, and maintenance	The upper limit of system performance improvement is predictable (limited by the upper limit of vertically scalable hardware capabilities)
	Data consistency is relatively easy to achieve	Single instance cannot achieve high availability. The system needs to be extended to master-slave, or more complex distributed mode to achieve high availability

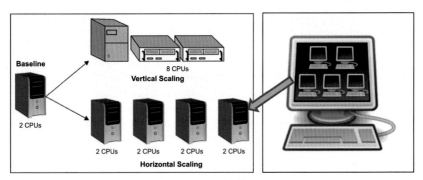

FIG. 6 Possible forms for low-cost horizontal scalability.

There are many forms of horizontally scalable systems. If divided according to the cost, a low-cost implementation method is shown in Fig. 6. From the original standalone 2-core CPU, horizontally expanding to four instances each with 2-core CPUs (8 cores in total)—the hardware investment is linear (4×), and there is even a possibility of sublinearity when there is a volume discount. In a vertical expansion scenario, upgrading from a 2-core CPU to an 8-core CPU may result in a price difference of more than four times (taking Intel X86 as an example, in its high-end server-class CPU, every 10% performance improvement means the doubling of the sales price, which is the main reason for Intel's high-profit margin. On the other hand, each multicore CPU is a set of complex distributed processing microsystems. With the increase in the number of cores, the complexity of intercore communication, collaboration, and consensus inches higher.). In comparison, vertical expansion may incur a superlinear cost increase though there is less hardware involved.

In fact, in the past 50 years of database system development, the first 20–30 years were dominated by miniframe architectures, and only in the past 20 years, with the evolution of distributed and PC server architectures, there have been distributed database systems built with dozens or hundreds of PC servers, and the overall price of the latter (PC cluster) can be lower than the former (miniframe), which is quite a spectacle. The Hadoop-like systems pioneered this spectacle. Their core idea is to use clusters of low-profile machines to realize multiple copy storage (HDFS) of massive amounts of data, and at the same time, it can perform nonreal-time processing (AP = Analytical Processing) of data in a divide-and-conquer manner (MapReduce). However, this concept is not suitable for graph databases, especially for scenarios that require online processing. What is more, with the vigorous development of cloud computing, many so-called horizontally scalable systems use virtualization technology to divide the resources that were originally from a physical server into multiple instances through virtualization, containerization, etc. (such as shown on the right side of Fig. 6), and clusters are formed between these instances. The construction of this so-called horizontally distributed cluster violates the core promise of distributed systems, which is to provide higher system throughput because these virtualized instances originated from the same underlying hardware, no matter how "cleverly designed," are not able to break through the performance bottleneck of the underpinning hardware. We can call this kind of horizontal scalability a typical distribution for the sake of distribution.

The core of distributed system design can be summarized as "distributed algorithm (logic)," which has four aspects:

- *Coordination* between multiple instances
- *Cooperation* between multiple instances
- Information dissemination and task decomposition (*Dissemination*)
- *Consensus*

The initials of these four aspects are C-C-D-C, which we call the CCDC of the distributed system. In the design and implementation of any distributed system, these four aspects must be addressed. All the knowledge we introduce in this section revolves around these four aspects.

There are many challenges faced by distributed systems, such as network issues, security issues, data consistency issues, state synchronization issues, system robustness issues, etc. We can generally summarize the following aspects:

- Local Execution vs Remote Execution
- State and data consistency issues
- Exception or failure handling
- Network-related issues (network partitioning, split-brain, etc.)

In the design and implementation of distributed systems, the first thing to consider is the difference between localized execution and remote execution of tasks. Both the former and the latter will involve the following three challenges:

1. *Decoupling*, that is, there is a time difference between a process (or thread) receiving a task, starting execution, completing execution, and returning, no matter how large or small the time difference is, decoupling (decomposition) helps to achieve modularization.
2. *Pipeline* or queue, the purpose of the pipeline (or queue) design is to avoid the situation of refusing to serve the next or multiple other tasks when the previous task is not completed.
3. Absorb and alleviate high loads (*Offloading*). As earlier, when high loads or extreme loads occur, even if the system takes longer to process each task on average, it is better than refusing to accept further loads, which is equivalent to downtime (or DoS = Denial of Service).

The previously mentioned three challenges are actually aimed at solving the problem of potential prolonged delays caused by remote task execution—the delay of remote tasks is naturally higher than that of local tasks—do not ignore two-way network communication and serialization, deserialization, interface calls, and other series of operations which bring about salient increase in time consumption!

The state and digital consistency of a distributed system is a very wide-ranging issue. In a distributed system, the concept of CAP has been deeply rooted in the hearts of many people, and it has even been raised to a theorem level—C.A.P = Consistency, Availability, and Partition tolerance (the three guarantees of CAP). The CAP theorem (or hypothesis) states that any distributed system (particularly data store) can only provide up to two of the three guarantees. Obviously, if the three guarantees of CAP are to be distinguished, the author thinks that availability is undoubtedly the first priority. If the system is not available, what is the point of having consistency and partitioning? The second is consistency, and the third is partition tolerance.

Based on the previous discussion, most systems will be implemented in a way that satisfies the A+C mode; individual systems will be implemented in accordance with the A+P mode, but some of the consistency will be sacrificed, that is, data inconsistency among multiple instances will appear in the cluster at a certain moment; rarely seeing the C+P type system, that is, data consistency and partitioning can be realized, but the system cannot provide external services (unavailable)—this kind of system probably only exists in the theoretical world. It is generally believed that Google Spanner was the first system to realize CAP, but the cost of its implementation is very high—Google uses GPS and atomic clocks to ensure the time synchronization of database servers around the world, while the consensus (data consistency) among instances is achieved through the Paxos algorithm. In addition, strictly speaking, Spanner implements sequential consistency—a weaker form of consistency than strong consistency (Strict Consistency), so it does not conflict with the core conclusions of CAP.

We need to understand that in a distributed system, the prerequisite for achieving consistency is consensus, which can be reached between instances first, and then state synchronization and data synchronization following the suit. In the following section, we use Google Spanner (collectively referred to as Cloud Spanner on Google Cloud) and its Paxos algorithm implementation as an example to introduce the core logic of the consensus algorithm.

Since its release in the late 90s, the Paxos algorithm has been used in a considerable number of big data platforms, cloud computing platforms, and NoSQL databases. Understanding Paxos' logic first requires understanding the three cluster instance roles it defines: proposer, acceptor, and learner, as shown in Fig. 7. The second is their interaction along the timeline, as shown in Fig. 8. Proposer initiates a proposal, the proposal will carry a unique value n, after getting the promise n of the acceptor, it will send the value n and a corresponding value, any proposal after this time point whose value is less than n will be ignored, and those greater than n can replace the previous value, but will not change the value. When this consensus is accepted and reached by the majority of instances, other roles in the cluster will automatically accept it, forming a consensus of all. The whole communication process is much like the TCP communication protocol in which three handshakes establish a connection and four handshakes terminate the connection, except that Paxos communication may require much more than four handshakes.

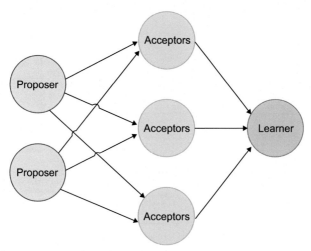

FIG. 7   Schematic diagram of Paxos consensus algorithm.

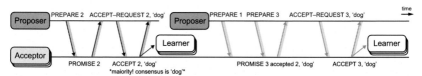

FIG. 8   Communications between Paxos roles along the timeline.

There are many variants of the Paxos algorithm. To be precise, it represents a large class of "consensus algorithms" (e.g., Classic-Paxos, Multi-Paxos, Fast-Paxos, Flexible-Paxos, Egalitarian-Paxos, etc.). The overall logic of the Paxos algorithm is relatively complex, the implementation complexity is high, and the communication is heavy, as shown in Figs. 7–9. In a large Paxos cluster of 110 instances, there may be a lot of network communication between Proposers(10), Acceptors(80), and Learners(20). The communication scale is about: (10 * 80 * $N + N * 80 * 20$), where $N \geq 4$—and this is just to reach a consensus, such as an election due to a role change. Network congestion, instability, or other reasons, such as clock errors, can cause this consensus-reaching operation to occur frequently, and repeat the same operation due to frequent failures, resulting in the network load pressure.

From the earlier Paxos consensus algorithm, we can see that the instances of the cluster are exchanging data according to a network communication mode whose complexity is close to that of the whole network broadcast—this mode is suitable for small and medium-sized networks (<30 nodes) but can create a lot of problems in medium to large networks. It is necessary for us to have a more comprehensive understanding of the possible modes of network communication and to choose the appropriate network communication mode wisely in

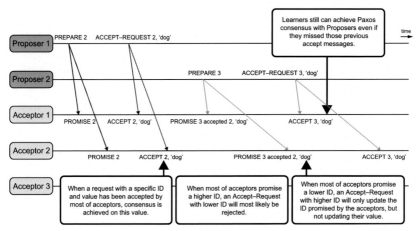

FIG. 9 Logic diagram of Paxos consensus formation between roles.

the design of distributed systems. Fig. 10 lists three common network communication patterns in distributed systems:

- Centralization (broadcast mode)
- Decentralization (local broadcast mode)
- Distributed (grid mode)

In 1987, Xerox Palo Alto Research Center published an article titled "Epidemic Algorithms for Replicated Database Maintenance."[a] In the thesis, three network communication methods are studied, namely: Direct Mail (centralized broadcast mode), Rumor Mongering (distributed network/grid mode), and Anti-entropy (decentralized, local broadcast mode). Note: The latter two decentralized modes can be regarded as generalized anti-entropy modes.

The concept of anti-entropy is particularly important in the design of distributed systems—entropy is an important concept derived from the field of thermodynamics, that is, the second law of thermodynamics, which states that there is irreversibility in the thermodynamic process, and any isolated system will spontaneously evolve toward the maximum entropy, also known as the maximum disorder state—if this concept is extended to a distributed system, multiple instances in the system will naturally produce more state inconsistencies over time, the data is inconsistent, so the anti-entropy communication mode is an important guarantee to achieve the state and data consistency between multiple instances through the distributed consensus algorithm.

There are many anti-entropy synchronization algorithms for distributed systems, which can be roughly divided into the following types:

- *Read synchronization*: Read-Repair, Digest-Read, etc.

---

a. https://courses.cs.washington.edu/courses/csep590/04wi/papers/demers-epidemic.pdf.

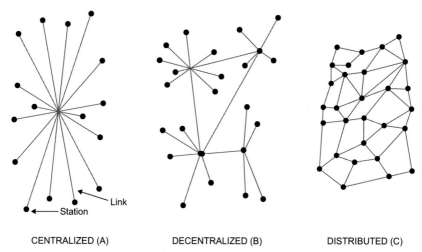

CENTRALIZED (A)   DECENTRALIZED (B)   DISTRIBUTED (C)

FIG. 10   Three network communication modes.

- *Write synchronization*: Hinted-Handoff, Sloppy-Quorum, etc.
- *Full data synchronization*: Merkle tree (by building a multilevel hash tree to track data in any range in the full amount of data and realize hierarchical synchronization), Bitmap Version Vector (by building and maintaining a bitmap version vector to ensure that instances in the cluster are recovering the latest data to be written after going back online).

The consensus algorithm is a very important part of distributed system design because other important features and components in this type of system rely on the distributed consensus algorithm. To name a few:

- *Fault-tolerance*: Distributed system achieves fault-tolerance through the reliability of data transmission and multiple copies, but multiple copies also mean that the difficulty of keeping data consistent increases with the increase in the number of replicas, even with exponential increase, that is, the complexity of maintaining $2*N$ copies is N^2 times as much as N copies (refer to the synchronization communication complexity in Paxos).
- *Reliability*: Reliability can be seen as a superset of fault tolerance, but its focus is slightly different, such as network reliability, storage stability, reliability of data lossless transmission, and network protocol design.
- *Availability*: High availability is the first problem that all distributed systems need to solve. Usually, in the design of distributed systems, a minimum number of instances must be online to provide complete services, such as >50% ($N/2$), and some systems have more stringent requirements, such as at least 2/3 of the nodes are online, or at least a complete 100% full amount of data is, and the specific percentage of online nodes to all nodes depends on the specific logic of data segmentation.

- *Integrity*: Data integrity includes physical integrity and logical integrity of data storage. The underlying implementation of a database uses locks to achieve data integrity at the logical level, especially in scenarios with high concurrent access. Although the use of locks is necessary in many cases, it should be avoided as much as possible, because once the lock is locked, all other related processes (or threads) will be in a waiting state, which can be regarded as a waste of resources. A good design will try to avoid locking. The design and appearance of latch(es) is to ensure data integrity at the physical level. It has a finer granularity than locks and is at the Page level, so it takes less time.

In a concurrent system, regarding the use of locks or latches, the following three situations are generally encountered:

- *Concurrent read-only*, which is the simplest concurrency scenario, can maximize concurrency completely under lock-free conditions.
- *Read-write mixed mode*: The optimization point is that we can separate the read-write tasks to ensure that multiple write tasks do not overlap each other, and the read tasks can be completely concurrent when there are no write tasks. However, the difficulty lies in that when there are simultaneous reading and writing, RW Lock (read-write lock) can help solve the contention problem.
- *Concurrent writing* (updating): requires locks, and lock optimization logic includes finer lock granularity. In traditional databases, there are database-level, table-level, and row-level locks; in graph databases, it depends on the logic of the underlying storage engine, no matter it is row storage, column storage, KV storage, or other storage modes, similar coarse-grained to fine-grained lock logic can be implemented. The main dimensions to consider include the scope of the data involved in the write operation itself—is it the entire graph or just a small portion of the graph on one instance—because of the characteristics of the data structure of the graph, we generally know that the optimal efficiency of multiple concurrent writes can be achieved according to vertex locks, edge locks, vertex attribute locks, edge attribute locks, or more refined lock logics for enhanced concurrent-write performance.

Most of the concurrency implementation logic is called Pessimistic Concurrency Control, such as the famous 2PC (Two-Phase Commit)—essentially a "blocking protocol," because coordinating processes within the cluster or node failures causing the node to go offline can trigger blocking. Therefore, someone proposed 3PC (Three-Phase Commit), which avoids blocking by introducing the Abort state mechanism and helps the system recover quickly by introducing the PreCommit (also known as the Prepare, preparation phase) phase—even when both coordinating and participating nodes fail during the commit phase. 3PC therefore brings higher fault tolerance to distributed systems than 2PC implementation.

Most of the lock implementations mentioned previously are pessimistic locks. In the field of distributed databases, a more efficient concurrent implementation method is "optimistic lock." Compared with the implementation mechanism of pessimistic locking (or pessimistic concurrency), it is to judge whether there are Transaction Conflicts and decide whether to block or terminate the task when the task is running. Optimistic locking (or optimistic concurrency) assumes that transaction conflicts are a small probability event, so transaction execution is split into three phases:

- Reading phase
- Verification phase
- Writing stage

Usually, the time consumption of the verification and writing phase is shorter than that of the reading phase. When the verification phase is successful in most cases, 3PC can obtain more efficient concurrency control than 2PC, although 3PC also contains critical areas (Critical Section), accessing this area under the condition of distributed transaction also needs to be serialized.

MVCC or MCC (Multiversion Concurrency Control) technology minimizes cross-task coordination at the storage level by allowing multiple versions of data records to be created as well as transaction IDs and timestamps—read tasks will not block write tasks and vice versa. Like traditional databases, graph databases can also use MVCC technology to support concurrent read-and-write scenarios. MVCC still uses locks, but reduces the use of locks (and mandatory serialized access) as much as possible, and therefore (with high probability) achieves better performance (shorter time consumption)—it can be counted as a strategy of exchanging space for time, as shown in Fig. 11.

If information is updated on the graph dataset, whether it is insertion, update, or deletion of vertex or edge data, it can be realized through MVCC. The logic of MVCC implementation at the micro level is similar to Copy-on-Write, that is, a write (update, delete, or insert) task in the graph database does not directly operate on the original data, but first creates a new version of the data item. In complex cases, multiple versions may exist at the same time, but the version that each concurrent task can see depends on the current isolation level—generally speaking, databases can implement four isolation levels from low to high:

- Read Uncommitted
- Read Committed
- Repeatable Read
- Serialization

The highest level of isolation is to avoid some possible problems, such as Dirty Read, Un-repeatable Read, Phantom Read, etc., but the higher the level of isolation, the lower the performance of the database, and the highest level is not used by default in most database implementations. Most databases' default

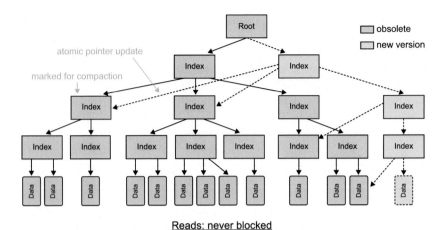

Reads: never blocked

FIG. 11 MVCC illustrated.

isolation defaults to read committed mode, and treats read uncommitted and committed equally.

When the write task creates a new version, other concurrent read tasks access the old version. In this way, as time goes by, many old data versions in the MVCC system must be cleaned up. Obviously, if the full amount of data needs to be scanned (e.g., table scan operations in traditional databases, in graph databases this is like scanning all vertex or edge data structures) to update the latest version of data into the database, and all the old version data is cleared (as shown in Fig. 11, the new data is linked to as the latest data, and the old data is deleted), which will be a very expensive operation (called Stop-the-world in relational database, e.g., in PostgreSQL, this operation is called VACUUM FREEZE[b]).

Another possible way to implement "multiversion concurrency" is to implement Snapshot Isolation which divides data into two parts at the corresponding storage engine layer: source data and undo-log—the source data keeps the metadata of the latest committed version, and the undo-log log can be used to create old versions of the source data. A potential problem with this design is that when the amount of source data is large and update operations occur frequently, the undo-log will be several times larger than the source data, and maintaining such a huge log will be a complex challenge. Log access slowdown and the loss of readability due to insufficient storage space will result in the inability to create snapshots, which in turn will cause the system to fail to work properly due to the inability to achieve snapshot isolation.

The biggest difference between graph databases and all other types of database systems is that the architecture design of traditional SQL databases

---

b. VACUUM FREEZE https://www.postgresql.org/docs/9.4/sql-vacuum.html.

is based on the storage engine of the storage layer. This design has its inevitability—the storage of two-dimensional relational tables naturally matches the storage engine architecture, and this design idea continues to the NoSQL field. Wide-table databases, column databases, document databases, and time-series databases all use similar storage logic. However, if the graph database also uses a two-dimensional relational table structure, the biggest challenge is that it is difficult to achieve high-dimensional graph queries. The general characteristic of graph data is that there is a high-dimensional relationship between data, and the challenges brought by this high-dimensional relationship cannot be solved by traditional distributed storage. Let us use a simple example to explain what is meant by high-dimensional association.

Suppose we have 200 vertices, each vertex has 100 edges associated with other vertices on average, and the whole graph has 10,000 edges and 200 vertices (Note: each edge connects two vertices). If we calculate the shortest path from any vertex in the graph to any other vertex, theoretically, the computational complexity of calculating the shortest path or the largest K-neighbor on this highly connected graph is: $(|E|/|V|)^k$, where E is the number of edges, V is the number of vertices, $K \sim = |V| * |V|/|E| = 4$, and the complexity is about 100^4 (about 100,000,000). If this graph is enlarged 1000 times proportionally (10,000,000 edges, 200,000 vertices), the computational complexity of the same query does not increase! Because starting from a certain vertex in the large graph, the number of neighbors in each hop is still only 100 on average, as the $|E|/|V|$ ratio is unchanged. This is the characteristic of graph databases. The complexity of graph data-oriented networked analysis is not directly proportional to the global data volume, but related to the topology, query logic, and query mode of specific datasets.

In actual realization of the shortest path query, if we adopt a horizontally distributed architecture (and data structure), any query for neighbors on the next hop can only find these neighbor vertices on multiple instances scattered across the cluster. Factors such as network delays, information synchronization, and even locks will cause the performance of this query to be exponentially slower than when all the neighbors are on the same instance, and the deeper you traverse, the slower the query becomes.

In SQL databases, a deeper query is equivalent to one more table participating in the JOIN calculation, and the complexity of the algorithm will increase exponentially. The horizontally distributed architecture can show its advantages only when facing metadata (shallow data) queries and calculations. For example, when traversing 10,000 vertices and 100 million vertices, the complexity of the latter is indeed 10,000 times that of the former, and the horizontally distributed architecture can greatly increase throughput through distributed computing. Suppose there are 10 instances to process the query, a near-linear $\sim 10\times$ performance improvement over a single-instance can be achieved. However, if these 10 instances are used to complete a multihop path query, the result may be that each hop (layer) of depth increase comes with

~10× performance degradation. This is where graph databases (designs) are more complex than traditional databases. In the next section, we will focus on the design of specific high-availability and distributed graph systems, and compare the pros and cons of different architectures.

5.2 Highly available distributed architecture design

In the previous section, we explored several possible ways to scale a system and the pros and cons of each. In this section, we can use specific architectural design schemes to evaluate the design considerations, ROI (Return over Investment) considerations, various indicators, and functionality choices.

When designing a high-performance, high-availability graph database, starting from a single instance and a single node, we generally have three architecture evolution options:

- Active and standby high availability
- Multi-Instance Consensus
- Large-scale Horizontally Distributed

We all know that the implementation complexity of these three systems is gradual from low to high, but this does not mean that the more complex system can obtain better results under different application scenarios, user needs, query patterns, query complexity, and data characteristics.

As future graph database architects, users, or enthusiasts, we hope that every reader can calmly and soberly analyze the challenges you face and identify the most suitable solutions you need when picking an architectural design. Fig. 12 shows the entire evolutional pathway of the graph database from standalone single-instance to horizontal scalability.

5.2.1 Master-slave HA

The simplest high-availability database is expanded from a single instance to a dual instance, but the two instances setup can differentiate into many possibilities for role-playing:

- One instance (A) is responsible for reading and writing, and the other instance (B) is responsible for backup.
- One instance is responsible for reading and writing, and the other instance can participate in the read-type load-balancing.
- Both instances support reading and writing, and they are mutual backups to each other.

In the first role-playing, instance A is responsible for carrying out all client requests, while instance B generally does not interact directly with the client, and it is only responsible for passively accepting backup requests from instance A. Only when instance A goes offline for some reason, instance B will go online and begin to carry customer load.

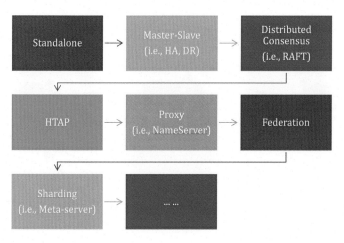

FIG. 12 From single instance to horizontal scalability.

In fact, even in such a seemingly simple active-standby mode, there are still many details worth considering, such as:

- How to ensure reliable communication between instances A and B?
- When one instance goes offline, how can another instance be brought online or stay online?

The search for the answers to the previous two seemingly simple questions will lead to the "Pandora's box" of networked and distributed system architecture design—unless we can be 100% sure that the network is 100% reliable, and the running program and data are 100% safe and reliable, otherwise, it is quite complicated to determine the reliability of communication and data from A to B or B to A. Because, when A sends the backup information to B, how to determine that B has received the information and completed the backup operation? Naturally, we want B to send a receipt or even two receipts, one to express receipt (ACK) and the other to express completion (ACK+DONE) to A. However, do we need to let B also know that A has received its reply? This communication process of replying and replying can become an endless cycle of dependence. Fig. 13 graphically illustrates this possible problem of unlimited communication (synchronization) between the two instances—this problem is famously called the "Two Generals' Problem."

The two-general communication problem is a simplified version (a special case) of the Byzantine Generals Problem, which expresses the existence of a possibility that the system cannot be consistent in the presence of arbitrary communication failures. In real-world engineering practice, we can only avoid the occurrence of extreme situations to a certain extent, such as the three-way handshake in the TCP protocol to establish a network connection, and the four-way handshake to terminate the TCP network connection scheme, we can only assume that in most cases, the communication network is reliable, and the

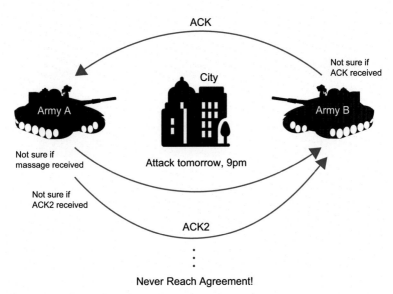

ACK

City

Not sure if
ACK received

Army A

Army B

Not sure if
massage received

Attack tomorrow, 9pm

Not sure if
ACK2 received

ACK2

Never Reach Agreement!

FIG. 13 Two Generals' Problem.

programs running on the A-B instances have integrity—the communication problem between the two armies tells us that any system will have unreliability, which is why we use several 9 s to measure the system reliability, such as an online rate of five 9 s (99.999%). We have seen some public cloud service providers claiming a reliability of eleven 9 s (equivalent to a 1-s offline failure every 3000 years), but simply unplugging one or two network cables or killing a process or two can bring the entire system offline. We are not even sure that any computer system created by humans can survive 50 years without failure. After all, no computer system has been used continuously for 50 years.

If the dual instance design continues to evolve, a cluster with at least three instances can be constructed, as shown in Fig. 14—such cluster can be called MSS (Master-Slave-Slave system) or MS-LBS (Master-Slave Load Balanced System).

When the MSS system has three instances (A, B, and C), the possible communication between them becomes more complicated, and there are at least 8 (2*2*2) possible ways of interaction. Usually, we will start with the simplest master-standby implementation, that is, only synchronize data from A to B and C in one direction. When A goes offline, select (manually or automatically switch) an instance between B and C as the new master. The master node is responsible for addressing requests from clients.

However, when the A instance goes online again, there will still be a need to synchronize data from the B or C back to A. While B becomes the primary instance, if instance C goes offline, only instance B in the cluster is online and can still provide services, although in this case, it is no longer a highly available system.

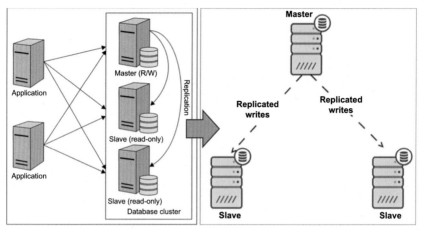

FIG. 14   Schematic diagram of master-slave backup system.

Another common implementation of the active/standby system with load balancing to a certain extent is that the active instance (A) carries all the read and write operations, and instances B and C load balance all the read-type operations from the clients and they synchronize with A constantly.

In the system architecture of active-standby mode, a big assumption is that in any time slice, at least one instance has full and latest data. If this premise cannot be guaranteed, the data consistency of the current system has been destroyed (another possibility is that the system is not running in the active/standby mode, which is a question we need to discuss later in this section).

The architecture of the active and standby systems can also evolve into modes such as intracity disaster recovery and remote disaster recovery. The remote disaster recovery model, as shown in Fig. 15, typically has only one cluster working online and the other cluster passively accepting synchronized data as a whole. To some extent, such a system is highly redundant, at least at the time of write operations, only 1/6 of the nodes are working, while the other 5/6 nodes perform data synchronization, and it is divided into two phases of data synchronization, that is, 2/3 of the instances in the primary cluster and 1/3 of the instances in the secondary cluster perform the first phase of synchronization, and the other 2/3 instances in the secondary cluster perform the second stage of synchronization. In the first phase of the synchronization process, there is a greater delay in the synchronization completion time of the primary instance of the secondary cluster due to the limitation of distance and network bandwidth, but often times we ignore this delay. In the actual 30-km intracity dual data center, it takes 0.0001 s (0.1 ms) over fiber optics for data transmission, if it is an operation requires ACK, it will take 0.2 ms (or 0.3 ms over three-way communication), which is already a nonnegligible delay in a real high-performance system design. Besides, such delay will be aggravated and amplified when passing through many processes within the system.

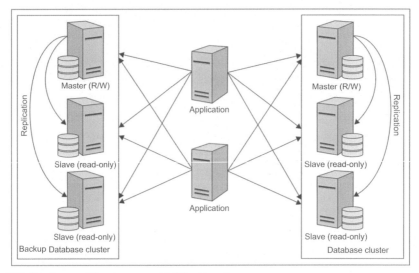

FIG. 15 Remote (disaster recovery) MSS system diagram.

This is why we as consumers may experience delays in many transactional scenarios because the transaction may go through multiple sets of business systems, such as multiple models of the antifraud system and in the full submission of transactional processing, a delay of 2 s is extremely normal. It is precisely because of these communication delays that the ability of graph databases to process massive data online (low latency) and high concurrency is particularly valuable, after all, the complexity of high-dimensional correlation, aggregation, and deep penetration calculations is significantly higher than that of low-dimensional and shallow calculations of traditional databases.

5.2.2   Distributed consensus system

We mentioned in the previous section that even in the simplest active-standby system architecture, the problem of inability to guarantee consistency in the system may occur. This is also because any distributed system is asynchronous and distributed in nature. All operations (network transmission, data processing, sending receipts, information synchronization, program start or restart) take time to complete. Any transaction processing is asynchronous even within one instance, and this asynchrony between multiple instances will be greatly magnified. One of the most important principles in distributed system design is: coexist with asynchronicity, do not pursue perfect consistency, but you can assume that the entire system can work normally most of the time, even if some processes or networks have problems, the entire system can still provide external services.

The distributed consensus algorithm (and system) came into being and is used to ensure that even if various problems occur in the distributed system, the overall service can still remain online. The distributed consensus algorithm has several core features:

1. Validity (Validated Proposal)
2. Unanimity (Reaching Agreement)
3. Quick Termination (Process Can be Terminated)

Validity means that the dissemination of instruction information between processes needs to be based on reasonable and effective data, and the effective set of processes can reach a consensus, and the time consumption for finally forming a consensus and terminating the synchronization process needs to be reasonable, completed in a short period of time. A high-performance distributed system should work at the millisecond level, while a large-scale cross-region distributed consensus system may need seconds to minutes to form consensus or require manual intervention. The specific termination time delay depends on the corresponding business need.

Reaching unanimity = forming consensus, similar to a democratic election among multiple processes, once they form a consensus, any process that participated in the election is no longer allowed to disagree with the result or perform tasks in a way that does not match the result—this is the "Byzantine general problem" we mentioned earlier.

In Fig. 16, the process of forming a consensus among multi-instances (multi-processes) is illustrated. This process is not fundamentally different from the behavior in our daily life. A, B, C, and D discuss where to go at night. At first, A proposes to go to the movies, and B agrees, but C quickly proposes to go to dinner. D agrees with C, and then B agrees with C's proposal, and finally, A also agrees to C's proposal instead. At this time, the four people reached a consensus to go to dinner in the evening. This is the simplest form of multitask consensus formation. In the actual distributed consensus algorithm, issues such as roles, stages, and how to terminate the consensus will also be introduced, which we will analyze one by one in the following.

The process of forming a consensus requires a clear termination algorithm, otherwise, there will be problems of unresolved and infinite waiting. For example, when a certain instance (process) goes offline, if the remaining processes wait for it to come back online indefinitely, or as shown in Fig. 16, the four processes of A-B-C-D have endless new proposals, so that the four parties can

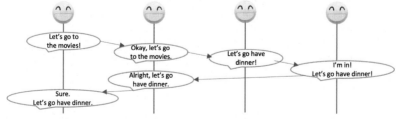

FIG. 16   How to reach consensus among multiple instances.

never reach a consensus. Consensus algorithms need to consider these factors and avoid their occurrence. Essentially, no matter what cross-process intracluster communication method is used, the algorithm needs to be as concise as possible, so that the cost of reaching (and terminating) consensus is as low as possible.

So, what kind of communication means should the distributed consensus system adopt? In the previous section, we mentioned the three communication methods of the network system (centralization = broadcasting, decentralization = hierarchical regional broadcast or multicast, and point-to-point distribution). For small distributed systems, the simplest and most direct way is the broadcast method, but there are many details in it because the interaction logic between the broadcaster and the information receiver determines whether this interaction process belongs to the one-way one-time transmission mode of the type of "doing-everything-according-to-fate" or other more reliable interaction modes. Understanding this process requires us to consider a possible situation, that is, if the instigator A sends a request to B and C and goes offline, but does not send it to D, then in the broadcast algorithm, we need to consider having B and C to continue broadcasting to D. From the perspective of communication complexity, such an implementation is a "reliable broadcast" mode. In a distributed system with N instances, its complexity is $O(N^2)$. Obviously, such a flooding communication mode is inappropriate for large-scale distributed systems, but it already has the characteristics of achieving system integrity and consistency through redundant communication.

Next, let us analyze how to ensure the order of messages in the process of distributed consensus communication, that is, at least two features need to be realized: multimessage transaction (atomic, indivisible), and sequence maintenance. A relatively simple implementation of the atomic broadcast algorithm is ZAB (Zookeeper Atomic Broadcast) in the distributed KV project Apache Zookeeper,[c] which divides all processes in the Zookeeper cluster into two roles: leader or follower, three states: following, leading, or election, and its communication protocol divided into four phases:

- Phase 0: Leader Election Phase
- Phase 1: Discovery Phase
- Phase 2: Synchronization Phase
- Phase 3: Broadcasting Phase

In the leader election phase (Phase-0), a process in the election state starts to execute the leader election algorithm to vote a process in the cluster to become the leader.

In the discovery phase (Phase-1), the process checks the votes and decides whether to become one of the two roles. The process that is elected as a potential leader is called a prospective leader, and then the process will communicate

---

c. ZooKeeper's ZAB http://www.tcs.hut.fi/Studies/T-79.5001/reports/2012-deSouzaMedeiros.pdf.

with other follower processes to discover the latest accepted transaction execution order.

In the synchronization phase (Phase-2), the discovery process is terminated, and the follower process is synchronized with the prospective leader's updated history. If the following process's own history is slightly behind the prospective leader's history, send an acknowledgment to the prospective leader process. When the prospective leader is confirmed by the quorum, it sends a commit message. From this point on, the prospective leader becomes the established leader. The pseudo-code logic is as follows:

```
/* ZAB Phase-2: Synchronization Phase (Pseudo-Code)*/
// Leader L:
Send the message NEWLEADER(e' , L.history) to all followers in Q
upon receiving ACKNEWLEADER messages from some quorum of
followers do
    Send a COMMIT message to all followers
    goto Phase 3
end

//Follower F:
upon receiving NEWLEADER(e', H) from L do
    if F.acceptedEpoch = e' then
        atomically
            F.currentEpoch ← e' // stored to non-volatile memory
            for each ⟨v, z⟩ ∈ H, in order of zxids, do
                Accept the proposal ⟨e', ⟨v, z⟩⟩
            end
            F.history ← H // stored to non-volatile memory
        end
        Send an ACKNEWLEADER(e',H) to L
    else
        F.state ← election and goto Phase 0
    end
end

upon receiving COMMIT from L do
    for each outstanding transaction ⟨v, z⟩ ∈ F.history, in
order of zxids, do
        Deliver ⟨v, z⟩
    end
    goto Phase 3
end
```

In the broadcast phase (Phase-3), if no new downtime problems occur, the cluster will remain in this phase. At this stage, there will be no two leader processes, and the current leader process will allow new followers to join and accept the synchronization of transactional broadcast information.

Phases 1–3 of ZAB are all asynchronous and detect whether there is a failure through regular heartbeat information between the follower processes and the leader process. If the leader process does not receive a heartbeat within the preset expiration time, it will switch to the election state and enter Phase-0.

Conversely, the follower process can also enter Phase-0 after not receiving the heartbeat from the leader process.

In the specific implementation logic of ZAB, the leader election is the core part. ZAB adopts a simplified but optimized strategy, that is, the process with the latest historical record will be elected as the leader, and it is assumed that all processes having the most recent proposed transactions have all committed transactions. The premise of this assumption is the serialization and sequential growth of IDs in the cluster, this assumption greatly simplifies the logic of leader election, and therefore it is called Fast Leader Election (FLE). However, even so, there are still many considerations in the FLE process, especially in various boundary conditions. The following is an excerpt of the FLE implementation code logic of the processes participating in the election process:

```
// ZAB Faster Leader Election (Pseudo-Code)
// Peer P:

timeout ← T0 // use some reasonable timeout value
ReceivedVotes ← ∅; OutOfElection ← ∅ // key-value mappings where
keys are server ids
P.state ← election;
P.vote ← (P.lastZxid, P.id);
P.round ← P.round + 1

Send notification (P.vote, P.id, P.state, P.round) to all peers

while P.state = election do
    n ←(null if P.queue = ∅ for timeout milliseconds, otherwise
    pop from P.queue)
    if n = null then
        Send notification (P.vote, P.id, P.state, P.round) tall
        peers
        timeout ← 2 × timeout, unless a predefined upper bound
        has been reached
    else if n.state = election then
        if n.round > P.round then
            P.round ← n.round
            ReceivedVotes ← ∅
            if n.vote ≻ (P.lastZxid,P.id)then
                P.vote ← n.vote
            else
                P.vote ← (P.lastZxid, P.id)
            Send notification (P.vote, P.id, P.state,
            P.round) to all peers
        else if n.round = P.round and n.vote ≻ P.vote then
            P.vote ← n.vote
            Send notification (P.vote, P.id, P.state,
            P.round) to all peers
        else if n.round < P.round then
            goto line 6

        Put(ReceivedVotes(n.id), n.vote, n.round)
        if |ReceivedVotes| = SizeEnsemble then
            DeduceLeader(P.vote.id);
            return P.vote
        else if P.vote has a quorum in ReceivedVotes
        and there are no new notifications within T0
```

```
        milliseconds then
            DeduceLeader(P.vote.id); return P.vote
        end
    else // state of n is LEADING or FOLLOWING
        if n.round = P.round then
            Put(ReceivedVotes(n.id), n.vote, n.round)
            if n.state = LEADING then
                DeduceLeader(n.vote.id);
            return n.vote
        else if n.vote.id = P.id and n.vote has a quorum in
        ReceivedVotes then
            DeduceLeader(n.vote.id); return n.vote
        else if n.vote has a quorum in ReceivedVotes and
        the voted peer n.vote.id is in state LEADING and
        n.vote.id ∈ OutOfElection then
            DeduceLeader(n.vote.id); return n.vote end
        end

        Put(OutOfElection(n.id), n.vote, n.round)
            if n.vote.id = P.id and n.vote has a quorum in
            OutOfElection then
                P.round ← n.round
                DeduceLeader(n.vote.id); return n.vote
            else if n.vote has a quorum in OutOfElection and
            the voted peer n.vote.id is in state LEADING
            and n.vote.id ∈ OutOfElection then
                P.round ← n.round
                DeduceLeader(n.vote.id);
                return n.vote
            end
end
```

Zookeeper and ZAB first started as a subproject out of Yahoo!'s Hadoop project, it was later spun off as a distributed interserver process communication and synchronization framework and became a top-level project in the Apache open-source community after 2010. From the pseudo-code as shown earlier, the algorithm logic and communication steps in ZAB are rather complicated. Similarly, the implementation of Paxos-like algorithms has strict requirements for clock synchronization between system instances across geographical regions, and the algorithm logic is very complex and hard to understand. In Google's Spanner system, an atomic clock is used to implement Paxos, but for enterprises that cannot design and manage such a complicated system architecture, Paxos seems to be a major hurdle to overcome.

After 2014, a simpler and explainable distributed consensus algorithm came into being, the most famous of which is the RAFT algorithm proposed in Diego Ongaro's doctoral dissertation called LogCabin.[d]

---

d. RAFTpaper http://web.stanford.edu/~ouster/cgi-bin/papers/OngaroPhD.pdf.

In RAFT, each consensus participating process in the cluster may take one of three roles:

- Candidate
- Leader
- Follower

Unlike Poxos, which uses clock synchronization to ensure the global order of data (transaction) is consistent but similar to ZAB, RAFT divides time blocks into terms. During each term (the system uses a unique ID to identify each term to ensure that there will be no term conflicts), the leader is unique and stable.

There are three main components (or stages) in the RAFT algorithm:

- Election phase
- Regular heartbeat phase
- Broadcast and log replication phase

Fig. 17 illustrates the interaction between the client and the server cluster in a 3-node RAFT cluster. The whole process revolves around the leader role process, which can be decomposed into 10 steps, and this only covers RAFT's broadcast and log propagation phase.

The three roles and their responsibilities are as follows:

- Followers will not actively initiate any communication, but only passively receive RPCs (Remote Procedure Calls).
- Candidate, who will initiate new elections, incrementally controls election terms, issues votes, or restarts the earlier tasks—in this process, only candidate with all submitted commands will become leader and notify other candidates of the election results via RPCs, and avoid split-vote pitfall (random election timeouts are used, e.g., randomly creating timeouts between 0.15 and 0.3 s to avoid problems caused by two instances simultaneously issuing candidate process resulting in a tie in the vote count). In addition, each process maintains its own set of logs, which is called LogCabin in the original RAFT algorithm implementation.
- Leader, periodically sends heartbeat RPCs to all followers to prevent expiration (and re-election) due to excessive idle time. The leader is usually the first to face the client process request, append to the log process and initiate replication, submit and change its own state machine, and synchronize the log to all followers (as shown in Fig. 17).

What RAFT describes is a general algorithmic logic. There are many kinds of specific implementations, and there is a lot of room for adjustment and tuning. For example, the original RAFT algorithm is similar to the MSS (One Master and Multiple Slaves) architecture, and only one instance is serving client requests at any time. If we contemplate the possible query scenarios of graph databases, we can transform and augment RAFT in stages (difficulty from low to high):

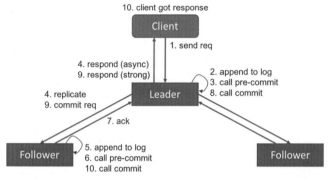

FIG. 17   RAFT cluster workflow decomposition.

- Multiple instances receive read request loads at the same time: writes are still implemented through the leader node, and read loads are balanced among all online nodes.
- Multiple instances simultaneously receive read-first and then write request loads: typically, for graph algorithms that require write-back, all nodes can carry the graph algorithm. In the write-back part, local write-back is performed first, and then asynchronously synchronized to other nodes.
- Multiple instances simultaneously receive and forward update request loads: write requests can be sent to any node in the cluster, but followers will forward them to the leader node for processing.
- Multiple instances process update (write) requests at the same time: This is the most complicated situation, depending on the requirements of the isolation level. If multiple requests change the same piece of data on multiple instances at the same time with different data values, it will cause data inconsistency. The most reliable way to achieve consistency in this case is to use serialized access to critical regions! This is what we have repeatedly mentioned in this chapter. It is critical for a distributed (graph) system to consider the need for serial processing when necessary.

At present, there may be more than 100 specific implementations of the RAFT algorithm, such as ETCD, HazelCast, Hashicorp, TiKV, CockrochDB, Neo4j, Ultipa Graph, etc. They are implemented in various programming languages, such as C, C++, Java, Rust, Python, Erlang, Golang, C#, Scala, Ruby, etc., more than enough to reflect the vitality of distributed consensus algorithms and systems.

In the highly available distributed system architecture based on the consensus algorithm, we made an important assumption that most of the time, each instance of the system has a full amount of data. Note that "most of the time" means that at a certain point in time, there may be data or state inconsistencies among multiple instances, and therefore it is necessary to pass a consensus algorithm in the distributed system to achieve eventual data consistency.

5.2.3 Horizontally distributed system

By now, the readers should have a certain understanding of the complexity of the design and implementation of distributed systems. In this section, we need to explore the most complex type of system architecture design in distributed graph database systems—horizontally distributed graph database systems (Note: This challenge can also be generalized to horizontally distributed graph data warehouses, graph lakes, graph platforms, or any other system that relies on graph storage, graph computation, and graph query components).

Before we start exploring, we should make it clear that the purpose of building a horizontally scalable graph database system is not to allow clusterized machines with lower configurations to carry tasks that can only be completed by a single machine, nor is it to reduce the machine configuration, then accomplish the same task with a larger cluster size. On the contrary, under the same machine configuration, a cluster composed of multiple machines can achieve larger-scale data throughput. It is best to achieve a linear distribution of system throughput with the growth of hardware resources.

The two routes we described earlier, for example:

- Route 1: Originally one machine with an 8-core CPU, 128 GB memory, and 1 TB hard disk, now it is necessary to replace with eight low-end machines, each with 1-core CPU, 16 GB memory, and 0.125 TB hard disk, and hoping to complete the same task, and achieving better effect.
- Route 2: Originally one machine with an 8-core CPU, 128 GB memory, and 1 TB hard disk, now uses eight instances with the same configuration to meet the challenge that is several times (theoretically ≤8) as much as on the previous one instance.

In route 1, if only very simple microservices are provided to the client, it is feasible to split 1 machine into 8 machines, and because of the potential higher IOPS provided by 8+ hard disks of 8 machines. However, the computing power of a 1-core CPU may be less than 1/8 of an 8-core CPU. The limitation of 1-core computing resources will only allow any operation to be performed serially, and network communications with other instances in the cluster will greatly reduce the efficiency of any task execution. This means that downgrading (low configuration) operations are only suitable for short-chain, single-threaded, and simple types of database query operations—in other words, storage-intensive or I/O-oriented operations can be solved with this "low configuration distribution." This is why in the past, in the development process of databases with relational databases being the mainstream, storage engines were "first-class citizens," while computing exists attached to storage engines is a "second-class citizen."

In the analysis of route 1, there is an important concept: short-chain tasks, also known as short-chain transactions in the Internet and financial industries, such as flash-sales operations and simple inventory inquiries—the operation logic involved is straightforward. Even on a large dataset, it only needs to locate and access a very small amount of data to return, so it can achieve low latency and large-scale concurrency.

Corresponding to short-chain tasks, long-chain tasks are much more complicated, and the data access and processing logic are more sophisticated. Even on a small amount of datasets, a relatively large amount of data may be involved in processing logic. If a traditional architecture (such as a relational database) is adopted, the operation time of long-chain tasks will be much longer than that of short-chain tasks, and each operation itself occupies a higher degree of computing and storage resources, which is naturally not conducive to the formation of large-scale concurrency. In Fig. 18, we graphically illustrate the differences between these two operations. In route 1, short-chain tasks may be satisfied, but long-chain tasks cannot be effectively completed due to the lack of computing power of each instance. Typical long-chain tasks include online risk control, real-time decision-making, attribution analysis, advanced data analytics, etc. The corresponding operations on the graph database include path query, template query, K-hop query, or a large query nested with multiple subqueries.

In view of this, in route 2, when expanding from one instance to eight instances, we do not reduce the instance configuration at all. The goal is to enable the cluster to solve the following three major challenges in a horizontally distributed manner:

- Challenge 1: Dealing with Larger Datasets
- Challenge 2: Providing higher data throughput (concurrent request processing capability)
- Challenge 3: Offering processing capabilities that consider both long-chain and short-chain tasks (computing power increase)

As shown in Fig. 19, in a cluster of multiple instances, each instance runs on a typical x86 system architecture, in which the northbridge computing components around the CPU are highlighted and the southbridge storage and other I/O devices are downplayed. In fact, each instance is a microsystem that scales concurrently, such as the multicore CPU, CPU instruction cache, CPU data cache, multilevel caches shared between multiple cores, system bus, memory controller, I/O controller, external memory, network card, etc. Taking one 8-core CPU and eight 1-core CPUs in route 1 as an example, under the premise of sufficient concurrency, the computing performance of the latter is

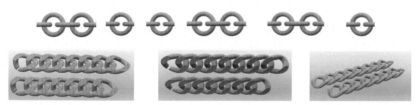

FIG. 18   Short chain tasks vs long chain tasks.

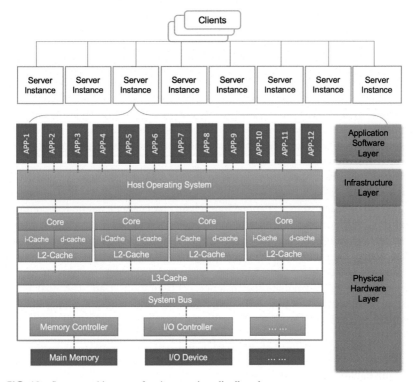

FIG. 19 System architecture of an instance in a distributed system.

significantly lower than the former, and the storage IOPS (I/O Operations Per Second) of the former is lower than the latter. Route 2 has one 8-core moving to eight 8-core units that may proportionally boost computing power and storage capacity. The first and second challenges of horizontally distributed systems are mainly solved by devices connected to the southbridge (storage, networking), while the second and third challenges are mainly solved by devices connected to the northbridge (CPU, memory).

In route 2, computing power finally becomes a "first-class citizen," which is crucial for graph databases. In the architecture design of the graph database, the computing engine is on an equal footing with the storage engine for the first time. This is very rare in other NoSQL or SQL-type databases. Only in this way can graph databases solve difficult problems such as real-time complex queries, deep traversal and penetration, and attribution analysis that cannot be solved elegantly by other databases.

Before we start designing a horizontally distributed (graph) database system architecture, there is a set of important concepts that need to be clarified: Partitioning vs Sharding. Generally, we define sharding as a horizontal partitioning mode, and partitioning is by default equivalent to a narrow vertical partitioning

Original Table

Customer ID	First Name	Last Name	City
1	Amy	Backhand	Dallas
2	Bobby	Dirt	Pleasanton
3	Carren	Swift	Denver
4	Davison	Tumble	Chicago

Vertical Shards

Customer ID	First Name	Last Name
1	Amy	Backhand
2	Bobby	Dirt
3	Carren	Swift
4	Davison	Tumble

VS1

Customer ID	City
1	Dallas
2	Pleasanton
3	Denver
4	Chicago

VS2

Horizontal Shards

Customer ID	First Name	Last Name	City	
1	Amy	Backhand	Dallas	HS1
2	Bobby	Dirt	Pleasanton	

Customer ID	First Name	Last Name	City	
3	Carren	Swift	Denver	HS2
4	Davison	Tumble	Chicago	

FIG. 20 Vertical partitioning vs horizontal partitioning.

(or vertical sharding) mode. In many scenarios, partitions and shards are treated equally, but the specific data-partitioning mode is modified by horizontal or vertical prefixes. Fig. 20 shows how a relational database table is partitioned vertically and horizontally.

In vertical partitioning (traditional partitioning mode), different columns are repartitioned into different partitions, and in horizontal partitioning, different rows are partitioned into different partitions. Another main difference between the two is whether the primary key is reused. In vertical partitioning, the primary key is 100% reused in two different partitions, but there is no reuse (of the keys) in horizontal partitioning. However, horizontal sharding also means that an additional table (or another data structure) is required to track which primary keys are in which shard, otherwise table-splitting queries cannot be efficiently completed! Therefore, the mode of partitioning or sharding often depends on the designer's own preferences or which method can achieve better results and cost-effectiveness in specific business query scenarios.

Although the partitioning or sharding of graph databases is more complex than traditional databases, there are still many ideas that can be used for reference. The author sorts out the following points:

- How to partition the vertex dataset.
- How to partition the edge dataset.
- How to partition attribute fields.
- After completing the earlier mentioned partitioning, how to optimize queries and boost performance.

In Chapter 3, we introduced how to cut vertices and edges in graph datasets (Figs. 20–22) to implement a horizontally distributed graph computing framework. Next, let us deduce a possible construction method for a horizontally distributed graph system that combines vertex-cutting and edge-cutting.

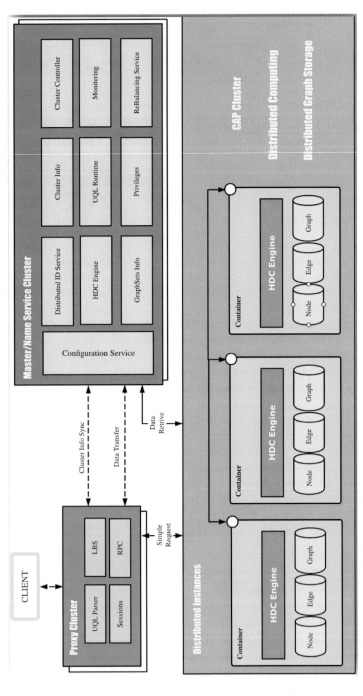

FIG. 21 Horizontally distributed graph database system.

FIG. 22 Original edge and reverse edge.

Fig. 21 illustrates a possible construction mode of a multi-instance distributed graph database system, in which each instance has its own vertex set, edge set, and other graph-related information, and the most important component is the distributed ID service (shown in the purple area on the upper right)—this ID service needs to be invoked to locate the instance (or instances) that contain specific vertex, and the ID server is only responsible for a few simple tasks (to allow for lowest latency):

- When inserting a vertex or edge, generate the corresponding ID and the corresponding instance ID.
- When querying or updating, provide the corresponding IDs.
- When deleting, provide the corresponding IDs and recycle the IDs of vertices or edges.

It may sound like a simple job, but let us just take the insertion of vertex and edge as an example to subdivide the possible situations:

1. Insert new vertex:
   a. On a single instance
   b. On active and standby instance pairs
   c. On multiple instances
2. Insert an outbound-Edge for an existing vertex:
   a. On the only instance
   b. On HA instances
   c. On multiple instances
3. Insert an inbound edge for a vertex:
   a. On the only instance
   b. On the active and standby instances
   c. On multiple instances
4. Add a new attribute to a vertex:
   a. On the only instance
   b. On the active and standby instances
   c. On multiple instances
5. Add a new attribute to an edge:
   a. On the only instance
   b. On the active and standby instances
   c. On multiple instances

Just inserting operations around vertices or edges are divided into 15 possible combinations, and other operations such as update and delete are more complex to consider. In graph databases, edges are attached to vertices, meaning that there must be vertices at both ends before there are edges, and edges have directions. The direction of the edges is very important, in graph theory, there is the concept of undirected edges, but in computer program implementations, because graph traversal involves moving from a certain starting vertex to the ending vertex passing through an edge, the edge is directional. Similarly, when storing one edge, you need to consider how to store the reverse edge, otherwise, it is impossible to reversely traverse from the ending vertex to the starting vertex. Fig. 22 illustrates the necessity of the existence of a reverse edge, when we add a unidirectional edge from the blue vertex to the red vertex, the system needs to add a reverse edge synchronously and implicitly by default, otherwise, the red vertex will not reach the blue vertex.

Similarly, when deleting edges, there is no need to delete vertices, but deleting vertices needs to start with the operation of deleting edges, otherwise, the system will immediately give birth to many orphan edges, which will cause serious problems such as memory leaks. In fact, deleting vertices and edges in batches in a graph database is a relatively expensive operation, especially in a distributed database, the complexity of this operation is proportional to the number of vertices and edges affected (note that edge deletion also needs to consider the reverse edge). A possible operation of edge deletion proceeds as follows:

1. Locate the instance ID of the deleted edge according to the edge ID (note that there may be cases where the starting and ending vertices of an edge are on different instances, and two or even multiple instance IDs will be returned).
2. (Concurrently) Search and locate the specific position of the edge ID in the data structure on each instance returned by Step 1.
3. (Concurrently) If the data structure of continuous storage of all edges of the vertex to which the current edge belongs is adopted, the located edge and tail edge switch positions, and the newly replaced tail edge is deleted.
4. If there are multiple copies of each instance, synchronize the copies.
5. After all the previously mentioned operations are successful, confirm and submit that the edge is deleted successfully, and the status of the entire cluster is synchronized.
6. (Optional) Recycle the edge ID.

Depending on the specific architecture design, data structure, and algorithm complexity, the time complexity of deleting an edge is $O(3 * \log N)$, where N is the number of all vertices ($\log N$ is the time complexity of locating an edge by attribute), and 3 means the times that positioning queries and synchronization operations between instances are repeated. Obviously, if it is the deletion of a super node with millions of edges, the operation may be very expensive. It is necessary to delete 1 million edges on one instance first, and then delete all the reverse 1 million edges. If these edges are distributed on different instances, which will trigger a lot of edge-locating queries. Of course, we can reduce

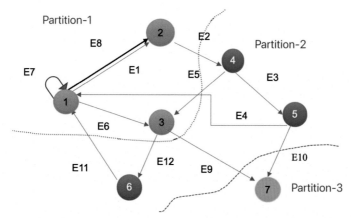

FIG. 23 Multigraph edge cutting.

the number of network requests by submitting the IDs of all edges to the server as a whole (or divided into multiple queries), but the complexity of this operation is similar to deleting large trunks of rows in large tables in traditional databases.

Let us take a specific example in the following to show how to segment a graph dataset and store the data on instances with different horizontal distributions. Taking Fig. 23 as an example, the following situations shall be sufficiently considered in system design:

- Support of multigraph, which allows multiple edges between pairs of vertices.
- Supports self-loops, that is, there can be edges starting from a certain vertex and returning to itself.
- Allow vertices and edges to carry their own attributes.
- Allows starting from any vertex to access adjacent vertices along the outgoing or incoming edges (in other words, support reverse edges and directed graphs).
- Allows horizontal sharding of vertices in graph datasets.
  a. (Optional) all vertex neighbors are stored in the same instance.
  b. (Optional) All edges associated with any vertex are stored in the same instance (regardless of whether the instance has multiple copies).
- (Optional) Cut the edges in the graph dataset, the cut edges will be repeated in two different instances.

In Table 3, we list a possible cluster-wide logical storage structure of all vertices from Fig. 23. If vertex-centric cutting were performed, every instance within the cluster would still hold 100% of vertices. However, if edge-centric cutting were conducted, as illustrated in Fig. 23, the vertices would be divided into three partitions—the orange one, the blue one, and the green one, and each would primarily occupy a different instance.

TABLE 3 Logical vertex storage.

Node UUID	Ext. ID	Porperty-1	Property-2	Property-3	...
1	...				
2	...				
3	...				
4	...				
5	...				
6	...				
7	...				
...	...				

A possible index-free adjacency edge storage data structure is listed in Table 4. Note that we arrange them in order of vertices, and all adjacent edges of each vertex are stored in forward and reverse aggregation. The sharding logic of the edge set is the same as that of the vertex partitioning logic in Table 3.

TABLE 4 Edge shards with vertex-cut.

Start Node UUID	End Node UUID	Edge UUID	End Node UUID	Edge UUID	End Node UUID	Edge UUID	End Node UUID	Edge UUID	SEP ARA TOR	R-End Node UUID	Edge UUID	R-End Node UUID	Edge UUID
The header above is for illustration purpose only, by no means a fixed setup													
1	2	E1	3	E6	2	E8	1	E7	SEP	5	E4	1	E7	6	E11
2	4	E2	SEP	1	E1	1	E8						
3	7	E9	6	E12	SEP	4	E5	1	E9				
4	5	E3	3	E5	SEP	2	E2						
5	1	E4	7	E10	SEP	4	E3						
6	1	E11	SEP	3	E12								
7	SEP	3	E9	5	E10								
...	...												

To verify whether this simple graph-cutting method has a significant performance impact on queries on the graph, we investigate the following situations:

- Node query (including read, update, delete).
- Edge query (ditto).
- Check all 1-hop neighbors of a node.
- Check the K-hop neighbors of a node ($K \geq 2$)
- Find the shortest path between any two nodes (assuming the same weight on the edges, ignoring the direction):
  a. Two nodes on the same instance.
  b. Two nodes on different instances.
- Other possible query methods:
  a. Graph algorithms.
  b. Template query, path query in other ways, etc.

Node queries can be subdivided into many situations:

- Search by ID:
  a. According to the global unique ID (aka UUID).
  b. Query by original ID (provided before ingestion/ETL).
- Search by an attribute.
- Fuzzy matching by text in an attribute field (full-text search).
- Other possible query methods, such as querying all IDs in a certain range, etc.

In the simplest case, the client first sends a query command to the ID Server to query for its corresponding UUID and corresponding instance ID through the original ID of the vertex. The complexity of this query can be $\leq$ $O(\log N)$.

The most complicated is the full-text search. In this case, the client request may be forwarded to the full-text search engine by the ID Server. Depending on the specific logic of the engine's distributed design, the engine may have two options:

- Continue to distribute to all instances, and assemble the vertex ID list corresponding to the specific search on each instance.
- The engine returns a list of all instance IDs and corresponding vertices IDs.

Edge query is similar to node query, but the logic is slightly more complex, as mentioned earlier since each edge involves a forward edge and a reverse edge. In addition, edges are attached to vertices, and there may be situations where you need to traverse some or even all of the edges of a vertex to locate an edge. Of course, we can optimize all possible traversal situations, such as using space-for-time tradeoffs, but still consider the cost and cost-effectiveness of implementation.

The biggest challenge for a horizontally distributed graph system is how to deal with networked graph data queries (during data penetration), such as finding all its neighbors from a certain vertex. Taking vertex 1 in Fig. 23 and Table 4 as an example, if you want to query for all its 1-hop neighbors, the complete steps are as follows:

1. (Optional) Query for its UUID based on the original ID.
2. Query the ID of the instance that contains it based on the UUID.
3. Locate the edge per its UUID in the edge data structure on the instance.
4. Calculate the number of all edges and associated vertex IDs.
5. Perform deduplication operations on associated IDs and return the result set.

In the previously mentioned operations, the core logic is step 4 and step 5. In step 4, if all outgoing and incoming edges of vertex 1 exist in the form of continuous storage, the sum of their incoming and outgoing degrees can be calculated with an extremely efficient O(1) complexity (value $= 7$); in step 5, because vertex 1 may be connected to other vertices multiple times, so the result of its 1-Hop is not 7 but 4, because the three edges associated with vertices 1 and 2 need to be removed to obtain the result after deduplication. Refer to Table 5.

In the previously mentioned data structure design, we can calculate all its 1-hop neighbors (implicitly including all its out-degree edges and in-degree edges) in the instance where any vertex is located, without any communication with other instances. In this way, we can keep the performance of shallow (<2 layers) query operations at the same level as that of nonhorizontally distributed architectures, but the ability to concurrently query can obviously be doubled.

Let us take a look at how to calculate 2-Hop neighbors of vertex 1. Still taking Fig. 23 as an example, the specific steps are as follows:

1. Obtain and locate the ID of the instance where vertex 1 is located.
2. Locate vertex 1 and its edge data structure on this instance, obtain all adjacent vertex IDs, and deduplicate IDs (obtain 2, 3, 5, 6).
3. "Divide and Conquer" the set of vertices from step 2:
   a. Query for the specific server IDs of the shard instance where the earlier deduplicated IDs are located, and send them to the corresponding instance service.

TABLE 5 1-Hop computing on one partition.

Start Node UUID	End Node UUID	Edge UUID	End Node UUID	Edge UUID	End Node UUID	Edge UUID	End Node UUID	Edge UUID	SEP ARA TOR	R-End Node	Edge UUID	R-End Node	Edge UUID
1	2	E1	3	E6	2	E8	1	E7	SEP	5	E4	1	E7	6	E11

    b. Vertices 2 and 3 are processed like step 2 because they are on the current instance.

    c. The instance holding partition 2 processes vertices 5 and 6, and the processing method is the same as previously mentioned.

4. Assemble and deduplicate the result set.

5. Return to the client.

Table 6 shows the process of calculating the 2-hop neighbors of vertex 1 across multiple shards. According to the earlier-mentioned steps, the following results can be derived:

1. Vertex 1 is on shard 1.

2. 1-Hop: The neighbor vertices after deduplication of vertex 1 in shard 1 are 2, 3, 5, and 6.

3. 2-Hop: Operate on each instance in a multithreaded concurrent manner:

    a. Next-hop neighbors of vertex 2: 4, 1, of which 1 needs to be deduplicated (because it has been visited at 1-hop and 0-hop).

    b. Next-hop neighbors of vertex 3: 7, 6, 4, 1, of which 6 and 1 need to be deduplicated (because they have been visited at both 1-hop and 0-hop, but if they do not communicate with other threads of the same instance or other instances at this time, impossible to know vertex 4 need to be deduplicated).

    c. Next-hop neighbors of vertex 5: 1, 7, 4, among which 1 needs to be deduplicated (same logic as mentioned earlier).

    d. Next-hop neighbors of vertex 6: 1, 3, among which 1 needs to be deduplicated (same logic as mentioned earlier).

TABLE 6 2-Hop computing of a vertex across shards (with deduplication).

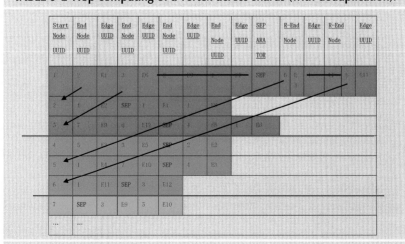

4. The results in step 3 need to be aggregated, the logic is as follows:
   a. All shard instances aggregate data to instance 1.
   b. The vertex sequence is as follows:
      i. Hop-0:1.
      ii. Hop-1: 2, 3, 5, 6.
      iii. Hop-2: 4, 1; 7, 6, 4, 1; 1, 7, 4; 1, 3.
   c. Hop-2 results after cross-hop deduplication: 4, 7.
5. Return the calculation result after deduplication to the client: vertex 1 has two 2-Hop neighbors, including vertices 4 and 7.

Note that in the specific operation steps mentioned earlier, the most complex and core algorithm logic lies in how to deduplicate data. Duplicated data brings several significant disadvantages:

1. Nondeduplicated data leads to wrong results, and this type of graph computing is a precise science, not a matter of probability as in AI/ML.
2. Nondeduplicated data, especially intermediate results, will lead to unnecessary downstream calculations, waste computing power, and increase network communication costs.

However, how to deduplicate the data is a very skillful thing and a challenging problem. As in the process described earlier, we deduplicated the calculation results in steps 2, 3, and 4c on the premise of not requiring interinstance communications in each hop of operation, and the deduplicated ID list of the previous layer can be listed in the process of rebroadcasting from the upper layer to the next layer. This "downward propagation" operation can be implemented in a logic similar to recursion.

   If we expand the depth of this K-neighbor query to 3-Hop or 4-Hop, or if the data magnitude of the graph increases exponentially, can our above algorithm still maintain high efficiency?

   Using Fig. 23 as an example, we start the K-hop query from vertex 1, and the maximum hop for it 2-Hop. At this time, if we calculate 3-Hop, the returned result should be 0 (empty set)! However, if we do not verify and deduplicate the data of each layer, we cannot know that there is no 3-hop nonempty result set relative to vertex 1 in the graph dataset before completing 2-Hop.

   There are two specific implementations of the deduplication algorithm described earlier:

1. During the downward propagation of the current layer (hop), deduplicate data and pass down the data passed down from the previous hop.
   a. Advantages: Partial deduplication can be achieved in each hop as much as possible, reducing the pressure of repeated calculations in the next hop.
   b. Disadvantage: A large amount of data may be transmitted, causing network pressure, and the current shard instance can only complete partial deduplication.

2. The data in the current layer is deduplicated, but it is not propagated to the next hop and waits for the final summary and deduplication at the last hop.
   a. Advantages: the logic is relatively simple.
   b. Disadvantages: There may be many repeated calculations (waste of computing power, waste of network bandwidth), and the amount of data assembled in the last hop is large, which does not conform to the principle of divide and conquer.

Using the same logic, we can also complete other queries and algorithms described earlier. The author only lists a possible horizontally distributed implementation here, and I believe that smart readers can come up with their own native and distributed graph architecture design.

Finally, we summarize some important concepts of distributed graph system:

- Distributed design is not trying to replace high-end equipment with low-end equipment and hoping to get better results. After all, a (high-performance) graph database system cannot be realized with the concept and framework of Hadoop—if you are still in the Hadoop era, your knowledge stack and cognition need to be upgraded greatly.
- The most important concept of distributed graph system design is to carefully decide which scenarios of graph computing need sharding and which ones do not. In other words, the scenarios that sharding can solve are shallow queries and shallow computing, while deep computing is naturally against sharding.
- With the development of computer architecture today, every standalone and single-instance system is a complete set of systems that can support high concurrency and large-scale concurrency at the bottom. It is the upper layer software that determines the system's capability! A high-concurrency underlying system does not mean that the software naturally achieves high concurrency!
- The characteristics and efficiency of the data structure determine the ultimate efficiency of the database system or the upper limit of the efficiency of the system.
- Programming languages are varied in efficiency. If there are still people who say that Python can write software that is much more efficient than C++, then vice versa, and the probability is much higher.
- The development of database systems requires an in-depth understanding of many components such as operating systems, file systems, storage, computing, and networks, and requires a lot of engineering tuning. It is very necessary to combine academic theory with daily engineering practice.
- Many features distinguish graph databases from traditional databases, the most important one being high dimensionality. Designing and using graph databases requires graph thinking.
- The high dimensionality of graph databases means that its challenges are not only storage but also computing!

Chapter 6

A world empowered by graphs

This chapter focuses on the application scenarios of graph databases. First, let us sort out the core capabilities that a competent graph database should possess.

- High-speed graph search capability: high QPS/TPS, low latency, real-time dynamic pruning (filtering) capability.
- Deep, real-time query capabilities for graphs of any scale (10+ hop).
- Support of intuitive GQL (graph query language).
- Robust, fault-tolerant, and highly available graph database system.
- Scalable computing (and storage): supports vertical and linear scalability.
- Highly interactive GUI with great explainability.
- A rich collection of graph algorithms for automated AI/ML chaining and data analytics.
- Convenient and low-cost secondary development capabilities (graph query language, API/SDK, toolchain, etc.).

With the above core competencies, a graph database can be widely applied and serve in the following fields:

- BFSI Industries: Smart marketing in the banking, financial services, and insurance industries, risk control, antifraud, antimoney laundering, risk assessment, asset liquidity management, smart auditing, an online recommendation system, and other scenarios.
- Supply chain finance: In the supply chain finance network, graph databases can empower and accelerate sophisticated management and monitoring of decentralized asset and circulation technology platforms.
- Telecom operators: Customer 360, intelligent recommendation, call fraud detection, network monitoring, graphical network management, etc.
- Internet of Things: IoT generates huge amounts of data and desires the ability to quickly and deeply extract value out of the data with networked analysis, for this purpose, making full use of graph databases is the most sensible choice.
- Internet: NLP, knowledge map, intelligent search and smart recommendation, chatbot, e-commerce, and other functions.
- Smart city, meteorology, information retrieval, criminal investigation, government affairs, military, and other fields.

The Essential Criteria of Graph Databases. https://doi.org/10.1016/B978-0-443-14162-1.00007-6

Based on many years of practical experience in the industry, the author has collected six application cases and user scenario white papers for readers' reference.

- Real-time decision system (online antifraud)
- Fraud prevention: monitoring of the bank's guarantee chain for corporate business companies
- Smart graph analysis based on industrial and commercial data (identification of ultimate beneficiary, analysis of holding relationship, etc.)
- Knowledge graph and real-time computing (real-time intelligent recommendation and other scenarios)
- The core indicator measurement in the liquidity risk management system is: the liquidity coverage ratio (LCR)
- Various other use cases.

6.1 Real-time business decision-making and intelligence

With the advent of the big data era, not only does the volume of data continue to grow, but the complexity and diversity of data continue to increase. More and more real-time business decisions rely on the understanding of how data is related and correlated. Traditional RDBMS database systems are not designed to solve this challenge, not even the newer big data frameworks, they may have good scalability, but often cannot process sophisticated data analytics in real-time. In credit card or loan applications, real-time decision-making is ideal, but it is very difficult to achieve without cutting corners, meaning, simplifying business decision-making processes and sacrificing precision, white-box explainability, and much more.

We will use two real-world use cases to illustrate how graph databases can help accelerate decision-making and boost intelligence. The first use case is a credit card or loan application decision scenario, and the second use case is online transaction fraud detection.

6.1.1 Credit loan application

During the process of applying for a credit card or loan, there are several scenarios:

1. Scan all loan or card application data to find all phone numbers (or other types of attributes, such as e-mail, address, etc.) shared by more than a certain number of applications.
2. Sift through the entire application data, and find applications that share any combinations of the following: Company, Referrer, Email, or device ID (phone number, IP address, etc.). The filtering rules can be strengthened to find out how many applications share all the above attributes.

3. Discover circles or loops. For example, application #1 uses mobile phone #2, which is used by application #3, which uses mailbox #4, but this mailbox is also used by application #1. The cyclic path looks like App [X] → Phone → App [Y] → Mailbox → App [X].

4. In more complex query scenarios, identify if there are deeper/longer loop paths with 10+ nodes (and minus-one edges).

5. Community recognition can intelligently classify applicants into different communities (customer groups), which will help credit card and loan companies to better understand customer behavior patterns.

To handle the above scenarios, we must first consider how to construct a data model that best handles the data dependency requirements in these scenarios. The following is a typical data modeling method to facilitate graph data patterns:

- Each application is considered a node.
- All attributes of the application, such as email, company, device, phone, and ID# are also modeled as nodes.

The graph schema described above is illustrated in Fig. 1, as you can see it is quite different from traditional SQL-style schemas and tables. The advantage of such graph schema design is that if we are trying to identify the correlations between any (number of) applications, we can intuitively find the associative (shortest) paths (or network, or subgraph) between them—such as two applications sharing a common property node, be it Email, Phone, Device ID, Referrer or Company.

In a financial institution, there are credit card/loan application data sets of more than 400 million applications and associated attribute nodes. Taking Scenario 1 as an example, the graph database (Ultipa) takes <2 s to identify more than 1.8 million phone numbers that have been used by more than five applications, while the Apache Spark system needs 780 s (13 min) to calculate (not including the ETL time which may double the overall time). The

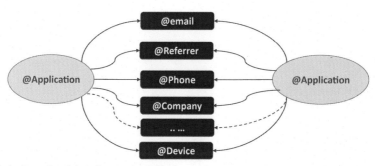

FIG. 1 Loan fraud detection graph schema.

performance difference is 458 times. The results were staggering, with 1.85 million phone numbers being reused in more than 45 million applications that could be considered potentially fraudulent. In the most extreme case, several phone numbers were used by 17 applications which indicate something is abnormal.

In Scenarios 2 and 3, the goal is to find applications that share common attributes (email, company, device, referrer, etc.) as illustrated in Fig. 2. Such scenarios are best described as graph data processing models, where all applications must be processed, so that the computing complexity is high, and have been traditionally considered large-scale time-consuming batch processing. The numbers captured in Table 1 show that while the Spark system may take an hour to finish the batch processing, the graph database can take less than 1-min.

It is possible and relatively easy to validate if the "batch-processing" results in Scenarios 2 and 3 are correct. As shown in Fig. 2, there are at least two ways to validate the results.

- Method 1: Simply run the shortest path query between the two found application nodes, the results would show six 2-hop paths with the six common attribute nodes in the middle.
- Method 2: Start from either one of the two application nodes, run the Jaccard Similarity algorithm, return the top-1 node as a result, and the other application node would be returned given its highest similarity score.

Louvain community detection algorithm is a recent addition to the family of graph algorithms (invented in 2008). It forms multiple communities after multiple rounds of iterative convergence on the full graph data, and closely

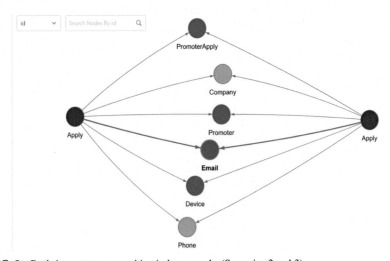

FIG. 2 Real-time pattern recognition in large graphs (Scenarios 2 and 3).

TABLE 1 Performance comparison of different graph systems[a].

Scenarios	Apache spark	System-U	System-N	Python NetworkX
OLAP (Scenario 1)	780 s	1.6 s	Not tested	N/A
OLAP (Scenario 2/3)	3600 s	44 s	Not tested	N/A
Loop discovery Scenario 4	N/A	10,000 QPS	30 QPS	N/A
Louvain community recognition (Scenario 5)	N/A	5-min (including disk write-back time)	3.5-h (including projection time)	Fail to complete in 24-h

[a]System-U stands for Ultipa, System-N stands for Neo4j.

connected data will be allocated in one community, which is very valuable in scenarios such as social network analysis, fraud detection, and marketing recommendations.

The challenge and disadvantage of the Louvain algorithm is that its original algorithm is serial, and once the amount of data becomes large (e.g., on the order of millions of nodes and edges), the algorithm will run very slowly. If you apply it in an antifraud setting, Louvain can take hours or days to run (or fail to complete at all). The other challenge with Louvain is that it is not easy to rewrite the algorithm in a fully parallel way, because a certain portion of the algorithm is inherently serial, if a forced complete parallelization is realized, the results can be wrong, therefore very careful engineering refactorization and optimization are needed (In the next Chapter, we will reiterate in expanded details on how to ensure accuracy while we accelerate things like graph algorithms). The numbers captured in Table 1 for Scenario 5 illustrate that the running time between different graph systems may vary hugely, from near real-time to $T + 1$ or longer. The significantly shortened latency is particularly valuable which means the data set can be updated more often and business insights can be extracted more rapidly.

Fig. 3 (left) illustrates how fraud users behave differently from normal users based on their networked behaviors, how fraud users tend to form their closely connected communities (just as normal users would also form their own tightly connected communities), and Louvain algorithm can help automatically identify these communities. It is also possible to sample the Louvain results and visually present such communities, as shown in the right portion of Fig. 3.

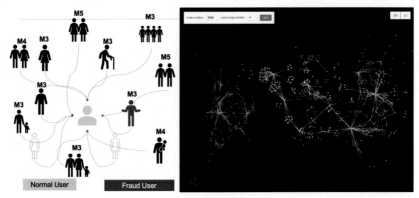

FIG. 3 Louvain community detection for fraud-detection.

6.1.2 Online fraud detection

Every year enterprises and governments are losing hundreds of billions of dollars due to fraud. As more and more transactions are conducted over the internet, online fraud detection, antimoney laundering, and decision-making have become a necessity.

Traditional RDBMS or Big-data-n-AI-based fraud detection or decision-making systems are not capable of online real-time fraud detection or decision-making. There are three major drawbacks holding them back:

- **High latency**: The average latency for transaction fraud detection on these systems is over 300 ms, and the highly concurrent nature of online transactions causes traffic to pile up and wait forever for a decision to be made. Online fraud detection must be real-time, every extra second taken to respond means degraded user experience and friction for trusted users.
- **Fail to handle supernodes**: Some transactional counterparties have tens of thousands of transactions a day, These counterparties are considered supernodes and their associated transactions have to be removed in the existing antifraud system, such as Apache Spark or even Neo4j, to allow for faster transactional behavior analysis, therefore causing the transaction network incomplete (and decision made inaccurate).
- **Fail to support real-time ETL**: Many existing antifraud systems cannot do real-time ETL (or update the data set as transactions flow in). Spark can only process static data after it is ETLed, but transactions keep on flowing in. Even though some vendors have claimed to use Spark as a real-time data warehouse, it is simply not, the latency is way beyond what a real-time fraud detection system can tolerate.

Taking a large commercial and retail bank's use case as an example. The bank understands that it needs a capable graph database to deal with the real-time fraud detection of card transactions. They had played with so-called in-memory databases (such as VoltDB), which were incapable of handling online antifraud with table-centric data modeling, and they also played with Neo4j Enterprise Edition and realized that though it is faster than Apache Spark by 30%, it still cannot handle supernodes and cannot match their expectation of 30 ms latency for each transaction fraud analysis (on average, 80% of the transactions can be analyzed within 300 ms, but Neo4j fails to respond when a transaction involves hotspot supernodes or when rush-hour traffic flows in and transactions pile up).

Let's be analytical on the key features a graph-powered real-time decision-making system (RTD) should possess:

- Real-time transaction update, note that each incoming transaction changes the graph dataset and its topology, with new vertices and new edges (or updating existing vertices or edges).
- 10+ graph models executed against every transaction to examine the transactional counterparties' networked behavior over a certain period (i.e., 3-month).
- Graph dataset contains 90-day transactional data.
- No hotspot supernode is left behind.
- Total latency is within 30 ms for each transaction's fraud analysis.

Table 2 captures the key differences between the three graph systems.

TABLE 2 RTD system feature and performance comparison[a].

RTD system comparison	System-U (v3.2)	System-N (v4.x)	Apache Spark + GraphX
Cluster size (instances)	3 (HTAP)	3 hot-standby	8
Vertices (million)	260	260	260 M minus all hotspots (e.g., POS)
Transactions (million)	1000	100	100 minus hotspot supernodes
Data volume (days)	90	7	7
Avg. latency (ms)	<20	>200	>300
Graph models processed (Per-transaction)	>10	<10	<5
Hotspot nodes?	Yes	No (traversal of supernodes will be extremely slow)	No (supernodes must be removed first)

[a]System-U stands for Ultipa, System-N stands for Neo4j.

For the first time, the bank's IT department sees a complete card transactional behavior network with all supernodes included, and no backlog is piled up when tons of transactions are rushing in during peak hours as they would experience otherwise with Spark or System-N. In this case, the ROI and performance gain with System-U is in the range of 100–1000 times, if you zoom into each transaction's decision-making, each model takes only 0.2–0.3 ms (200–300 μs), that is the beauty of real-time graph computing.

6.2 Ultimate beneficiary owner

In the world of business, many owners have chosen to hide their identities from the surface, They designed schemes to use layers of intermediary companies to gain anonymity or tax benefits or simply for fraud or money-laundering purposes. During the KYC (Know Your Customer) or law-enforcement investigation processes, it is possible to identify the ultimate beneficiary owners (UBOs), quantify a certain beneficiary's % of ownership to a certain business entity, and explore sophisticated investment controlling paths and networks by certain parties with appropriately modeled business data—Industrial and Commercial Graphs that capture how business entities and stakeholders are correlated.

Here is one such challenge faced by local, state, and federal authorities in San Francisco, In the past decade San Francisco's real-estate properties that were supposedly allocated to low-income local families have been bought up by LLCs (Limited Liability Companies) that are hard to trace their UBOs. The issues were so prevalent that this has become a major concern, government agencies like IRS (Internal Revenue Services) and local law enforcement are interested in understanding what parties are hiding behind these LLCs, the manual exploration process can be very labor-intensive and time-consuming, because the eventual UBO parties may hide many hops (layers) behind the surface LLCs, and often these UBOs intentionally hide and cross-owning their shares, making the overall ownership structures highly complicated (Fig. 4).

The three main pain points with UBO identification are:

- Data presentation: Traditional databases all use columns and rows to store data and present the data using tables which are counterintuitive in many real-world applications. For instance, to understand a business entity's ownership structure, the most intuitive way is to present the relevant data in a graphical and correlated fashion. If a person acts as the legal representative of a company, in a graph setup, this is an edge connecting two nodes, one node being the person, pointing at the other node which is the company, and the edge (relationship) is labeled "Legal."
- Data connectivity: Assuming we have collected all data of a business entity, starting from the business entity node, recursively, all entities that are linked with it can be retrieved and form a subgraph. Note that the resulting collection of nodes and edges may form a graph instead of a tree, A tree does NOT have circles, but a graph may, and this reflects real-world scenarios better

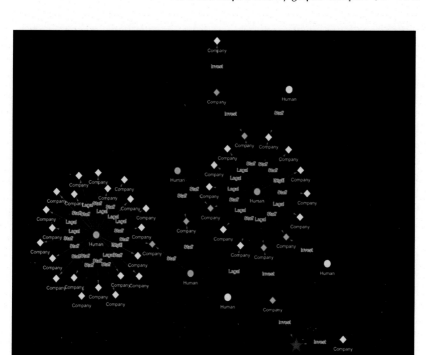

FIG. 4 A company's UBOs may be hiding layers behind surface.

because business entitles and persons may form circles or cross-investment/ owning business structures, which can NOT be represented with a hierarchical tree structure.

- Data exploration (deep traversal): In real real-world setup, it is NOT uncommon to see the eventual owners (aka Ultimate Business Owners, or majority-stakeholders) that are many hops away from the business entity that are being examined. A traditional RDBMS or document database (or even most graph databases) is NOT capable of addressing such exploration (deep penetration) in a fast and timely fashion due to the joining of multiple tables will inevitably create the "cartesian product" problem, therefore making the query complexity exponentially difficult and slow.

The computational complexity of finding the UBOs may be very high. Let us do a simple calculation here:

- Assume that each company has 25 owners (investors, equity holders)
- Time complexity to dig 5-hop deep: $25 * 25 * 25 * 25 * 25 = 9,765,625$ ($\sim$10 M)
- If we dig 10 layers deep: $10^{14} = 100$ trillion.

If we do not have more efficient data structures, algorithms, and architectural systems, it will be impossible to solve these challenges only by relying on relational query methods and system architectures. Fortunately, many of the above challenges can be solved through the optimization of the system architecture, especially the low-latency, high-density concurrent data structure and real-time graph computing engine, so that users can find the ultimate beneficiary in real-time, even in microseconds.

Now, we know that deep graph traversal can identify the selective shareholders of a business, no matter how far they are apart from the starting business entity, sometimes even among thousands of shareholders. Fig. 5 illustrates the dramatic overall differences between SQL and GQL in identifying UBOs—the differences lie with not only performance but also the code complexity—while SQL takes a long time to go recursively traversing, GQL is instant, while it takes GQL just a one-liner, SQL needs many lines of hard-to-understand code to achieve the same.

In a large stock exchange, the auditing department (per regulatory mandate) needs to examine all possible links between the 4000 publicly listed companies and 10,000 key persons. The computational complexity is incredibly high, because one query between one public company and one key person may have 1000 links (or shortest paths) not to mention the same kind of queries have to be conducted 40,000,000 times, the resulting paths could be in the range of 40,000,000,000. The stock exchange tried to solve the problem with System-N only to find out that on average each query takes $\sim$0.5 s, and finishing the entire query would take months ($\sim$4000 h or 5.5 months). By adopting System-U, they were able to finish this large-scale batch processing in a $T+0$ fashion ($<$1 h), and on average each path query takes only 0.46 ms! System-U on average is 1000+ times faster than System-N, in complex queries, System-U tends to be 10,000s of times faster. The table below (Table 3) captures systemic differences between different graph systems.

Presenting entity relationships in a networked and animated fashion is both intuitive and productive. Fig. 6 shows how the UBO query results can be shown visually and intuitively, and users can interact and explore the graph in real-time to expand on their findings, such as:

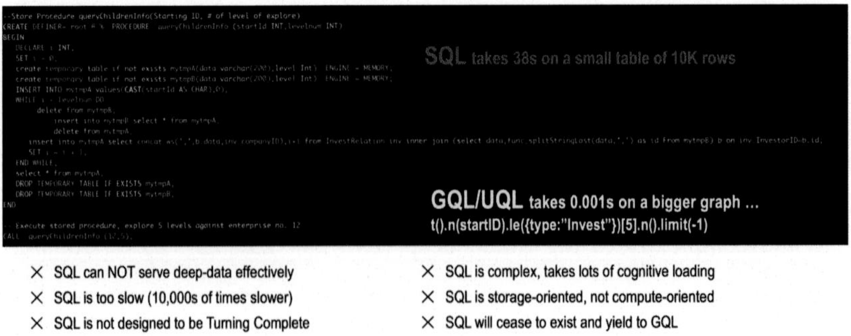

✕ SQL can NOT serve deep-data effectively	✕ SQL is complex, takes lots of cognitive loading
✕ SQL is too slow (10,000s of times slower)	✕ SQL is storage-oriented, not compute-oriented
✕ SQL is not designed to be Turning Complete	✕ SQL will cease to exist and yield to GQL

FIG. 5 The problems with SQL vs the benefits of GQL.

TABLE 3 UBO graph systems feature comparison.

Smart enterprise graph	System-U v3.x	System-N 4.x
Cluster size (instances)	3 (HTAP)	3 (hot-standby)
Data volume (nodes and edges in millions)	1000	1000
Max concurrency (threads or vCPUs)	80	4
Max query depth (hops or layers)	>32	≤7
Avg. query processing time (ms)	0.46	500

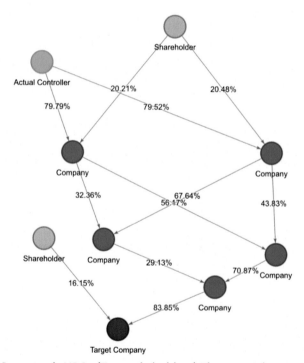

A Company's UBOs (Top stakeholders) That are 4-hop Away

FIG. 6   UBOs hops away from target company.

– Finding connections (paths) from an existing entity to any other entity.
– Finding the most similar entities across the entire network.
– Edit/Update any entity or relationship.
– Show, hide, or delete any entity/relationship.
– Customize the graph in any way that is possible.

6.3 Fraud detection

Globally, most of the profits of large commercial banks come from corporate loans. Banks require companies to use collateral to hedge against potential bad debt risks. In many cases, such collateral can also exist in the form of a guarantee issued by another company. We call this Corporate Guaranteed Loan in corporate business. When an entity (a company or a person, called A) applies for a loan from a bank, the bank needs another entity (usually another company, called B) to make a financial guarantee that if A cannot repay the loan, then the party B will pay off the loan for A.

A commercial bank typically processes thousands of corporate loans. Front-office staff disburse a large number of loans daily, and back-office staff identify potential risks associated with each company guarantee. This process has traditionally been time-consuming and labor-intensive.

There are some typical types of fraud, intentional or not, related to guarantees:

- The simplest form of fraud: A guarantees B, B guarantees A, which is a direct violation of the precondition for any bank to issue a loan (although many times banks fail to exercise this regulatory self-check responsibility).
- Form a chain or circular guarantee: A, B, C, D, E, A, This circular guarantee chain is difficult to detect because you have to dig deep enough to E, and then find that E has provided a guarantee for A, thereby violating bank guarantee risk avoidance rules.
- A more complex topology, for example, may involve many loans + guarantees between multiple entities, thus forming a sophisticated guarantee network (and potentially including many loops).

Fig. 7 shows the red-dot company (1121) guaranteed two other companies (600, 1120), and the two companies further guarantee three other companies (239, 1120, 1122), eventually forming four loan guarantee loops, specifically, three triangles (triangle is the simplest form of guarantee ring) and one loan guarantee quadrilaterals (involving four parties).

The example above is a zoomed-in survey of a specific company, typically a bank runs all the loan guarantee data in a batch fashion to see how many breaches there are. This process can be very time-consuming without the

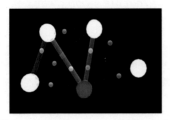

Ring-Path-1: [1121, 600, 239, 1121]

Ring-Path-2: [1121, 600, 239, 1120, 1121]

Ring-Path-3: [1121, 1120, 1122, 1121]

Ring-Path-4: [1121, 600, 1120, 1121]

FIG. 7 Guarantee loops by multiple entities.

support of real-time graph computing because a large number of companies (usually tens of thousands or more) are involved. Through real-time graph computing technology, the overall processing time can be shortened exponentially (from days or hours to seconds, or even milliseconds). Fig. 8 is a three-dimensional panorama of a bank's current corporate loans. All loan guarantee chains and guarantee circles formed between enterprises are visualized in real-time, which is very intuitive and easy to understand.

The guarantee-chain structure in the center of Fig. 8 involves as many as 100 companies, Such a structure is difficult to sort out manually, but it is very common in real business loans—many hidden illegal guarantees that violate bank risk control rules, the back-office business personnel need to be reminded for further investigation.

According to some market research reports in 2019, nearly 40% of corporate loans in Wenzhou City, Zhejiang Province, China is involved in illegitimate guarantees. Before the loan is issued, it can be very beneficial to check if any guarantee chain or circle exists to help the bank understand the legality (compliance) of each loan application. Through such a system, a bank can also continuously monitor the flow of loans and take preventive measures when the loan risk exceeds a certain limit, and help banks take preemptive measures to protect loans and avoid large-scale domino effects.

With the help of information aggregated from various channels, such as court documents, litigation judgments, social media, and sentiment analysis, we can accurately and preemptively identify high-risk companies (and their loan guarantees, capital flows, etc.), so that banks can decide whether to early recover loans to protect bank assets from loss. Fig. 9 illustrates a company found to be convicted in connection with six lawsuits involving large sums of money. At the same time, this corporate client has provided multiple loan guarantees for

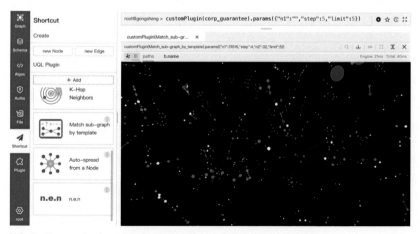

FIG. 8 Panoramic view of guarantee chains and rings in corporate banking.

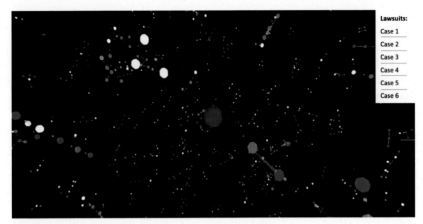

FIG. 9 Identification of high-risk loan guarantees and entities.

other entities. This situation can be judged as a potentially high risk, and banks should investigate and revoke these loan guarantees as soon as possible to prevent imminent losses.

6.4 Knowledge graph and AML

Knowledge graph should be credited to Google for gaining popularity since its inception in 2012. In the process of constructing the search engine, to improve the search relevance, not limited to the weight ranking of keywords, Google has invested considerable energy in building a network of knowledge entities, Different entities are associated with specific relationship(s) among them, and this association may be very extensive, ranging from logical tree classification to causal relationship, etc. The knowledge graph of a general search engine can be regarded as a general-purpose knowledge graph, and its boundary is endless. To some extent, this is a coupling between the computing power of the machine and the knowledge set of humans. Of course, we believe that it will take a long time for this matter to be perfected. Perhaps only when human beings fully understand how our brains work and how humans work together can we announce that the graphs of human knowledge have attained a closed loop.

If we are to draw an analogy between the mechanism of how human brains function and how knowledge graph's function. We will see ourselves comparing the human left brain and right brain to knowledge graphs and graph computing engines while the former is logical but shallow and slow, the latter is high-performance, intuitive, and instant.

Building a general knowledge graph is extremely complex, and so far, no company, government, institutional team, or individual has declared to have completed this "mission impossible." Given this, a divide-and-conquer strategy and approach have been adopted to build knowledge graphs with more controllable scales and for vertical industries such as health care, insurance, finance, or supply-chain knowledge graphs. Even with such a limited scale, there are still countless challenges, from data collection, cleansing, NLP (Natural Language Processing), and data modeling, to computing, data penetration, and visualization.

Fig. 10 shows a knowledge graph of investment and cooperation relationships. There are only five types of vertices in the graph:

- GP investors
- LP investors
- Companies (investment projects of LPs)
- Company executives
- Universities or institutes

This is a relatively simple graph, but it can be convenient to implement some human brain-like intelligence on it, such as querying the associating network between two investment institutions (LPs) or between executives of two or more companies. Query operations on these graphs can be used to assist in investment and financing decisions. If there is no graph system with adequate computing power, users may still use complex Excel spreadsheets to analyze the above relationship network, and that will be unimaginably complex, quite a torment for analysts per se.

The knowledge graph, as illustrated in Fig. 10, not only allows users to enter the micro world to interact with specific data points and their relationships, but also gives a panoramic view to understand the spatial topology of the knowledge network—after all, compared with machines, human beings always have an advantage in having a global view, and graphs seem to be able to better assist humans in this direction. Another characteristic of the graph is that you can operate recursively on it, such as expanding from a vertex in breadth or depth, discovering its neighbors from an impact scope perspective, or finding appropriate linkage with (paths to) specific destinations.

It is this large-scale and intricate data-point connection that leads to the complexity of building general knowledge graphs. A side effect of constructing knowledge graphs is the potentially overwhelming computational complexity of traversing hotspot supernodes—which are entities with significantly higher than average connectivity (degree). For example, when we were trying to identify the "causal linkages" from Genghis Khan to Isaac Newton (or vice versa), both entities are well-connected supernodes, and traversing from either entity to reach the other would be computationally challenging. The matter can be complicated when there are more parties involved or the query depth is prolonged….

Graph databases provide more than one method to truly solve the above problems, such as point-to-point deep path search, network search between multiple entities, template path matching search, or fuzzy full-text-based path or network search. The complexity of fuzz-matching path search is exponentially

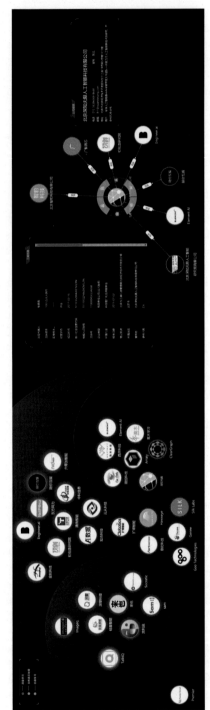

FIG. 10 Interactive knowledge graph for investment research.

higher than that of existing search engines—the specific logic is broken down as follows: first, fuzzy keywords search a set of entities, then start from these entities to find the target entities that can finally be reached on the graph, and these target entities also meet the fuzzy keyword search results. This kind of search is similar to human beings' inferences, rapid divergence, and then convergent fuzzy matching and search again. However, traditional search engines based on a single keyword do not have this capability, to be precise, they cannot perform fuzz-matching path search between two entities on the graph. Network search is also very interesting. For example, the association network formed between five criminal suspects (or five knowledge points) through an ad hoc network. Its computational complexity is as follows:

- The pairwise path queries to be run are: $C(5, 2) = 10$
- Under the condition that the path depth is five, it is assumed that each entity has 20 association relationships (average degrees)
- (Theoretically) Computational complexity: $10 * (20 ^ 5) = 32,000,000$

When the number of vertices in the network increases, the depth becomes deeper, or the density of the graph increases, the resulting computational complexity will only increase exponentially.

A knowledge graph without the support of graph computing power is like a car without an engine. It is dead, offline, and cannot create much value. Graph computing power is the core of a knowledge graph, the strength of its performance directly determines the effect and capability of the knowledge graph. We will showcase two scenarios to illustrate how graph computing can help augment and accelerate the knowledge of graphs.

6.4.1   Antimoney laundering scenario

Antimoney laundering is a widespread problem worldwide. It is the most common method used by organized criminal gangs to launder illegal income in huge financial cash flows. Governments around the world are looking for ways to identify and filter out these "black money" streams.

Graphs are considered to be the most natural way to express money flows through mathematics and graph theory. Especially when there are multiple hops in the process of cash flow—criminals intentionally construct multistep and multilayer cash flow models to launder money. They use the graph method (shown in Fig. 11), and if regulators are still in the era of relational database or shallow graph computing, then any of these money laundering paths through deeper camouflage will not be able to be identified. Worldwide, graph technology is being increasingly used to address a range of challenges in antimoney laundering. Capabilities such as deep graph search, real-time, white-box explainability, and stability are useful for identifying suspicious funds in antimoney laundering scenarios, especially in huge capital networks, for real-time

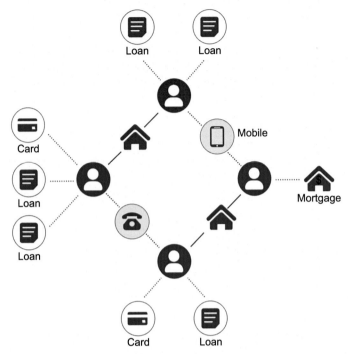

FIG. 11 Typical AML loop formed by multiple parties.

screening of money laundering behaviors, for discovering complex transactions between suspicious persons, and for rapid risk assessment.

Common antimoney laundering scenarios share the following symptoms:

1. Starting from multiple linked accounts (transfer-out accounts) to find paths to other multiple linked accounts (payees). Or start from one initial account and go back to the account or the associated account of the initial account after going through multiple hops of fund transfers (literally forming a money-launder loop).
2. Accounts involved in money laundering usually have a very high proportion of the transfer amount in the average daily balance of the account ($\sim$100%), the residence time of the transfer amount in the account is short, multiple accounts may share the same device, phone, WiFi APID or have other same attributes, etc.

There are usually millions of accounts in the banking system. We can assume that most of the accounts are law-abiding, and should be quickly filtered out, so as not to waste time and computing power on these good accounts and behaviors. The recent behavior indicator of the account is the primary filter condition to be concerned about, such as abnormal multidimensional relationships between accounts, strange capital exchange activities, and so on. Any money laundering that has occurred must have a starting point and an ending point (the ending point will absorb most of the amount transferred from the starting point), and it is not difficult to find them on the graph. How to block ongoing money laundering in real time is the goal and focus of many financial institutions. Money laundering does not happen instantly, it usually goes through many steps involving many accounts and spans across a certain period—From an AML detection perspective, a pattern will form first after several batches of transactions are executed, and many more will follow suit—being able to identify the patterns as early as possible and prevent the same pattern from repeating, ideally in real-time, is the key of graph-powered AML.

Fig. 12 shows that starting from one account (the green node), through layer-by-layer transfers, the money is finally transferred back to itself. This mode of loop transfer itself does not necessarily mean that it is 100% illegal, but at least it is worth raising alarms to let professionals pay attention to this pattern and identify whether it is money laundering. In addition, for financial institutions, the scale and frequency of money laundering by organized crime is usually much higher than that of individual retail customers. Therefore the focus of AML should be the large scale of money laundering, high frequency of occurrence, many accounts involved, and cross-bank and cross-border money laundering. In the figure above, operations such as filtering and analyzing transfer paths, locking fund collection entities (or converging accounts, usually are nodes with in-degree much greater than the out-degree), and high-frequency large-value transaction accounts can efficiently identify antimoney laundering participating accounts, and provide insights for real-time blocking or secondary analysis.

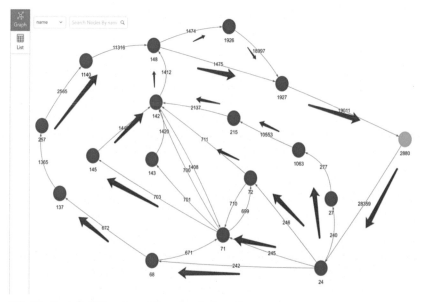

FIG. 12 Typical AML pattern of deep transfer loop discovery.

FIG. 13 AML fund converging after 10 hops.

Fig. 13 illustrates an AML scenario after starting from an account and going through ~10 layers of decentralized transfers, the funds gradually converge to a final account. If the mandates by the regulatory agency are to do a shallow 3-hop inquiry, and the max query depth of financial institutions is 5 hops, then it is impossible to find such a deep money laundering model. This kind of deep traversal and network behavior recognition is almost impossible to achieve without the support of native and high-performance graph computing engines.

Big-data framework-based graph systems, such as Spark+GraphX, Janus-Graph, ArangoDB, or the earlier generation of native-graph systems such as Neo4j, will appear extremely slow or even unable to return when performing deep queries of more than 5 hops. For the financial systems, speed and performance must come first. Time is money, and every second wasted on traversing links that cannot lock down the criminals is appeasement and connivance of crimes and damage to financial assets.

High-performance graph computing empowers the financial industry to achieve real-time in-depth antimoney laundering and has the characteristics of high visualization, ease of use, and convenient integration. We firmly believe that the next generation of antimoney laundering IT infrastructure will have a place for real-time graph database platforms.

6.4.2 Intelligent recommendation

If we want to better understand the value of recommendation system solutions based on graph computing systems, we need to first understand the status quo and problems of existing, traditional types of recommendation systems, most of which have the following commonalities:

- Require preprocessing, making real-time recommendation difficult if not impossible.
- Latency for data updates is often measured in hours or days $(T+1)$.
- A large amount of redundant data will be produced, wasting storage space (storage cost is high).
- Multiple heterogeneous models, coexist and it is difficult to reach a consensus.
- Client-side integration work is required.

Fig. 14 illustrates how to realize real-time and intelligent recommendations on the knowledge graph, the specific logic is as follows:

- User A browses (purchases, favorites, adds to the shopping cart, etc.) product A.
- Product A is browsed (or purchased, favorited, added to the shopping cart, etc.) by other users (such as B, C, D, etc.).
- Users B and C (and other users) also favorite product B; users C and D (and other users) browse product D and other products.
- By integrating and sorting other user behaviors in step 3, we know that products B and D receive the most attention—this process is the simplest implementation of "Collaborative Filtering" (CF hereafter).

Now, if we want to calculate whether there is a correlation between products B and D and product A, the knowledge graph of products comes into play. For the recommendation system implemented by traditional rule-based schemes,

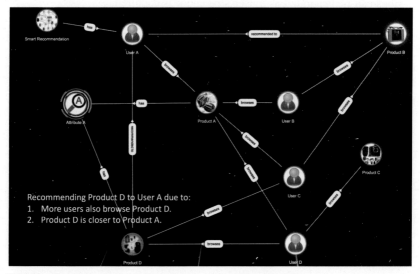

FIG. 14 Knowledge graph-based smart and real-time recommendation.

ignoring the correlation relationship between products and user behavior and expectations will recommend ridiculous results. For example, user A just bought a refrigerator, and the recommendation system found that the refrigerator was purchased at a high frequency recently, so it continued to recommend the refrigerator to user A—this is unwanted system behavior. If a knowledge graph of how products are related to each other is created, and integrated with customer shopping behavior, we will be able to evolve from the previous rudimentary recommendation to more sophisticated yet intelligent product recommendation logic—for example, the recommendation can be upgraded from another refrigerator-to-refrigerator magnets, ice cube boxes, fresh produce.... Only the following two types of product data are needed to achieve the above intelligence:

- Commodity classification data: Commodity SKUs may be on the order of tens of thousands to millions, but classification data is usually on the order of hundreds to thousands. The classification data is like the metadata of commodities, and it is essentially a tree, at the leaf level of the tree, commodity SKUs can be attached, eventually forming a larger graph (Note the difference between tree and graph is that graph has loops, because when an SKU is connected with two metadata points, a loop is formed). Products of the same kind (i.e., within the same category or having the same attribute) are usually closer on the graph.
- Label data: Different types of commodities can also be related through multidimensional labels, such as the "refrigerated" tags (labels) assigned to fresh food. These labels can form another layer of heterogeneous metadata in the product knowledge graph so that smart linkage-based recommendations can be generated as a user's behavior is captured.

In graph databases, it is intuitive and fast to implement collaborative filtering. Unlike big data and machine learning, there is no sophisticated data sampling, training, or black-box steps to achieve it. The reason is that the logic and process of collaborative filtering are described in natural language, and it can be implemented in simple template path queries.

As illustrated in Fig. 15, the steps for CF are:

1. Starting from the user (red dot), find 1-hop related items (via edge-type: browse, add to cart, purchase).
2. Find all users associated with the above items (via edge-type: browse, shopping cart, purchase, etc.).
3. Find other products associated with the above users.
4. Perform operations such as classification, sorting, and identification of the above products for the initial user, and recall the final set of products to be recommended.

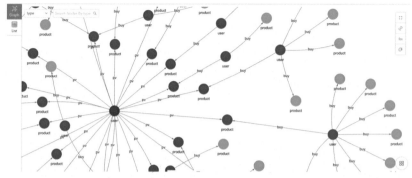

FIG. 15   Real-time collaborative filtering based on template path query.

The first three steps from above can be implemented with one simple GQL statement:

```
n({_id == "StartUserID"}).re({behavior == "pv"})
.n({type == "product"}).le({behavior == "buy"})
.n({type == "user"}).re({behavior == "buy"})
.n({type == "product"} as results)
return(results) limit(10000)
```

Of course, there is some extra work to be done in step 4 for collaborative filtering in real-world applications. However, it is undeniable that what is delineated in the first three steps is the maximum range of products that can be recalled for recommendation. In some e-commerce scenarios, the above query is equivalent to searching for three layers (3-Hop) on the graph in the form of a path template and filtering by node and edge attributes. In a graph with high connectivity, the products to be recalled may be as many as tens of thousands, additional filtering and sorting based on both user behavior and product association strength are necessary to narrow the scope of recall.

We devised two methods for such purpose:

- CRBR, which stands for Community Recognition Based Recommendation
- GEBR, which is a Graph Embedding Based Recommendation

CRBR can be regarded as a relatively basic collaborative filtering implementation, but it is much more efficient, agile, and lightweight than the traditional recommendation systems that are built on top of big data frameworks such as Hadoop, Spark, and Flink. Its core concepts are as follows:

- Run Louvain community detection for all product items (and users)
- Group by user's shopping-activity behavior (browsing, carting, purchasing, etc.)

- Locate the highest-ranked Louvain communities, but exclude those superhot products, such as toilet paper, bottled water, masks, and hand sanitizers during the early COVID-19 epidemic (these products may be recommended in a specialized column, but not suitable for being returned within a generic recommendation system)

The above three steps can be regarded as the only data training on a graph. In a time, series graph, training can be done for certain time ranges, and validation can be conducted on other newer time ranges.

Among the above-grouped users, users with lower activity can be assigned higher scores (this logic may sound counterintuitive, but it is very simple: in the same community, the behavior of a less active user is more valuable than a superactive user, because the latter's coverage is too broad to attract the attention of other regular users (the targets of a focused recommendation), and it is difficult to achieve recommendation convergence.

The above logic can be further optimized. For example, leveraging the product classification and association data, environmental information, spatial–temporal information, user attribute information, and more to fine-tune the CF process and results.

In Fig. 16, we sampled 5000 vertices from the whole graph to draw the spatial topological network constituted by the top Louvain communities in real-time. It should be pointed out that the Louvain community recognition is calculated iteratively for the whole graph, while the computing complexity is high, it is possible to accelerate as explained in Chapter 4, so that near real-time

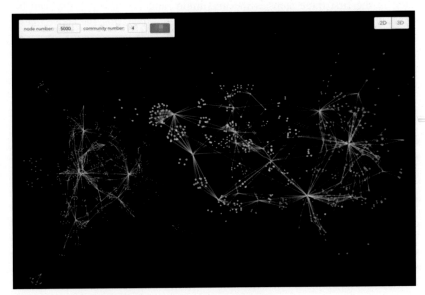

FIG. 16 Results visualization of louvain community detection.

performance can be achieved even on large graphs. The visualization part is usually sampled with vertices with a much smaller order of magnitude, usually in the range of hundreds to thousands of nodes and edges due to network-bound data transmission and on-screen rendering restrictions which we explained in Chapter 4 (noting the gigantic time–cost difference of handling many small files vs few large files).

The core idea of CRBR is to first find the communities (closely related communities) formed by all users and products, and then optimize the recommendation logic of collaborative filtering based on additional information such as user and product attribute information. For the recommendation system, our goal is not to design a brand-new algorithm, but to achieve higher efficiency and cost performance through graph data modeling. Relatively speaking, the core values of a graph-powered recommendation system lie in:

- Flexibility: The graph query template can be adjusted very easily.
- Speed: Much more efficient than traditional collaborative filtering, with a performance improvement of 100 times or more.
- XAI: The graph-centric data processing is white-box explainable.

CRBR is not the only smart recommendation solution. GEBR is another implementation. It utilizes graph-based deep learning which we call graph embedding. On the surface, graph embeddings may look gray-boxed, because the embedding process may involve deep random walks which is to generate low-dimensional vector-space values that represent the high-dimensional topo-spatial structures (by the products and user behaviors). For example, deep random walks are used in the data sampling process as an integral part of graph-embedding algorithms such as Node2vec, Struc2vec, GraphSage, FastRP, etc.

The deep random walk process is equivalent to the data sampling process in the machine learning process for massive data, and very efficient data sampling can be achieved through high-speed graph computing. Compared with traditional machine learning models, the performance of data sampling achieved by highly concurrent graph computing walks can be improved by more than 100 times. Fig. 17 shows that the graph data set is firstly deeply randomly walked based on the Node2Vec (or Struc2Vec) mode in a near-real-time batch processing method, and after obtaining the graph-embedded features of each user, the product recommendation is made according to the graph embedded features of the "subactive users." Note that this recommendation logic adopts a kind of "reverse thinking," that is, the behavior of the most active users and the most purchased products (low-value products, such as mineral water, and paper towels) are not necessarily the most in need of recommendation, but the behavior of those subactive users is more valuable for recommendation.

Another advantage of graph embedding (algorithm) is that if we are to make recommendations for one user, we do NOT have to traverse his/her neighborhood for other users' behaviors to assemble the list of recommendations, we can use his/her embeddings to compare with other users across the entire graph, so

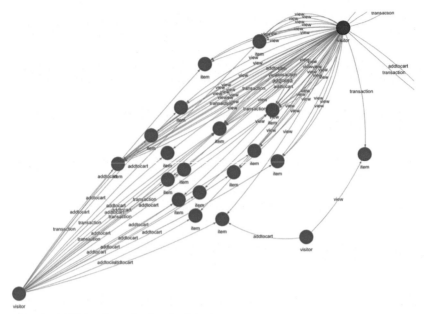

FIG. 17 GEBR based on subactive user behavior.

that the results can be quite diversified avoiding tunnel-vision type of recommendation which may be too limited. The same concept can be used in wholesale banking, for instance, the bank may have developed a whole set of bespoke services and products for one corporate, and the bank can replicate the services and product matrix to other corporate users having similar embeddings as the original corporate, even though they are not exactly connected on a graph data set!

For deep learning on graphs, many methods are still in the stage of continuous iterative transformation from laboratory to industrial applications, but they have shown considerable efficiency, accuracy, and explainability, which represent a near-term breakthrough in the development of AI.

Graph-based search and recommendation has the following advantages:

- Truly smart search is possible, be it multiple keywords correlation or linkage search, causality search, or some other types of searches that traditional search engines cannot handle.
- Real-time recommendation is possible as real-time data refreshing is made possible.
- Working with Knowledge Graphs, such as Merchandise Knowledge Graphs, the recommendation is very much human-like—100% intelligent, instead of relying on aggregated statistical data results!
- Recommendation Graph = Real-time Merchandise Graph + Customer 360-degree Graph, it offers unified all-in-one recommendation solution.

If you are going to build a search and recommendation system that is fast and smart, with low TCO and high ROI, adopting a graph-powered solution makes every sense, welcome to the rapidly evolving world of IT—graph is on the way to becoming mainstream. We are amid a trillion-dollar IT infrastructure and technology upgrade with the database being the core, and the graph is one promising technology most likely to win.

6.5   Asset and liability management and liquidity management

All tasks and services of a bank are part of Asset and Liability Management (ALM). Every single deal or transaction creates liquidity and risk. These risk positions need to be bundled, made transparent, and the risk quantified and managed. ALM requires the IT department to supply all relevant data in a consistent, complete, and online fashion, and allow back office and customer departments to drill down on any deals to know their contribution (attribution analysis or back-tracing), ideally in real-time (online) instead of $T+1$ or $T+N$. These requirements have made ALM particularly challenging. In this case study, we will address such challenges with real-time high-density graph computing which offers superior agility, flexibility, and unprecedented A&L management and monitoring features that help the ALM and Treasury department stay on top of their business data with deep insights.

Financial institutions have long been suffering from the pain points of traditional ALM systems, particularly in the following three areas:

- High Computing Complexity and Long Waiting Time
- Poor Flexibility and Coarse Granularity
- Lack of End-to-End Solution (siloed systems)

The main reason why traditional ALM systems built on top of relational databases or big-data architectures are slow and work in $T+1$ fashion is that these systems have high computational complexity when conducting ALM processes. Taking what is illustrated in Fig. 18 as an example, the attribution analysis of a key ALM indicator over RDBMS tables is very complicated and time-consuming due to many tables join, and the complexity can be much lower over a graph—the difference is exponential. The chart in Fig. 19 illustrates that as the query depth (comparable to the number of tables joined) grows, the performance gap between RDBMS and graph database enlarges exponentially, for relational databases, as the query depth reaches 3-hop and beyond (equivalent to four tables joining each other), things quickly go beyond control.

For example, in a typical financial application scenario of tracing the money flow of credit loan funds, the goal is to find accounts with a maximum transfer depth of 5 hops. The experimental data set includes 1 million accounts, and each account on average has about 100 transfer records (this makes the data set with 1 million nodes and 50 million edges). The performance results by MySQL and a

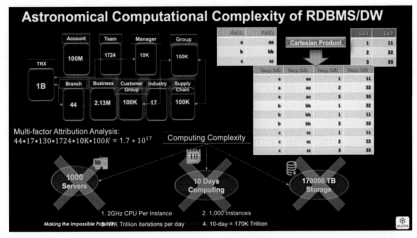

FIG. 18 Overwhelming computing complexity with RDBMS.

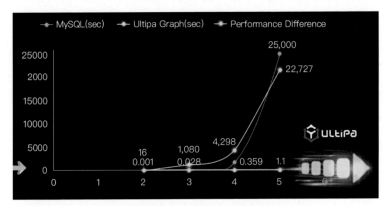

FIG. 19 Performance gaps between RDBMS and graph database.

graph database (System-U) on money-flow tracing are captured in Table 4 and charted in Fig. 18.

Due to the high complexity, traditional ALM systems tend to avoid directly dealing with transaction-level data, and handle aggregated or statistical data instead. This coarse granularity limits these ALM systems to provide fine and deep insights.

Another problem with legacy ALM systems is the fact that the diversified functionalities are offered in multiple siloed systems, as shown in the left portion of Fig. 20. This gives ALM users a fragmented and inconsistent user experience and makes it difficult to conduct deeper and correlated analyzes among many ALM indicators.

TABLE 4 RDBMS vs real-time graph database.

Query depth	MySQL execution time(s)	Graph database execution time(s)	Number of records (accounts) traversed	Number of edges (transactions) traversed
1	0.001 (1 ms)	0.0002 (0.2 ms)	≈10	N/A
2	0.016 (16 ms)	0.001 (1 ms)	≈1000	≈1000
3	30.267	0.028	≈100,000	≈10,000,000
4	1543.505	0.359	≈900,000	~45,000,000
5	Cannot complete	1.1	≈1000,000	50,000,000

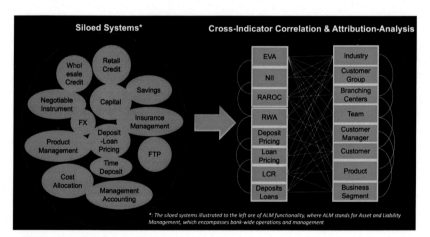

FIG. 20   ALM systems siloed vs unified.

What is illustrated in the right part of Fig. 20 is the cross-indicator analysis spanning multiple financial indicators that ALM cares about, together with multidimensional metrics for attribution analysis. Having such capability in the ALM system would give ALM users a one-stop-shop place to centrally analyze ALM data and save much time. The comparison between the two types of ALM systems is captured in Table 5.

In the second part of this case study, we will use liquidity risk, specifically an internationally regulated financial indicator LCR (Liquidity Coverage Ratio)

TABLE 5 Graph powered ALM vs traditional ALM.

ALM comparison matrix	Graph-powered ALM	Traditional ALM systems
Typical latency	<2 s	$T+1$ or longer
IT architecture	HTAP graph computing	RDBMS/big-data or lakehouse
Data granularity	Finest, single deal/trx	Aggregated
# of indicators	2000	Fewer and segregated in siloed systems
Development procedure	No-code	SQL + Excel
Explainability (white-box)	Yes	No (Black-box)
Attribution/contribution analysis	Yes, real-time	No, or $T+N$
Back-tracing	Yes, real-time	No, or $T+N$
Stress-testing/simulation	Yes, real-time	No, or $T+N$
Visualization	2D and 3D	Headless, Excel-based
Cross-indicator analysis	Yes	No (impossible to implement)

to illustrate how a graph database can help improve ALM (and LRM, short for Liquidity Risk Management, is a subset of ALM).

Liquidity risk may sound remote to some readers, but the collapse of Silicon Valley Bank (SVB), America's 16th largest bank, in the first half of 2023 should have made many people aware that the illiquidity and insolvency are the direct cause of the fall of the bank. During the international financial crisis in 2008, many banks and financial institutions were in trouble due to a lack of liquidity despite their apparent capital adequacy, and the financial market also experienced a rapid reversal from excess liquidity to shortage. After the crisis, the international community has paid unprecedented attention to liquidity risk management and supervision. In 2008 and 2010, the Basel Committee successively issued the "Principles of Robust Liquidity Risk Management and Supervision" and "The Third Edition of the Basel Accord: International Framework for Liquidity Risk Measurement, Standards, and Monitoring", which established the liquidity risk as one of the major risks to be monitored by the banks around the world. In January 2013, the Basel Committee announced the "Third Edition of the Basel Accord: Liquidity Coverage Ratio and Liquidity Risk Monitoring Standard", which revised and improved the liquidity coverage ratio (LCR) standard published in 2010.

LCR is an important financial indicator stipulated in the Basel III treaty. Major banking institutions in all sovereign countries around the world are required to monitor this indicator by January 1, 2023. It is designed to increase liquidity for banks while strengthening capital requirements.

Following the bankruptcy of SVB, we cannot help but ask: building a powerful combo of risk-control toolkits has already become the consensus of modern risk control management, so why a bank as large as SVB, with 200 billion in assets, has not used a set of "financial weapons," particularly the liquidity risk monitoring tools to help the bank executives to stay on top of their routine, and even intraday, liquidity risk management? Something must have been overlooked from both the bank and regulator's perspectives.

Since the last global financial crisis in 2008, regulations and theories on liquidity risk management have matured, but there have been few breakthroughs in technology empowerment—traditional SQL-type databases, big data, and data lakes frameworks cannot realize real-time liquidity risk management, quantitative attribution analysis, scenario simulation and other core business demands for the whole bank with the full amount of data. With the vigorous development of the digital economy, the digital transformation of commercial banks has become the only way forward. Traditional liquidity risk management systems are also facing the reality of digital transformation. LCR is a complex and difficult-to-control "new species" for many commercial banks. Even for banks that have deployed LCR systems, solutions based on traditional relational databases (such as Oracle) have the following problems (pain points):

- Black-box: The existing LCR indicator calculation systems are all implemented in a black-box (uninterpretable) way. The entire operation process of the system is opaque, and there are no detailed and quantitative indicators to track elements such as rate of change and conduction path. This limitation limits the bank's understanding of the liquidity coverage ratio to a % value, and cannot deeply understand the impact of business changes on the liquidity coverage ratio.
- No backtracking: Due to the lack of graph computing support in the past, the liquidity coverage ratio indicator cannot be backtracked. It is impossible to backtrack from the LCR indicator and trace it back to the business, account, or other factors that have the greatest contribution to the indicator. This untraceability means that banks can only use an LCR indicator to cope with supervision, but they cannot deeply understand their core business performance and adjust business development indicators according to specific business conditions.
- No detail-oriented simulation: The ability opposite to backtracking is forward simulation, that is, starting from a certain branch, a certain industry, a certain region, a certain type of account, and certain transactions according to the way of transmission along the path, from leaves to root, to simulate the impact of changes in certain factors on LCR. The lack of this ability makes it

impossible for banks to intelligently and comprehensively predict, evaluate, and design their products and adjust business direction to potential impact on their liquidity status.

- Lack of visualization: Graph visualization is an important means of calculating LCR indicators transparent and interpretable (see Fig. 21). The liquidity coverage ratio without the support of these means is just a simple indicator, which is of no help to achieve internal efficiency through a comprehensive analysis of assets and liabilities.
- Nonreal-time: The loading and calculation of LCR-related business data takes a long time and cannot be calculated in a $T+0$ or real-time manner, let alone real-time simulation, backtracking, quantitative calculation, and other operations.

In recent years, graph computing and graph database technology have developed rapidly. We can use the latest graph technology to rebuild the liquidity risk management system. The liquidity risk management information system shall at least realize the following functions:

- Monitor the liquidity status and calculate the cash inflow, outflow, and gap in each set period.
- Calculate liquidity risk regulatory and monitoring indicators, and increase monitoring frequency if necessary.
- Support the monitoring and control of liquidity risk limits.
- Support the monitoring of large capital flows.
- Support the monitoring of high-quality liquid assets and other assets without liquidity barriers.
- Support the monitoring of finance information concerning collateral (pledge) type, quantity, currency, location and institution, custody account, and others.
- Supports real-time stress testing under different hypothetical scenarios.

With commercial banks facing "strong supervision" and "internal efficiency," a high-performance graph computing system can efficiently empower banks to upgrade their business models, adjust their asset and liability structures, optimize resource allocation, and pursue the transformation of light banks with light capital consumption, while improving profitability and capital efficiency.

Compared with LRM solutions built on traditional architectures, a graph database can clearly and efficiently reveal complex relational patterns, process massive amounts of data in real time, and visualize results and conduction paths in real-time. These are the core demands of LCR's external supervision and internal efficiency. The LRM system based on a real-time graph database has the following advantages (as shown in Fig. 22):

- White-box interpretability through high-performance, easy-to-operate 3D visualization.

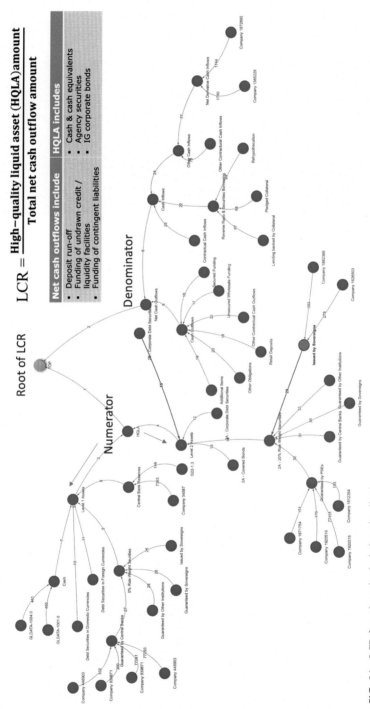

$$LCR = \frac{\text{High-quality liquid asset (HQLA) amount}}{\text{Total net cash outflow amount}}$$

Net cash outflows include	HQLA includes
• Deposit run-off • Funding of undrawn credit / liquidity facilities • Funding of contingent liabilities	• Cash & cash equivalents • Agency securities • IG corporate bonds

FIG. 21 LCR formula and calculation visualized.

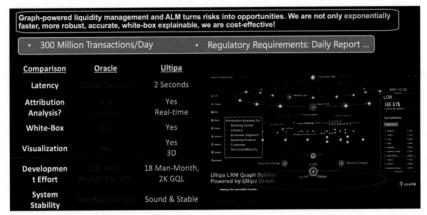

FIG. 22 Oracle LRM vs Ultipa LRM.

- Real-time traceability allows banks to locate and trace the main factors and transmission paths of LCR (and other liquidity indicators as well) changes in real-time and visually.
- The ability of real-time simulation allows banks to conduct quantitative analysis based on scenario simulation for core assets and debt products and businesses.

The core of the LRM system is to achieve fast computing and real-time visualization of liquidity indicators by connecting the business data of the whole bank, completing data development, and building a graph computing framework.

$$LCR = \frac{High - quality\ liquid\ asset\ (HQLA)\ amount}{Total\ net\ cash\ outflow\ amount}$$

Taking the development of LCR as an example, the development tasks can be listed in four groups as outlined in Fig. 23. Note the four sequential groups can be viewed as Data Preparation/Collection, Data Cleansing and Modeling, Foundational Feature Implementation, and Application Development (Visualization).

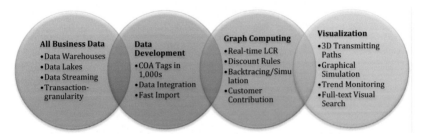

FIG. 23 Liquidity risk management development tasks.

Development Phase

Testing Phase

FIG. 24 Liquidity management system development process.

Fig. 24 shows a typical engineering approach to LRM system implementation which can be further broken down into the development phase and testing (and verification) phase.

To glue everything together, an LRM system would have an architecture comparable to what has been sketched in Fig. 25.

The main functions of a graph-powered LRM system include:

- Visual display and breakdown of LCR (other liquidity ratios follow similar design ideologies). Illustrated in Fig. 26.
- Attribution Analysis, particularly the real-time backtracking feature (illustrated in Fig. 27) when comparing multiple days' LCRs and quantifying which factors contribute how much to the change of LCR.
- Full-text Visual Search (illustrated in Fig. 28), as a key complementary to the formulated user interface and allows users to search freely.
- Interactive Analysis (Fig. 29), examining how a customer, a banking service or product, an industry, a banking-center brank, a business customer, or a client manager and his/her team has contributed to the bank's overall liquidity status—such requirement involve both historical data analysis and attribution analysis.

Different from the pain points of poor timeliness and slow calculation in traditional liquidity systems, the graph-powered liquidity risk management system can perform real-time data penetration on massive and complex data and accurately measure the reasons for changes, helping business parties to quantify risk changes in the first place and fulfill regulatory requirements, adjust industry business decisions in real-time, formulate business rules, and realize the equilibrium between safety, profitability, and liquidity in the banking business.

After the financial crisis in 2008, it has gradually become a consensus of the industry and regulators to pay attention to liquidity risk management. Industry experts have found in their research that risks have the characteristics of correlation, mutual transformation, transmission, and coupling, and the risk transmission channels are complicated.

Basel III puts forward new and more refined requirements for the management and capital measurement of credit risk, market risk, and operational risk,

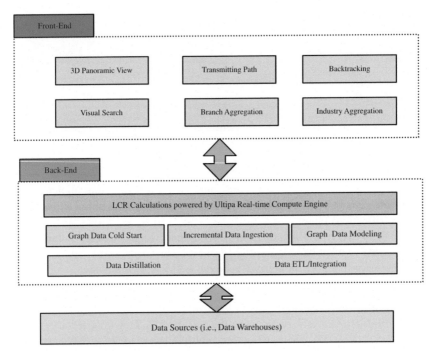

FIG. 25 Graph-based LRM system architecture.

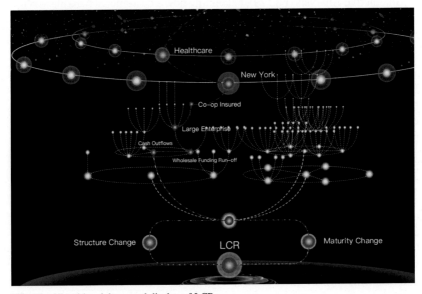

FIG. 26 Visual breakdown and display of LCR.

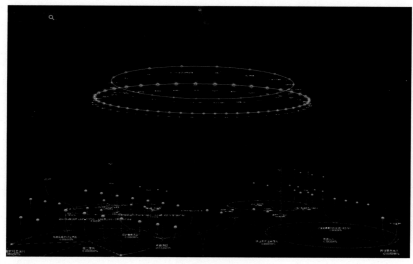

FIG. 27 Real-time LCR back-tracing with transmitting paths.

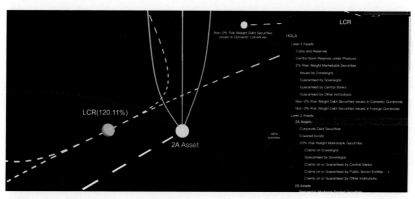

FIG. 28 Visualized full-text search.

and makes more detailed regulations for risk information disclosure. Special guidelines have also been issued for facility construction, risk data collection capabilities, and risk reporting. Commercial banks took the opportunity of the Basel III reform to take into account both new regulatory compliance and internal management efficiency needs, laying a solid foundation for the implementation of the Basel III system. Fig. 30 shows a graph-powered ALM system that covers the monitoring and management of all Basel III indicators such as AUM (Asset Under Management), LCR, IR (Interest Rate), Capital, RWA (Risk Weighted Asset), NII (Net Interest Income), RAROC

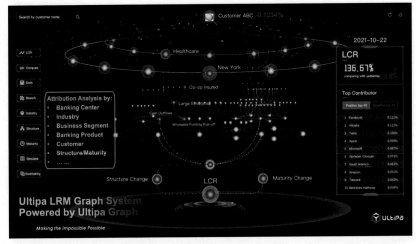

FIG. 29 The full picture of LRM.

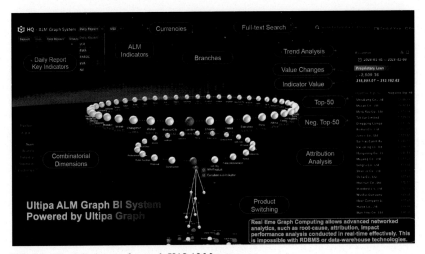

FIG. 30 The full picture of a graph-XAI ALM system.

(Risk-Adjusted Return On Capital), EVA (Economic Value Added), varied leverage ratios, and offers an off-the-shelf end-to-end solution to the banks to help comb through their infrastructure, data, and reporting mechanism, and fill in the gaps between with white-box explainability, highly visualization, and highly intuitive GUI. It goes way beyond the basic needs of regulatory compliance, but also satisfies business growth needs via deep correlation and 360-degree quantifiable insights of the crucial ALM data.

6.6   Interconnected risk identification and measurement

6.6.1   Innovation of graph computing in interconnected financial risk management

1. What are interconnected financial risks?

The butterfly effect is a concept in chaos theory. It refers to a dependent phenomenon of sensitivity to initial conditions——Small differences at the input end may transmit along the path, and magnify by multiple folds at the output end. In the 1970s, a meteorologist named Lorenz in the United States explained that the occasional flapping of the wings of a butterfly in the Amazon rainforest might cause a tornado in Texas 2 weeks later. By analogy, the butterfly effect also exists in the financial market. In the interconnected (interdependent) financial market, any "butterfly" flapping its "wings" may cause cross-market risk contagion, and a single individual's risk problem is likely to cause problems in the entire market. At present, interconnected financial risks are the types of risk most likely to cause systemic risks at the national or even global scale.

The 2008 financial crisis made the industry realize that several risks do not exist in isolation, but are interrelated, and interactive and eventually lead to systemic risks, and there is a chain effect between different types of risks. Counterparty credit risk is an important channel for individual volatility risk to spread to peers and eventually magnify into systemic risk. Liquidity risk is the amplifier of financial market volatility, and market risk will eventually evolve into systemic risk through risk contagion. Cross-market risk is not necessarily a domino-like serial or chain transmission, but a network transmission (parallel, multilevel transmission) that affects the whole body of multiple markets.

Taking Silicon Valley Bank as an example, its problem is not only liquidity risk, but the intertwining of at least four risks: credit risk, reputation risk, liquidity risk, and market risk, which ultimately constitute interconnected financial risks. Interconnected financial risks have three characteristics:

- Chain effect: Each risk, if observed singularly and independently, is transmitted in the form of a chain or chains.
- Network propagation: When all risks are considered interdependently, the propagation of risks shows a network pattern. Taking the SVB's March 2023 collapse as an example, the bank held a lot of long-term debt (facing interest-rate risk which is a type of market risk) that had declined in market value as the Fed raised interest rates to fight inflation, and as a result, the bank faced huge losses (via market risk) when it had to sell those securities to raise cash to meet the waves of withdrawals (liquidity risk and reputation risk) from the customers who are predominantly from the technology sector (sector-wide credit risk) which has been hit badly by the downturning market since late 2022.

- Propagation of risks may cause systemic risk. As the shockwaves triggered by the risks continue to spread, the entire market will be in a panic mode (that is why the U.S. government intervened swiftly to prevent the bank-run panic from spreading).

Interconnected financial risk cannot be simply measured by PD and LGD (probability of default and loss given default). It is necessary to trace, examine, quantify, and predict how risks are transmitted to each other in different markets, and it is clear that traditional, linear, or low-dimensional technologies are insufficient for such purposes. High-dimensional graph technology potentially and naturally can provide effective solutions for such purposes.

2. How to identify and measure interconnected financial risks: what key issues to solve?

Interconnected financial risks have features such as cross-market, cross-product, and cross-department, involving many counterparties, with complex risk types, having long chain of infection, relatively weak management, and broad scope of coverage. The root cause of the difficulty in managing interconnected financial risks lies in the strong contagion, and commercial banks have not yet formed a complete risk management system. At present, interconnected financial risk management urgently needs to solve the following three aspects:

- Scoping the audience for risk contagion: Commercial banks must form a panoramic view of the spread of interconnected financial risks and know which customer groups the risks are spread to.
- Measuring risk propagation paths: Commercial banks must identify the possible transmission paths and scale (depth, breadth, speed) of risks, and be able to evaluate the final result of the superposition of multiple risks on the transmission paths (a network eventually).
- Identify critical nodes in the risk contagion process. Once commercial banks find key nodes (products, customers, industries, regions, etc.), they can take action to alleviate or prevent further spread of risks.

The above aspects and characteristics of Interconnected Financial Risks are illustrated in Fig. 31.

3. Application of graph computing in interconnected financial risk management

Graph computing can calculate the relationship between data through deep penetration and mining of massive and complex data points, and solve complex multilayer nested relationship mining problems. This kind of calculation can be compared to reverse engineering of the working mode of the human brain—this is why graph computing is also called "neuromorphic computing", it can be used as a magical weapon to empower financial risk management.

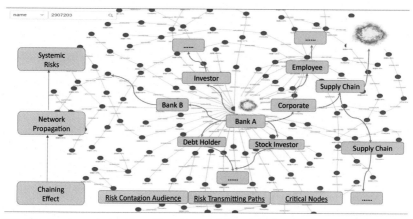

FIG. 31 Scoping and measuring of interconnected financial risks.

Taking SVB as an example, immediately during and after its crisis, the risk first spread to SVB's customers (mainly technology companies and tech investors), which is the first layer; if something goes wrong with these companies, the first to be affected are their employees and suppliers, which constitute the second layer; if the suppliers stop supplying goods and employees refuse to resume work, these companies will not be able to fulfill their contracted businesses with their customers, then the risk will spread to the market at large, which is considered the third layer—a gigantic network is therefore formed—for the sake of simplified discussion, we have left out other financial institutions that are potentially impacted by the downfall of SVB, but the point of discussion is that risks are spread hop by hop, layer by layer, and the "chaining, networked and interconnected effect" is obvious, as shown in Fig. 31.

In terms of measuring risk transmission paths, graph computing can help identify financial risk transmission paths through four types of paths (or connectivity):

- Equity: Recall the UBO scenario we discussed earlier, equity ownership can reveal how the bank and its customers are controlled on the one hand (for their management and investors) and how their investment landscape looks on the other hand (for the companies they have put money in).
- Guarantee: Recall the Corporate Guarantee Chain scenario, this will tell the investigator which other companies are likely to be affected quickly with the fall of the bank or the troubled corporate customer.
- Capital: Knowing the money flows would help identify the counterparties that are to be affected during the risk contagions.
- Supply chain: Same as above, but potentially in a broader spectrum, we are living in a world full of supply chain networks, and the fall of a major corporate customer or a bank may cause a ripple effect affecting tens of thousands of companies on the supply-chain network.

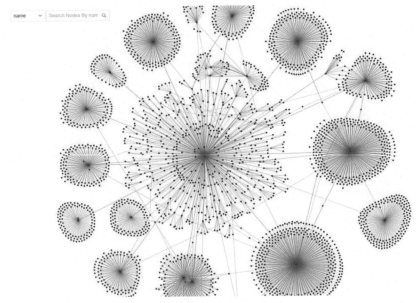

FIG. 32 A Corporate's interconnected risk customer groups.

In terms of customer groups for measuring risk transmission, after all the paths of risk transmission are identified by graph computing, all customer groups affected by risks are also identified, with collapsing bank or corporate as the "center" and the risk transmission path as the "radius". The risk-affected customer group is a "circle," and various "circles" are overlapped and intertwined to form a "network" from a global-view perspective. The risk-affected customer groups include suppliers, employees, investors, affiliated companies, stakeholders, and more as shown in Fig. 32.

In terms of finding key nodes in the process of risk propagation, graph computing technology can establish network relationship graphs, determine key entities (customers, products, or something else) for risk prevention and control, and introduce customer access, risk limits, and other means to effectively prevent risk spreading.

Due to the correlated nature of interconnected risks, it is difficult for traditional relational databases to achieve fast data modeling and computing of multilevel relationships. Graph computing technology can swiftly find out the key nodes, risk factors, and risk transmission paths, and empower the overall management of interconnected financial risks.

6.6.2 The broad prospects of graph computing in the financial field

Graph computing (graph database) is considered to be a robust third-generation AI technology implemented with graph-augmented intelligence. The third generation of artificial intelligence requires the collaboration of four elements: data, knowledge, algorithms, and computing power. Compared with the previous generation of artificial intelligence technology, the third generation of artificial intelligence pays more attention to the white box explainability of algorithms and the significant improvement of computing power. Digitally transforming companies can use the driving force brought by graph computing technology to build a new AI risk monitoring system based on complex networks, accommodate the architecture of complex network topology, and meet the requirements of high dimensionality, high visibility, high performance, strong security, and real-time-ness on the business side.

With the rapid development of economic globalization and the market economy, all things (enterprises and individuals) are connected through certain relationships to form a complex network, such as enterprises, banks, trust companies, insurance companies, guarantees, etc. Economic entities such as companies have formed an intricate graph of related relationships through equity, mutual guarantees, related transactions, financial derivatives, supply chain relationships, and multiple identities of management.

Compared with relational databases which are based on two-dimensional table schemas, graph databases are essentially high-dimensional databases, and their core and most unique capabilities are high-dimensional data computing capabilities (graph computing). Banks can use a graph database to build a customer relationship graph, pay attention to the correlation between various types of customer information, realize the leap of customer insight from local to network-wide, from static data to dynamic intelligence, build a customer-wide relationship graph, discover potential risks, and predict risk transmission paths, probabilities, and impacted customer groups.

Graph computing technology has revolutionized existing credit risk management. By combining graph computing (computing power) with business logic (relationship graph or knowledge graph), an artificial intelligence bank risk control brain with real-time online computing and analysis capabilities can be built. Transforming the subject of financial risk prevention from a single customer to risk customer groups and the timeliness of prevention from post management to advanced prediction. Typical application scenarios include: identifying hidden group-related risks, identifying guarantee circle risks, insight into customer group risks, and real-time monitoring of loan capital flow:

1. Hidden group-related risks

Graph computing technology can help unravel a company's complex equity structure by penetrating it layer by layer to meet the financial institute's (i.e., bank's) KYC (Know Your Customer) or Customer-360 requirements for the identification of the beneficial owner of the customer.

Banks use corporate relationship graphs to identify shadow groups and invisible groups, achieve unified credit granting to actual controllers, group customers, or single legal person customers, effectively identify high-risk customers, prevent multiple credit granting, excessive credit granting, credit granting to zombie companies or shell enterprises, and prevent financial frauds.

2. Identify the circle of guarantee risks

Graph databases can help identify the main risk companies and their complete guarantee paths in all guarantee circles and chains of a bank. The bank can use graph technology to carry out data modeling and timely identify and quantify the default risk of guarantee circle (chain) enterprises, conduct efficient checks on guarantee circle (chain) loans, analyze the causes of guarantee risks, and take preventive measures promptly. For example, by using graph algorithms to analyze the size of the enterprise guarantee circle and the intensity of the guarantee relationship, identify the enterprise guarantee circle with relatively large guarantee risk in the structural sense, carry out focused processing and analysis, monitor the most complex guarantee relationship, the largest amount involved, and the greatest risk in real-time, and then focus on real-time monitoring of the core enterprises in the guarantee circle or chain.

3. Insight into customer-group risk

Based on the customer relationship graph, comprehensively consider the customer supply chain, fund-flow chain, capital and guarantee circle, and other relationships to form a customer risk transmission path, to calculate the risk conduction probability between the enterprise to be assessed and the related enterprises within N hops (layers), and in-depth exploration of all the risk conduction paths (and impact factors) from the subject enterprise to "enterprise known to be falling." When a risk event occurs in an enterprise, the risk exposure of all credit customers of the bank is calculated in real-time—the same can be generalized to the central bank or other regulatory bodies to evaluate the potential impact of the collapse of any financial institute.

4. Real-time monitoring of loan capital flow

Based on graph technology, tracing the flow of loans can be conducted in a penetrative manner. After the loan is disbursed, the post loan management personnel of the handling agency and the risk management department should do the following:

- The flow of loan funds meets the agreed purpose, pays attention to the flow of funds in the bank, and reports suspicious matters in the process of supervising the flow of credit funds promptly.
- After disbursement, it is necessary to track and monitor whether the credit funds flow into enterprises that do not have a supply chain relationship with the borrower.
- Track and analyze the source of repayment funds of the borrower, and verify whether the interest repayment funds are regularly remitted by a third party and whether the principal repayment funds are centralized transferred by a

third party, before the repayment date, to judge the misappropriation of loan funds.

Graph computing technology tracks which account each loan fund eventually flows into, to determine whether the loan fund has been embezzled and whether it has flowed into areas of regulatory focus such as bit-coin, real estate, stock market, or other suspicious channels.

With the deepening of digital transformation, it is becoming more and more important to deeply explore the value behind the relationship. The traditional relational database can no longer meet the requirements of in-depth search, relationship discovery, and business priority. The graph database organizes data through the graph structure, which overcomes the challenges of deep relational data analysis that cannot be overcome by other databases, and provides important technological support and technical guidance for building enterprise knowledge graphs, building AI decision engines, and realizing in-depth business knowledge and value mining. At present, the predominant application of graph technology in the field of risk management is still in its infancy. In the future, graphs will dig deep into data and blossom in the fields of business innovation, intelligent marketing, smart customer service, advanced investment, business forecasting, and index measurement, and empower enterprises to smoothly tread the waters of digital transformation.

Chapter 7

Planning, benchmarking and optimization of graph systems

In software engineering, we may spend 20% of our time developing a system and 80% of our time testing, evaluating, and optimizing it. The 20-to-80 principle is especially true for database systems. Usually, the high-intensity code development of a database system may only take 1 year, but testing and optimization (iteration) may take several years. There is a saying in the industry: the customer is the best teacher. Customer needs and scenarios are the best whetstones for polishing a database product. Especially for emerging technologies such as graph databases, we believe that it is not difficult to make a wise choice between the two alternative routes of building a car behind closed doors or iteratively developing based on real business needs and data.

This chapter contains three parts:

- How to plan a graph system
- How to benchmark graph systems
- How to optimize a graph system

7.1 Planning your graph system

Graph System has a very broad meaning. It may be a graph database system, a graph computing framework, a knowledge graph platform, or an end-to-end business layer solution or graph product built on any of the above underlying systems. Compared with traditional databases, the graph system is a brand-new technology. Even for those big data frameworks that can only realize certain specific functions, business departments, and IT departments have uneven awareness of them. Therefore, when your department needs to deployment a graph system, it means that the business has encountered a very interesting challenge.

The Essential Criteria of Graph Databases. https://doi.org/10.1016/B978-0-443-14162-1.00002-7

The author has encountered various business personnel, customers, developers, peers, and partners in the past few years, and found that a consensus has already been reached that the graph system shall be planned for launching, however, people came to the consensus from different perspectives. Specifically, the awareness of the graph system can be roughly divided into three stages: the stage of problem or solution aware, the stage of briefly tapping into it, and the stage of in-depth application.

In fact, there may be an "unaware" stage before the "aware" stage, but customers at this stage still have a long way to go before the planning of a real graph system. However, their own business challenges and pain points may not be clearly summarized, however, they are still wandering in the circle of traditional architecture and have not transformed. Most of the customers who are aware of graph technologies have heard of the magic power of the graph database, but they have not had the actual opportunity to use it. Some of the customers who have tried it have already installed and tested at least one graph system, but they have not really let the graph system run in the production environment or serve the business in depth. Customers who are in the in-depth application stage usually have already begun to plan how to use the graph system on a large scale in a wider range of scenarios. After all, the meaning of the graph system is not as an optional icing on the cake subsystem, it will gradually blossom and bear fruit in empowering the applications of new scenarios, and eventually replace traditional types of databases.

According to the general law of development of things, everything goes from small to large, from slow to fast, from static to dynamic. If this set of rules is put on the graph system, it can also be confirmed in the development process of worldwide graph systems, as shown in Fig. 1—the magic quadrant for graph database systems.

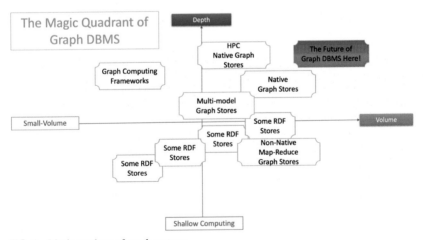

FIG. 1 Magic quadrant of graph systems.

There are multiple genres of graph systems, they are:

- RDF stores
- Graph computing frameworks
- Multimodel graph stores
- Nonnative Map-Reduce type graph stores
- Native graph stores

Combined with specific business demands, the systems in the above quadrants can be classified into three modes, in terms of how they would be able to handle the following data-processing requirements, ranging from simple to complex, nonreal-time to real-time, and high-latency to low-latency:

- Offline system, batch processing tasks (traditional BI mode).
- Online system (OLAP-centric), background batch processing tasks, shallow data processing capabilities (≤ 2 to 3 hops).
- Online system (HTAP-capable), online batch processing capable, with in-depth (>3 hops) online query processing capabilities.

The above three modes may be regarded as the main line of building a graph system solution. Along this main line, we will combine data, modeling, capacity, system coexistence and other dimensions to analyze how to fulfill the planning and construction of your new graph system.

7.1.1   Data and modeling

In the era of big data, any system without data flowing in and out will be water without a source and a tree without roots. Where does the data of the graph system comes from is a matter of different opinion. If we look at the life cycle (stage) of data flow, there are roughly three stages. Note that data may flow from the previous stage to the next stage, or may appear and exist directly in a certain stage:

- Data is generated from other existing systems and flows into the graph system.
- Data is generated in the graph system, or is retained in the graph system after flowing into it.
- Data flows out from the graph system to other systems.

The data types and source systems that can be connected by (and imported into) a graph system are abundant. Taking the data categories with graph-inclusion value in the enterprise IT environment as an example:

- User information data, account information data, etc.
- Product information data.
- Market information data.
- Rule data, policy data, configuration data, etc.

- Asset data information, etc.
- Third-party datasets.
- Any type of data that can be imported (such as ETL) into the graph.

What data can be handled (imported) in the graph is a very basic yet interesting question, but its answer is not set in stone. If we change our perspective, from the structure-of-data point of view, the data currently imported into the graph is predominantly structured data. For semistructured and unstructured data, they need to be converted into structured data through ETL tools before they can be picked up by the graph system. Many readers may not understand this passage well. We need to revisit the meaning and purpose of the existence of the graph system—to find the correlations amongst multidimensional and related data, or to find the micro or macro characteristics of these data, and perform accelerated and augmented value extraction. Compared with semistructured or unstructured data, it is most efficient to find associations between structured data. The emergence of big data systems introduces the concepts of structured, semistructured, and unstructured data. For example, multimedia data, documents, and most text files are typical unstructured data, while semistructured data and structured data are generated after different levels of "structural" processing. From the perspective of the most fine-grained storage (bit), all data is structured, but from the perspective of operating system, file system, data processing systems and upper-level applications, we artificially define different data types and processing means. From a macro and forward-looking perspective, any type of data can enter the graph system, but different types of data may experience varied utilization rates and the way of processing may be quite different. In other words, unstructured or semistructured data usually need to go through a "structurization" process, which is a necessary step in the "data modeling" process.

How does the data scattered across different siloed systems, as illustrated in Fig. 2, enter a graph system? We need the help of some tools (as shown in Fig. 3):

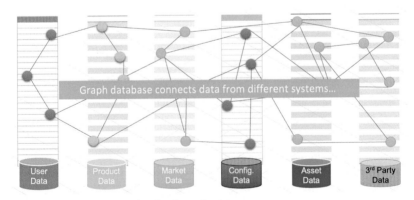

Graph database connects data from different systems...

| User Data | Product Data | Market Data | Config. Data | Asset Data | 3rd Party Data |

FIG. 2 Graph system connecting data from siloed systems.

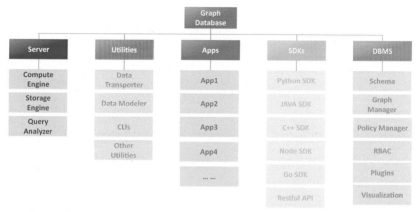

FIG. 3   Graph system toolchain.

- Data import tool (Data Transporter or Data Importer)
- Cross-system connector (Data Connector)
- Data Modelling and Schema Design tools (Graph Data Modeller)
- Graph Data Management and Visualization Tools (Graph Manager or DBMS)
- Other utilities like Command Line Interface tools (CLIs)

The data import tool (which can also include data export capabilities) reads and parses the generated graph data and loads it into the graph system. It can be divided into different types based on their working mechanism, such as local import or network/online import. The performance of network import is usually lower than that of the local import mode, and distributed/parallel import tends to be faster than singular/serial import. A cross-system data connector can be regarded as a more advanced import tool, which can directly connect to other systems (e.g., data warehouses, big data systems, or other database systems) and read the data fields of the system, importing them into the graph system online. The connector can be seen as a network-based cross-system graph data import tool. The graph data modeling, schema design and visualized data management tools are powerful for cross-table and cross-database data modeling, schema-management, and graph data management, especially with the visually interactive operation process that allows for more convenient data and model management.

Regarding data modeling, we have mentioned many times in the previous chapters that there are usually more than one way of data modeling in graph systems, and it can even be said that to explore the relationship between data in multiple modeling methods, you can do "All roads lead to Rome," however, the time, difficulty, and cost of each road to Rome may vary widely. The author has sorted out the differences in graph modeling in different modes, which can be roughly divided into the following three categories:

- Traditional graph computing and graph database systems
- Social graph (or graph dataset) and financial graph system
- Academic and theoretical graphs and industrial graphs

If we sort out and compare the commonality between the modeling methods in the above three groups of comparisons, it can be summarized in Table 1.

TABLE 1 Static graph computing vs dynamic graph databases.

Comparison matrix	Traditional graph computing Social graph Academia graph dataset	Graph database Financial graph Industrial graph dataset
Simple graph?	Simple graph	Multigraph
Isomorphic?	Isomorphic graph (entities tend to be of the same type)	Heterogeneous graph (entities may be of different types)
Static data?	Static	Need to support dynamic data
With attributes?	No attributes or very few attributes	Multiple attributes
Data volume	Small, or synthetic (simulated) large datasets	Generally larger
Multiple graph sets?	Single graph set (pursuing a single and large graph)	Multigraph set or linkage between multiple graph datasets
Architecture design	Focus on in-memory computing	Computing + storage (database)
OLAP/OLTP	Mainly offline processing mode, or AP system	AP → TP Or AP + TP (HTAP)
Data import method	File-based import	Need to support multiple import methods
Real-time?	No specific requirements. Some frameworks pursue real-time algorithms, such as GAP	May pursue real-time performance and low latency to address real-world decision-making and data analytics challenges
Visualization	Visualization is mostly used for statistical analysis	Visualization is used to provide interpretability and guide business
Industry characteristics	Academia, social SNS	General industry, financial industry, government and enterprises, etc.

Obviously, the graph data in the industry is dynamic and heterogeneous, and the correlation between data entities is diverse. From a data modeling perspective, the diversity is reflected in scenarios like some data can be modeled as entities, while they can also be modeled as edges or even attributes for certain vertices or edges. To help illustrate this, taking supply chain management as an example, if we are to model port data as entities, how would you model the freight and vessel data? There are multiple ways—model the freight and vessel as separate (of heterogeneous type) entities, or model the freight as relationships between the port entities, and vessel data as attributes associated with the freight, and so on.

The gist here is that we may adjust the data modeling according to the needs of the business to better solve the challenges of the business! Overall, the development trend of the graph systems is to shift from simple-graph to multigraph, from isomorphism to heterogeneity, from static to dynamic data, from no attribute, single attribute to multiple attributes, from single graph data set to multiple graph data sets, from AP to TP …

The above passage can be regarded as the core concept in the planning and design of any graph system. The graph is high-dimensional, and it can use flexible and diverse methods to achieve functions that are difficult (to realize) for low-dimensional (such as two-dimensional SQL) systems.

We can use a concrete example to illustrate the flexibility of graph modeling. Taking the financial industry as an example, the wire transfer network between accounts (as shown in Fig. 4) can take the account as the vertex and the transfer transaction between accounts as the edge. Naturally, the transfer network is formed.

From the graph theory point of view, Fig. 4 is a typical isomorphic graph, but because there may be multiple transfer transactions (edges of the same type) between two accounts, it should naturally be a multigraph. If the simple-graph mode is used, each transfer transaction needs to be modeled as a vertex, plus two additional edges to link the accounts of both counterparties, as shown in Fig. 5.

The two illustrations from Figs. 4 and 5 on the financial transfer/transaction network can be augmented and introduce other types of entities, such as POS machines, merchant accounts as well as different types of association

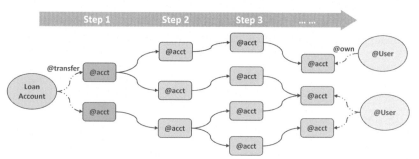

FIG. 4 Transaction (account transfer) network.

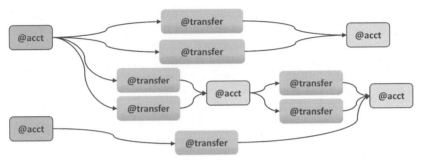

FIG. 5 Transaction (transfer as an entity) network in simple graph model.

relationships, such as holding relationships, investment relationships, affiliation relationships, kinship relationships, supply and marketing relationships, etc. In this way, the networked graph becomes a heterogeneous graph (a mix and match of different types of entities and different types of relationships), as shown in Fig. 6.

Taking the heterogeneous graph data set in Fig. 6 as an example, the borrower is a type of entity, that can have multiple attributes, such as gender, age, account level, balance, email address, work unit, phone number, etc. But there can also be another modeling method, which splits the above attributes as separate entities, as shown in Fig. 7, it becomes a typical credit or loan anti-fraud data modeling method.

In Fig. 7, to check whether each credit application (treating each application as an entity) has a quantifiable risk, check whether it has an associative relationship with other credit applications. If there are a common company, device, phone number, email, and other entities in any two credit applications, there is a high potential for fraud (duplicate applications). The Jaccard Similarity algorithm introduced in the fourth chapter can also quantify such risk efficiently—when two applications are having abnormally high similarities, they should be identified for further risk investigations. Finding the associative relationship between two vertices can also be realized by path query, such as loop query (or detecting cycles), starting from a credit application, if there are multiple 4-step circles, the application has a high risk of fraud. For example, each application shown in the figure has 15 loops with a depth of 4 ($C_6^2 = 6 * 5/2$).

Fig. 7 can be seen as a typical simple graph, and does not require complex attributes except IDs. But it should be pointed out that a simple graph is a special case of a multigraph, and the effect of a simple graph can be achieved with a multigraph, but not vice versa.

7.1.2 Capacity planning

The capacity (predominantly maximum allowed storage volume, supplemented with max client connections, etc.) of graph systems has always been a very interesting question. The author tries to help readers clarify and grasp the core

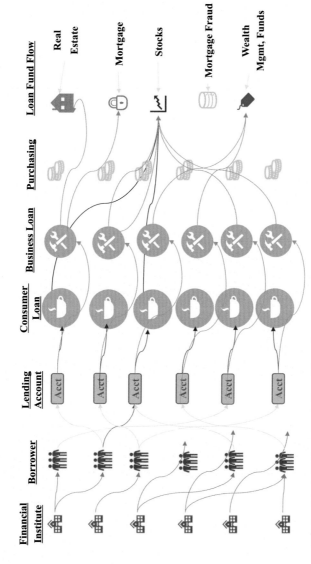

FIG. 6 A typical heterogeneous financial graph.

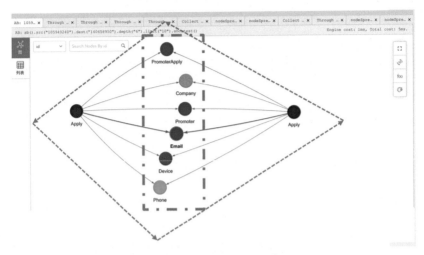

FIG. 7 A typical loan application antifraud graph data modeling.

of system capacity planning from two perspectives (mysteries per se), and once that is addressed, the rest of the planning problems can be naturally solved.

- Influenced by the big data concept, the urge to pursue excessively large (such as trillion-scale) graph systems.
- The capacity counting method is per cross-table-join as in relational databases.

Baptized by the wave of big data, when many technicians first come into contact with the graph system, it is easy to apply the concept of MapReduce/HDFS to the graph system, and directly set the expected data capacity of the system to trillion-scale, although the data at hand now, and in the foreseeable future, will not even be reaching 100 million mark.

Regarding the maximum data capacity of a graph system, it essentially depends on the specific business requirements and the kind of data and modeling methods used to address real-world challenges, rather than solely focusing on whether it is possible to build a trillion-level graph system. Putting the cart before the horse, if we consider using the Hadoop (or Spark) model to construct a graph system, it can indeed store trillions of data, as we have mentioned in previous chapters. However, such a system may encounter serious challenges when it comes to graph data processing. In other words, the primary objective of a graph system is to handle the association, traversal, drill-down, and aggregation of multidimensional data, accomplishing these tasks in less time and with lower resource consumption (hardware). These tasks heavily rely on computational capabilities, with computation as the core. The author has observed a trend in the industry in recent years, where there is a focus on the separation of storage and computing but a neglect of the timeliness of computing. There is a tendency to equate computing power solely with the enumeration and superposition of underlying hardware, without considering whether the software can fully leverage the computing power of the hardware. In short, regardless of the amount of data that can be stored, it will be futile without software that fully utilizes the underlying computing power through concurrency and efficient execution. The graph system must address computational challenges first and then align them with the appropriate storage capabilities. Ultimately, a complete system must integrate both aspects.

Furthermore, it's important to recognize that the scalability and performance of a graph system extend beyond just its storage and computing capabilities. The design and implementation of efficient algorithms, data structures, and indexing techniques also play a crucial role. Optimizing these aspects can significantly enhance the system's ability to handle larger datasets and complex queries. Additionally, considerations such as fault tolerance, data replication, and distributed processing are vital for building a robust and scalable graph system. Therefore, when evaluating the capacity of a graph system, it's essential to take into account a holistic approach that encompasses not only storage and computation but also algorithmic efficiency, system architecture, and fault tolerance mechanisms.

However, even if a trillion-scale data set exists, if any system treats trillion-level data indiscriminately, there are only two possibilities:

- The performance of the system is terrible (low performance)
- Very high system cost (high performance)

The former refers to a system operating in HDFS mode, whereas the latter would involve the implementation of costly hardware devices. When dealing with datasets on the order of trillions, a sensible system architecture design would involve processing the data in layers, rather than relying solely on sharding. It's important to note that sharding alone cannot effectively address performance bottlenecks. For readers seeking further clarification on this topic, Chapter 5 of this book on Distributed System Architecture Design provides valuable insights.

For instance, in a hierarchical storage design, data can be categorized into different tiers based on its usage patterns. For example, 10 billion records can be considered hot data, 90 billion records as warm data, and 900 billion records as cold data. This design aims to strike a balance between cost and performance. While storing all data in memory could potentially offer significantly better performance compared to storing it all on hard disks (possibly up to 100 times faster), the cost of memory is much higher than that of hard disks. Therefore, a compromise must be made. One of the reasons why the Spark system has gained popularity and is faster than Hadoop in recent years, apart from architectural and code efficiency improvements, is that it achieves higher memory utilization. Additionally, the cost of memory has gradually become more affordable over the past decade. Fig. 8 illustrates the concept of typical hierarchical data storage based on data access or usage patterns.

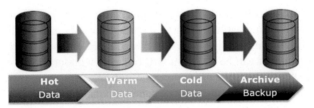

FIG. 8 Hierarchical data storage.

It is important to acknowledge that having trillions of data classified as hot data is both unrealistic and not aligned with real-world business situations. Additionally, the "modeling and composition" of such a graph must not be fully optimized. In reality, many so-called 100-billion and trillion-scale graphs often consist of entities that are actually smaller in scale, with a significant number of entities that can be better represented as attributes of the nodes or edges.

In social network scenarios, taking the world's largest social networks like Facebook and WeChat as examples, Facebook has around 3 billion users, while WeChat has over 1.2 billion users. If we were to design a graph that captures every message as an edge, it could indeed reach trillions or even more. However, such a composition would be inefficient and would overlook the core value of a graph system, which lies in extracting value from metadata and the associated network formed by metadata.

What data can be considered as metadata? In traditional databases, data corresponding to primary keys, foreign keys, or fields that can be matched in a one-to-one relationship can typically be used as metadata, such as cardholder information, SNS user IDs, or account IDs. However, data like chat sentences, pieces of text, voices, or pictures have limited value as metadata. Instead, they are better suited as auxiliary data, such as attribute fields of metadata, or as time series or column data stored in files and auxiliary databases (with pointers linking to the graph data set). These auxiliary data can be considered as warm or even cold data and stored in column databases or data warehouses that handle massive amounts of data at very low cost, while the graph system serves as the storage and compute engine for online hot and mission-critical data.

At present, the world's largest financial institution has a user scale of about 300–400 million, with accounts and card numbers topping 1 billion, merchant POS machines in the range of tens of millions, and transactions in the range of billions per year. Assuming we are also integrating other heterogeneous types of data of business interest into the graph systems, such as e-mail, kinship, and UBO data, the largest graph of this kind is on the order of tens of billions. If we store all the data for 10 years, then we can expect the scale of such a graph to reach hundreds of billions. However, this way of thinking is based on data warehouses or even data lakes. According to this storage logic, once data enters the warehouse or lake, it will "sink to the bottom" immediately, and it is unlikely to be used as an online system!

If the 10-year maximum 100 billion data in the above specific example is layered to provide services, a decomposition idea is as follows:

- The first layer: The data of the past 1–3 months enters the online graph system, the data scale is about 1 billion, and supports low-latency, high-concurrency query operations.
- The second layer: The data of the past 4–12 months enters the graph data warehouse system, which can provide rapid migration to the online system and perform near real-time batch processing operations.
- The third layer: The data of the past 13–120 months enters the graph lake system, which can provide fast migration to the data warehouse and large-scale batch processing operations.

Another approach to decomposition is to employ the concept of graph sharding or partitioning, allowing all 100 billion data over 10 years to be accommodated within an online graph system. Such a system would adopt a large-scale horizontal distributed architecture, where its scale is at least 40–120 times greater than that of a first-layer system. The complexity and cost of building and maintaining this "horizontal distributed" system are exponentially higher compared to the previous idea of constructing a layered graph data system.

Another misconception in capacity planning involves misjudging the scale of the graph that may actually need to be built. This misunderstanding often arises from the issue of "Cartesian Product" when querying data from traditional databases or data warehouses through multitable associations (JOIN). During a discussion with a knowledge graph company regarding the scale of the "impressive" graphs they serve their customers, it was discovered that they simply multiplied the number of rows in several SQL database tables, resulting in a figure of 432 trillion. However, in reality, the graph consisted of less than 20,000 entities, including 4000 companies, 3000 financial indicators, spanning 30 years (120 quarters), 30 industries, and 10,000 executives. When all these numbers were multiplied together, it resulted in 432 trillion, creating an "astronomical figure." In truth, all these items can be considered as entities (vertices) in the graph, amounting to only 37,030 entities. What is truly remarkable is that when these entities are interconnected, they theoretically can generate relationships in the trillions. However, labeling this graph as a trillion-level graph is essentially a case of exaggeration—in reality, these entities may form no more than a few million relationships which is considered a reasonably sized enterprise graph.

Let's summarize the gist of the capacity planning of the graph system:

- According to the actual business needs, decide which data is suitable as an entity, which is a relationship, and which is an attribute.
- The above data modeling scheme is not static. In different scenarios, an entity may be transformed into a relationship or an attribute, and vice versa.
- The capacity planning of the graph is based on: the current data level (the actual number of nodes+edges) * $(1+g\%)^N$, where $g\%$ is the estimated annual data volume growth %, and N is the estimated service life (years) of the system.

If a system is expected to serve for 10 years with a 20% data growth rate each year, starting with 100 million data in the first year, there would be 500 million data after 10 years. In reality, the majority of graph adopters have not yet reached a billion-level data volume. Therefore, planning for a trillion-level graph system may seem overly ambitious. The primary challenge lies in efficiently storing and processing data while keeping costs low. However, it is worth noting that existing market architectures claiming to support trillion-level graph system solutions are still primarily focused on data warehousing, with management and storage as the central considerations (and implicitly ignoring the business-oriented needs of deep/graph data processing).

A graph system without the ability to support in-depth computing lacks a fundamental reason for its existence. One of the elements that distinguishes graphs from traditional systems is their capacity to perform deep drill-downs, penetrations, and correlation analyses. Without the ability to delve deep into the data, the true power of graphs cannot be realized. Deep data processing is essential in painting a complete picture and uncovering meaningful insights from the underpinning data.

7.2 Benchmarking graph systems

The benchmarking of a graph system is a crucial aspect of verifying its functionalities and capabilities. Typically, there are three benchmarking methods:

- Self-benchmarking: This involves the system's own assessment of its performance, features, and capabilities. The graph system's developers or providers conduct internal benchmarking to measure its effectiveness and suitability for various use cases.
- Benchmarking by academic organizations (nonprofit or standardization bodies): Academic institutions or organizations dedicated to research and standardization may conduct evaluations of graph systems. These evaluations focus on objective analysis, benchmarking, and comparison of different systems based on predetermined criteria.
- Industry-bound graph benchmarking: Within the industry, organizations may perform evaluations of graph systems based on their specific requirements and use cases. These evaluations consider factors such as scalability, performance, ease of integration, and overall suitability for the specific industry's needs.

Self-assessment is an essential task that every graph system builder must repeatedly undertake. Through comprehensive self-assessment, builders can identify any omissions, fill gaps, and gain a better understanding of their own strengths and weaknesses. However, without third-party inspections, the results of self-assessment are often prone to questioning in terms of accuracy, fairness, comprehensiveness, and other factors.

Academic organizations conduct benchmarking that provides valuable insights into graph systems including initiatives such as the Linked Data Benchmark Council (LDBC), the GAP Benchmark from the University of California, Berkeley, and evaluations conducted by the Big Data Institute of Information and Communication in China.

Within the industry, the evaluation of graph systems is typically referred to as Proof of Concept (POC) testing. This internal evaluation and benchmarking process allows organizations to assess each graph system's performance and suitability based on their specific requirements and use cases.

By employing these evaluation methods, users and organizations can gain insights into the capabilities and performance of graph systems, helping them make informed decisions about their graph-system adoption and implementation.

The methodology, specific steps, and focus of the above three types of benchmarking are different. This section aims to introduce the reader to the commonalities between them, as well as the characteristics of each.

A complete graph system benchmark system must include at least the following four aspects:

- Functional testing
- Stress testing (or long-term performance/system-throughput testing)
- Interface testing
- Secondary development testing

Among them, interface testing and secondary development testing can also be regarded as a subset of functional testing, and they are listed separately because they take into account the emphases of different dimensions. Functional testing can be subdivided into the following subitems:

- Data import/export
- Metadata operations (CRUD or CRAP)
  - One-time large-volume
  - Batch and incremental
  - Single record (i.e., transaction)
- Graph query operations
  - Path
  - K-hop
  - Template
  - Variable calculation
  - Other complex queries
- Graph algorithm
  - General Graph Algorithms
  - Complex/bespoke graph algorithm

- API/SDK toolchain
- Graph tooling
  - ○ Visualization tools
  - ○ Data modeling
  - ○ Data management and processing tools
- Stress testing
  - ○ Sustained system performance (throughput, latency, max concurrency, resource consumption, stability, consistency, etc.) covering all the above functions under different loads (concurrency scale, task complexity)

In this section, we will introduce the three important parts of the benchmarking process in more detail:

- Benchmark environment (the testing beds)
- Benchmark content (tasks)
- Results/Correctness verification

7.2.1 Benchmark environment

The benchmark environment can be simply divided into two parts: hardware environment and software environment. The scope covered by the former includes (as shown in Table 2) server resources, network resources and virtualization environment, etc.

TABLE 2 Common graph system POC hardware environment.

Category	Configuration	IP & notes
Server	Quantity: 3 (or more) CPU: Intel 32 vCPU or 16-core Memory: 128 or 256 GB Hard disk: 1.2TB * 2 (RAID) Network card: 10 Gbps	The IP addresses of the three servers should be allocated within the same network segment, for example: 192.168.*.* 192.168.*.* 192.168.*.*
Network	10 Gigabit Network	$\geq$5 Gbps
Storage	NAS (mounted network disks)	Other network storage architecture

Table 2 presents the typical X86-based hardware environment commonly used in graph systems. While some graph systems may utilize RISC-based instruction set architectures, such as ARM processors or IBM's Power ISA architecture, or even GPUs (for accelerated matrix operations), it is important to note that these architectures are relatively uncommon in the context of graph systems. Their versatility and capabilities have not yet been extensively tested by time and the market. Consequently, this book does not cover graph systems based on these alternative architectures.

It is worth noting that, as depicted in Fig. 9, 97% (485 systems) of the top 500 supercomputing systems utilize the X86 system architecture. However, in the latest list released in June 2023, there is only 1 system (No. 10) built using Intel architecture from China, the No. 2 system was built by a Japanese company based on the ARM architecture. The No. 5, No. 6 supercomputing systems were built by two American laboratories using the Power architecture, and the seventh-ranked system was China's Sunway TaihuLight, which employs a RISC-comparable architecture. Additionally, half of the Top-10 systems utilized a heterogeneous hybrid approach of X86+GPU, leveraging the high concurrency capabilities of Nvidia's GPUs for X86 acceleration. It is important to note that the majority of systems on the Top-500 list still utilize the X86 architecture, specifically implementations based on Intel's (384/500) or AMD's (101/500) 64-bit CPUs.

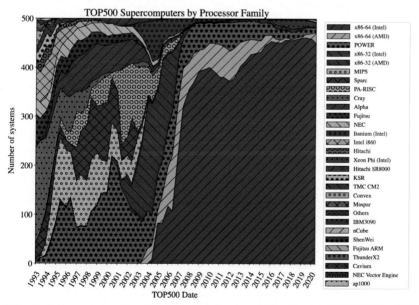

FIG. 9 Supercomputer (Top-500 List https://www.top500.org/lists/top500/2021/11/) CPU composition trend.

The choice of software system is closely tied to the hardware environment, and the graph system is typically built to accommodate the required operating, development, and testing environments. Each software component has its own unique dependency stack, calling methods, version naming, configuration parameters, and resource consumption. Therefore, the software environment of a graph system primarily focuses on key software categories, as outlined in Table 3.

Using supercomputing systems as an example, as illustrated in Fig. 10, since the end of 2017, all systems listed have been based on the Linux kernel. This suggests that the benchmarking of graph systems is predominantly conducted within a Linux operating system environment. However, with the increasing popularity of technologies such as virtualization and containerization, graph systems can be easily embedded in other types of operating system environments as well.

In the process of testing, it may also be necessary to provide a complete list of third-party software that a graph system integrates and relies on and provide a security vulnerability report according to the needs of security compliance. The relevant content is beyond the scope of this book and will not be expanded here. In most benchmarks, the software and hardware configuration of the test environment are heavily influenced by the dataset being tested. Particularly, certain hardware indicators are directly associated with the dataset's size. For instance,

TABLE 3 Common graph system software environments.

Category	Configuration	Notes
Operating system	Centos 7.5+ Ubuntu 18.4+	Linux OS distros
Virtualization environment	VMW ESX (Type I Hypervisor) Linux KVM Hypervisor	In some benchmark environments, the virtual machine monitor (Hypervisor) will be used as the hardware environment, especially type I
Container environment	Docker 19.x+ K8S Platform	Docker is the container runtime, and K8S is a container management platform
Other software	Each company formulates the required compilation environment, operating environment, etc.	
Interface	Data import format and method, export method, API and SDK calling method, etc.	

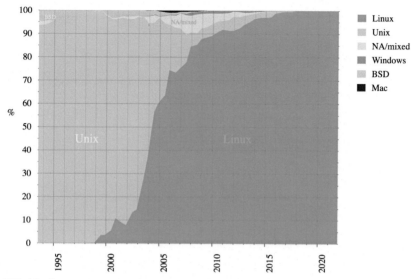

FIG. 10 Supercomputer operating system composition.

small datasets may only necessitate low-configuration servers or virtual machine environments, whereas larger datasets typically require higher hardware specifications. While we will delve into the specifics of dataset characteristics in the data loading section, it is important, for the sake of completeness, to provide an overview of some typical test datasets in this section for readers' reference.

Test data sets are generally divided into two genres and styles:

- Academic style: including social SNS graph, WWW network data set, road network information data set, artificially synthesized data set, etc.
- Industrial style: enterprise knowledge graph, financial graph, transaction network, credit antifraud data set, NLP data set, general graph data set, etc.

The data sets of the academic genre generally belong to the category of simple-graph, isomorphism, and with no or very few attributes of vertices and edges. This kind of data set has evolved from the fundamentals of graph theory, operations research, path planning, social psychology, SNS social network, and NLP research in the second half of the last century.

Industrial-grade graph datasets in the real-world context have emerged relatively recently, with the majority of them becoming prominent in the second-to-third decades of this century. These datasets typically exhibit characteristics such as multigraph, heterogeneity, and unique attributes. Transaction networks and knowledge graphs within the financial industry are examples of common industrial-grade graph datasets. These datasets reflect the complexity and interconnectedness of real-world industrial systems and provide valuable insights for various applications and analyses.

TABLE 4 Comparison of two genres of graph data sets.

Dataset categories	Academic and social networks Typical testing dataset	Typical industrial POC dataset
Very small dataset (10,000 level)	Various artificial datasets (# of nodes <10,000) Mostly used for publishing papers and proving graph algorithms	Very rare
Small data set (Million level)	AMZ dataset (edges: ~4 million)	For example, NLP data sets, credit data sets of small and medium banks
Medium dataset (Tens of millions)	Graph500 (edges: ~70 million) US road network dataset (80 million)	Alimama (edges: 105 million) Credit and risk control data sets of large banks, etc.
Large dataset (Billion level)	Twitter (points: ~1.5 billion) GAP-Web (2 billion), GAP-Kron (2.2 billion), Urand (2.3 billion)	Business graph (0.3–1 billion), large bank transaction data set (3 months, 1 billion trxes)
Very large dataset (>3 billion)	Synthetic datasets, for example 100× augmentation of Graph500 ...	Long-term transaction data sets of large banks (~10 billion), measurement data sets of various regulatory indicators (10–100 billion)

Note: The LDBC test data set is between the two types of data sets in the above table. So far, the test content has focused on SNB-type test data and scenarios, that is, Social Network Benchmarks. With the rapid growth of demand for graph data testing in the industry represented by the financial industry in recent years, the LDBC has been working on launching the Financial Benchmarks (FB hereafter) since 2022, The first FB draft was released in June 2023, with the full official FB standard to be ironed out by the beginning of 2024.

In Table 4, we comprehensively compare the two types of graph datasets. Although we use the size of the data set in Table 4 to divide the benchmark classification, the test complexity of the graph data is not directly proportional to the size of the data. This is also the core difference between the graph system and the traditional database systems. We explained this in previous chapters, and to reiterate here, the (average) complexity of each operation in a graph system depends on the topological structure(s) of the graph dataset. Some key aspects to examine include:

- Vertex-edge ratio (E/V): This ratio represents the relationship between the number of edges (E) and the number of vertices (V) in the graph. It provides insights into the density of the graph and can influence the performance of graph algorithms.
- Self-loops and loops: Self-loops occur when an edge connects a vertex to itself, while loops involve a sequence of edges that form a closed path. The presence of self-loops or loops can complicate various graph operations and algorithms.
- Connected components: Connected components refer to groups of vertices that are mutually reachable within the graph. Evaluating the number and sizes of connected components helps understand the connectivity and structure of the graph.
- Presence of supernodes: Note the density of a graph does not directly reflect the existence of supernodes, though the likelihood of high density usually comes with a higher probability of supernodes.

These characteristics of the topological structure of a graph dataset play a significant role in determining the efficiency and effectiveness of graph processing algorithms and operations. It has little to do with the size of the total data but depends on the connectivity characteristics of the graph data. Some graphs seem to have a large size in terms of orders of magnitude but are very "sparse" (very low vertex-to-edge ratio or density), and the computational complexity is low; some graphs have a small number of vertices and edges, but the density is high, and the computational complexity for certain operations can be very high (that's why you will find some graph databases time out on deep traversals against small data set).

When planning benchmarks or proof-of-concept (POC) tests, it is *not* necessary to solely rely on large datasets to evaluate the capabilities of a graph system. Here are two examples:

1. Benchmark with small to medium datasets: In many cases, small to medium-sized datasets are sufficient for a benchmark to reveal a graph system's performance and functionalities. For instance, the AMZ0601 dataset, which consists of approximately 400,000 vertices and 3.4 million edges in one connected component, can be used to randomly access any vertex and find all its deduplicated 1-hop, 2-hop, and deeper-hop neighbors, or find all shortest paths between any random pair of nodes—such queries, plus some graph algorithms, can reveal a system's deep-traversal performance as well as accuracy in the results.
2. On large datasets, the point of benchmarking a graph system is to identify if it has "dynamic pruning" capability which can trim unneeded data and accurately return with desired results. Taking the UBO scenario as an example, the underpinning industrial and commercial graph may reach a scale of hundreds of millions of nodes and edges, while queries starting from specific

nodes, such as companies or directors, supervisors, and senior executives, can be conducted to find their associations and holding paths. Instead of brute force calculations, intelligent graph systems should utilize query filtering by applying attribute filters on nodes and edges and setting thresholds to accurately retrieve specific information, such as a company's external investment network or shareholding paths. Note that this kind of graph queries over a large data set is very different from AI/ML, which tends to generate probabilistic results, unlike graph systems' fact-based laser-accurate results.

Large datasets are suitable for verifying a graph system's processing ability for large amounts of data, but it may not thoroughly test its in-depth query capabilities. For example:

- For graph algorithms that work against the entire data set, the time consumption and amount of data to write back are directly related to the data size.
- Incremental data processing capabilities, such as insertions and deletions, can detect how the system's time consumption changes as the data size increases.
- Operations like path queries, K-hop queries, and some graph algorithms are not always affected by the data size and can be considered as "Dataset Size Agnostic."

These examples demonstrate that the evaluation of a graph system should consider various factors, including dataset sizes, query requirements, filtering mechanisms, and algorithmic performance, to provide a comprehensive assessment.

7.2.2 Benchmark content

There is no standardized set of benchmark criteria for graph systems, but there are some typical test content and procedures that can be used as a reference. Here are some of the key aspects to consider when evaluating a graph system:

- Data modeling
  - Assess the system's data modeling capabilities, including the support of heterogeneous and homogeneous types of data, and connectivity support of third party data systems.
  - Whether the modeling can intuitively and conveniently reflect business needs, the flexibility of the modeling schemes in the face of business/data changes, the resource consumption rate, etc.
- Data loading and integration:
  - Evaluate the system's data ingestion capabilities, including the ability to import data from various sources, handle data transformation, and maintain data integrity.

- ○ Test the system's performance and efficiency in loading and integrating large-scale graph datasets.
- Performance (compute and storage):
  - ○ Measure the system's response time for common graph operations, such as node/edge retrieval, traversal, and pattern matching.
  - ○ Test the system's ability to handle increasing data sizes and query loads.
  - ○ Evaluate the system's performance under different concurrency levels.
  - ○ Assess the system's data storage capabilities, such as its ability to efficiently store and retrieve graph data, handle data partitioning and sharding, and support data replication and fault tolerance.
- Graph algorithms:
  - ○ Assess the system's implementation and efficiency of common graph algorithms, such as PageRank, Shortest Path, Connected Components, and Community Detection.
  - ○ Measure the execution time and accuracy of these algorithms on representative datasets.
  - ○ Assess the system's support of algorithm extension, customization, and feature richness.
- Query language and expressiveness:
  - ○ Evaluate the graph query language provided by the system, its expressiveness, and ease of use.
  - ○ Test the system's support for complex graph queries involving pattern matching, aggregations, filtering, path finding, as well as integration of graph algorithms.
- Data management and visualization:
  - ○ Assess the system's data management capabilities, such as a one-stop place (DBMS) to manage meta-data, schemas, indexes, access control and policies, cluster management and monitoring, plug-ins, query results, etc.
- Scalability and high availability:
  - ○ Evaluate the system's fault tolerance and availability features, such as data replication, clustering, and load balancing.
  - ○ Test the system's ability to scale horizontally with increasing data and query loads.
- Integration and ecosystem:
  - ○ Assess the system's compatibility and integration with other tools, frameworks, and data processing systems commonly used in the organization's data ecosystem.
  - ○ Evaluate the availability of connectors, APIs, and libraries that facilitate data integration and analysis.

These are some of the typical test content and procedures that can be considered when evaluating a graph system. The specific evaluation approach may vary depending on the requirements and goals of the graph-adopting organization.

Regarding graph data modeling, many people consider modeling abilities as a secret weapon and believe that the graph model should not be made public. However, I believe this idea is highly undesirable. If the model is black-boxed, opaque, and unexplainable, it can ultimately harm everyone involved and lead us back to the path of second-generation artificial intelligence systems based on deep learning and neural networks. The core of the augmented-intelligence AI system, representing the third generation of artificial intelligence, is graph augmented intelligence, which emphasizes white-box interpretability, transparency in the modeling process, and openness.

Furthermore, a graph system should not aim for algorithm monopoly. As a general-purpose underlying system, its focus should be on providing customers with superior computing power and transparent algorithms, rather than relying on black-box models and algorithms.

In summary, transparency and interpretability are crucial aspects of graph systems and the third generation of artificial intelligence. By embracing white-box models and algorithms, we can build more trustworthy and explainable systems that benefit both developers and users.

Taking the application of industry data as an example, one is a fixed data modeling mode, that is, the prospect proposes a specific data modeling method, and the graph system realizes and satisfies it; the other is an open modeling mode, that is, the prospect describes the business scenario, and the graph system is to design a satisfying data modeling mode. Let's take a look at the benchmark contents of these two different modeling modes.

Data modeling mode 1: Fixed mode modeling

Taking a transaction data set as an example, as shown in Table 5, the data modeling requirements are clearly defined. Node and edge data files have been generated, allowing for a direct assessment of whether a graph system can fulfill the specific data modeling requirements. This evaluation process will also involve judging and scoring the effectiveness of the system.

The judging rules related to data modeling mainly focus on the following points:

- Is it 100% consistent with the demand?
- Ease of modeling (tools are easy to use, fewer steps, and faster modeling shall be given better score).
- Comparison of source data size and storage size (data expansion ratio).
- Does the graph system provide other useful functions?

Data modeling mode 2: Flexible modeling

If there are no specific data modeling requirements and only the business requirements at the application layer, then any graph system can improvise and go free-style on data modeling. When one data-modeling method is superior to another for a specific business scenario, particularly in terms of resource

TABLE 5 Fixed mode modeling requirements.

Requirements	Requirement description	Remark
0	The node is the transaction party ID (card number), the edge is the transaction, and the attributes of the edge are used as the filtering condition in each query	Edge attributes include at least transaction time, transaction amount, etc.; it is required to apply attribute filtering in the query or graph algorithm
1	Data volume: 1.2 billion Nodes: 200 million Edges: 1 billion	The corresponding file size is 110 GB Nodes and attributes: 20 GB Edges and attributes: 90 GB
2	Data file server address	192.168.XX.XXX
3	Data file directory	/var/data/POC2022-12/
4	Node file: account.csv	Node (entity): card number
5	Edge file: transaction.csv	Edge (relationship): transaction, the order of each counterparty in each transaction A, B means: card (A) pays card (B)

utilization, timeliness, and system stability, we refer to this superior approach as the "Best Practice." The process of POC and benchmarking often involves discovering these best practices.

In Chapter 1 and this chapter's Figs. 6 and 7, we discussed the potential for flexible data modeling in graphs. It's important to note that the capabilities of heterogeneous graphs and multigraph can be seen as a superset of isomorphic graphs and simple graph. The former offers backward compatibility, while the latter lacks forward compatibility. This means that a system supporting multigraph has greater "modeling" abilities compared to simple-graph systems. However, in some specific scenarios, simple-graph has its advantages, such as in fraud-detection, as illustrated in Fig. 7, it is typical simple-graph data modeling, and it can be more effective in identifying frauds by finding specific topological structures via graph traversals which are hard to be achieved with multigraph data modeling (as in Figs. 4 and 5).

In Table 6, we present a typical requirement for heterogeneous data modeling. Unlike the explicit node and edge allocation modes outlined in Table 5, Table 6 only specifies the existing source data situation and the desired outcome and evaluation criteria in the precondition section. The specific modeling method, however, is entirely determined by the graph system waiting to be benchmarked.

TABLE 6 Flexible data modeling requirements.

Testing items	Load heterogeneous data into graph
Testing purposes	Investigate heterogeneous data, including modeling of mixed-mode graph data of various types of nodes and edges, and the ability to load and generate graphs, and evaluate the loaded data size (storage space occupation), loading speed, and loading time
Preconditions	o The data used are company equity and employment data o About 200 million corporate entities and shareholders-supervisors-executive entities; about 2 million companies in the supply chain; there is also some loan-guarantee information; account transfer information between corporate accounts, etc.
Expected outcome	Equity (investment) data and supply chain data are fully loaded into the graph, and it is possible to derive and calculate all shareholders and shareholding ratios of a certain company or explore its investment network or upstream investors and quantitative shareholding ratios from a natural person or company; Identify the ultimate beneficiaries, actual controllers, and find out whether there are compliance issues such as cross-shareholding and illegal benefit transfer, etc.
Benchmark standard	o Can the requested data be 100% loaded (ingested)? o Rank by loading data time and size o Ability and effectiveness in penetration depth and time, precise computing of shareholding ratios and identification of UBOs, etc.

Indeed, in the context of a business graph, it is more appropriate to represent relationships such as investments, transfers, guarantees, or supply and marketing relationships using edges to visually depict the connections. However, it is not uncommon to come across systems that represent transactions through vertices along with redundant pairs of connecting edges. The advantage of the former approach is its intuitiveness and lower computational complexity, while the latter requires more complex and potentially inefficient computational models to conduct similar tasks. Throughout the iterative process of designing the architecture for a graph system, as well as any database system or big data framework, such detours or suboptimal approaches can be encountered. It highlights the importance of carefully considering data modeling choices and optimizing computational models to ensure efficiency and effectiveness in performing graph-related tasks.

Next, we need to evaluate the data access capability of the system, usually, we will evaluate the following aspects:

- Multisource data access, multiformat data access capability.
- Ability to import and load massive amounts of data.
- Data output capability.

In the previous section, the data import (export) tools, cross-database connectors, and graph modeling tools we introduced can be used in the data access benchmark. Table 7 lists a typical data import benchmark requirement, and Table 8 lists possible data output benchmark requirements.

In the data access part, it is indeed important to test the data loading boundaries of a graph system. This involves determining the maximum amount of data that the system can store and handle efficiently under current hardware and software conditions. While these tests are not as common as other benchmark methods, they are often conducted as part of self-assessment tests. Developers of graph systems have a vested interest in understanding and evaluating the limits of their own systems, as it provides valuable insights into system scalability and performance. By pushing the boundaries and assessing the system's ability to handle large volumes of data, developers can gain a deeper understanding of their system's capabilities and identify any limitations or areas for improvement.

TABLE 7 Multisource data access requirements.

Testing items	Multisource data access
Testing purposes	The graph system needs to load various data files into the graph database. This item examines the data access capability: whether it supports different types of data formats to seamlessly access (import) the graph database, to flexibly respond to different data input scenarios
Preconditions	Source datasets (data exporter): defined by the benchmark initiator
	Access method: The data access scenario is customized by the graph system vendor
Expected outcome	Support access to at least three different data formats, for example: ○ File access (e.g., CSV) ○ Hive access ○ Spark access ○ SQL database access ○ Third-party graph database system access ○ Small-batch import, incremental data import, and item-by-item data import through API/SDK calls ○ Whether to support visual data access
Benchmark standard	○ Functionality completeness ○ Speed (i.e., number of records ingested per second, total time of ingestion, online-ingestion vs offline-ingestion) ○ User experience (ease of operation, celerity, system robustness)

TABLE 8 Data output (export) benchmark requirements.

Testing items	Various data output methods
Testing purposes	Investigate whether different types of data output (export) methods are supported
Preconditions	The data output scenario is customized by the party under test (graph system builder)
Expected outcome	Support at least three different data output access methods, for example: • Export files such as CSV • Export to SQL Compatible Database • Export to other types of databases or big data systems • Batch export, incremental export, and item-by-item export data through API/SDK calls • Visualized data export
Benchmark standard	• Functionality completeness, feature richness, etc. • User experience (ease of operation, system robustness, data export speed, etc.)

Mass data loading testing or limit testing (Table 9) is crucial for identifying the storage limit of a graph system and can provide valuable insights for optimizing the system architecture. By pushing the system to its limits, developers can uncover potential bottlenecks and areas for improvement. Some key points to consider during limit testing include:

1. Architecture optimization: By optimizing the system architecture, such as distributed architecture for partitioning logic, data backup, log management, and intermediate data optimization, developers can enhance the system's efficiency and store more data at a lower unit price.
2. Data structure optimization: Evaluating and optimizing the data structures used for storing graph data can improve performance and storage capacity. This includes techniques such as indexing, compression, and efficient data representation.
3. Cache optimization: As the data volume increases, cache performance becomes crucial. Limit testing can help identify potential issues with cache scalability and efficiency, enabling developers to optimize caching mechanisms and improve query performance.
4. Database design flaws: Limit testing with massive data sets can expose design flaws in the database schema and query patterns that may not be apparent in smaller-scale data. Identifying and addressing these flaws can improve overall system performance and data loading efficiency.

TABLE 9 Mass data loading requirements (stress testing).

Testing items	Massive data loading into graphs
Testing purposes	The actual application scenario of the graph database needs to load a massive amount of data (such as full transaction flow data) into the graph database and examine whether it can load, process, analyze, and manage data. The test indicators include the maximum loadable amount of data, loading time, and others. Note that, for those multimodel/NoSQL graph systems, there is no upper-limit for a loadable amount of data which is only bound by available hardware storage. In such cases, the benchmarking should focus on loading efficiency measured by latency and expansion factor
Preconditions	o The data used in the test loading is transaction flow data, with a data volume of 200–300 million per day, loaded into the graph on a daily basis, with a maximum of 2 years (730 days) daily data, that is, about 100 billion records o The test data is ready (node and edge files have been prepared) to be loaded in batches on demand
Expected outcome	o Data can be loaded correctly to generate a graph(s) o Minimum data sets to be loaded are 30 days (or the number of days allocated according to business requirements and specific hardware indicators) o Gauge graph query performances (average time consumption, maximum time consumption, concurrency scale, etc.)
Benchmark standard	o Failure to load the minimally required data will result in no credit for this test o Comprehensive scoring and ranking based on data volume per unit time of loading data, maximum loading volume, and graph query performances

5. Hardware limitations: The underlying hardware configuration imposes constraints on the maximum amount of data that can be stored. It is important to consider factors such as storage capacity, data caching, and incremental data storage to ensure optimal utilization of available resources.

6. Qualitative changes: When pushing the system to its limits, there may be qualitative changes in system behavior. For example, inserting a significantly large number of edges may result in unexpected issues like memory exhaustion (OOM). These stress scenarios help uncover stability-related problems and are essential for designing a robust and reliable system.

By conducting thorough limit/stress testing, graph system developers can gain a deeper understanding of their system's capabilities, uncover potential issues, and optimize the architecture and data handling processes to achieve better performance, scalability, and stability.

Metadata processing operations play a fundamental role in graph systems. These operations involve managing and manipulating vertices, edges, and their attributes. Table 10 focuses on batch processing of the metadata. By conducting batch processing on metadata, we can assess the system's performance, scalability, and robustness in handling common operations that are essential for managing graph data. This allows us to identify any bottlenecks, optimize the system's metadata processing capabilities, and ensure the smooth and efficient execution of daily business operations.

Deep graph query is the most distinctive feature of a graph system, setting it apart from other database systems and big data frameworks. The benchmarking of deep query operations varies depending on the context. For instance, the GAP benchmark presents all queries in the form of graph algorithms, including the breadth-first query (BFS), which is otherwise considered a regular graph queries in many contexts. In this book, we categorize in-depth graph queries into the following groups to conduct more focused benchmarks. We treat benchmarks related to graph algorithms as a separate category.

TABLE 10 Metadata benchmarking requirements.

Testing items	Meta-data processing
Testing purposes	Test the metadata processing capability of the current graph system, evaluate the storage space occupation, and processing timeliness, including the total processing time, the average processing time of individual metadata, and other indicators
Data characteristics	o The data used in the test loading is heterogeneous transaction-type data with a data volume of 1 billion o The vertex schema contains 20 attributes including account type, account level, account balance, etc. o The schema of the edge contains 30 attributes including start vertex, end vertex, transaction type, transaction amount, timestamp, etc.
Benchmark content	o Insert 10 million edges (transactions) incrementally o Batch update vertex attributes (create new attributes) o Batch update edge attributes (create new attributes) o Update an attribute field of vertices that fall in a specific degree range o Delete all vertices and their associated edges that match vertex attribute to a certain value (and interval value)
Benchmark standard	o % of completeness (functionality) o Rank is given by latency, with the lowest latency offered the highest ranking

- K-hop Queries: Refer to Table 11.
  - Kth-hop query.
  - M-to-N hops query.
  - 1-to-K hops query.
- Path Queries: Refer to Tables 12–15.
  - Breadth first query.
    - Shortest path (Shortest-path).
    - Full path query (All shortest paths).
    - Expansive query (Spread).
  - Depth-first query.
  - Cycle query (Finding circles).
- Template Queries: Table 12.
  - Template K-hop.
  - Template Path.
- Complex combinational queries: Refer to Table 16.
  - Path or K-hop query based on full-text search.
  - Multinode to multinode network query (network formation).

The graph system's support for secondary development is also a very important capability, such as the programmable interface (API/SDK), custom model

TABLE 11 Multihop neighbor query.

Testing items	K-hop (BFS) query benchmarks
Testing purposes	Investigate the graph system's deep traversal capabilities under conditions of massive data and batch (or online) processing. Run K-hop queries on the loaded graph data, verify the correctness of the query results, and measure the time consumed for these operations
Preconditions	o The test data set can be 30-day transaction data as in Table 9, and the transaction data has been loaded in the graph system o Provide 10 million vertices (can be randomized) as the starting vertex set for this query
Test content	o Count the 1-hop neighbors of 10 million vertices (1st hop) o Count the 2-hop neighbors of 1 million vertices (2nd hop) o Count the 3-hop neighbors of 100,000 vertices o Count the 4th hop neighbors of vertices in 10,000 o Count 5th hop neighbors of 1000 vertices o Count the maximum-hop neighbors of 100 vertices
Benchmark standard	o The results must be correct first, before gauging by performance o Ranked by overall and average time o Note: If duplicated neighbors appear in results, mark them as incorrect

TABLE 12 Template path query (UBO graph).

Testing items	Template path query in heterogeneous graphs
Testing purposes	Investigate the function and performance of breadth-first query in a heterogeneous graph formed by multiple types of entities as depicted in Table 6. Designate a type of vertex (personnel or enterprise) and query for all the other personnel and enterprises associated within the depth of 1–20 hops. The types of relationships include investment, shareholder, employment, etc., verify whether the returned results of the query are correct and count the time-consuming (Total time-consuming, Average time-consuming, Maximum time-consuming, Minimum time-consuming, etc.)

Test content	o Load the business graph data (UBO graph per se)
	o [Query 1]: Starting from a company entity
	  ■ Enterprise List:
	    ● Name: XXX Co., Ltd., YYY Company, ZZZ Company
	    ● UUID: CBDA, CXYZ, YYDS
	  ■ Query depth and filtering logic as follows:
	    ● Layer 1: Filter by certain types of relationships
	    ● Layer 2: Filter by certain types of relationships, such as outbound investment, calculate % of ownership
	    ● Layer 3: ditto
	    ● Layer 5: ditto
	    ● Layer 10: ditto (If there is any)
	    ● Layer 15: ditto (If there is any)
	    ● Layer 20: ditto (If there is any)
	o [Query 2]: Starting from a person entity
	  ■ Personnel list:
	    ● Name: Director-A, Director-B, Executive-C, etc.
	    ● UUID: ABCDEFGH, BCDEGHIJ, ABDCNNDFG, etc.
	  ■ The query logic is comparable to Query 1 but tracks:
	    ● Layer 1: All relationships (as manager or investor)
	    ● Layer 2 and beyond: All outbound investment & ownership

Expected outcome	The query result is correct Query results should be written to a disk file(s) for validation
Benchmark standard	o If the query is not supported or the returned result is incorrect, no points will be awarded o If some queries cannot be returned or finished due to excessive resource consumption of the graph system, it's an area for improvement o Ranked by response time

TABLE 13 Shortest path queries.

Testing items	Shortest path query
Testing purposes	Investigate the shortest path query ability between any two entities in the graph composed of multitype entities and edges. Specify two nodes (i.e., pairs of personnel or enterprises as in the UBO graph), and query for ALL shortest paths
Preconditions and test content	○ Successfully load the business graph data ○ Specified query input data: ■ List of starting entities: enterprise A name or id ■ Ending entities list: enterprise B name or ID (one-to-one correspondence with the starting point) ■ Query depth: Query the associated shortest path within six (or deeper) degrees (inclusive) ■ Query direction: Bidirectional (ignoring direction)
Expected outcome	The query result is correct Returning all paths instead of one or an incomplete list
Benchmark standard	○ If the query is not supported, or the returned result is incorrect, no points will be awarded ○ During the query process, if there is a graph system failure, mark it as a point for improvement ○ Rank by response time

TABLE 14 Complex path query requirements.

Testing items	Complex path (subnet) query
Testing purposes	Test the graph system's path query capabilities under complex logic conditions based on the loaded graph data, specify transactions that meet certain conditions (time, place, amount, transaction direction, etc.), and return the complete subgraph (network) form by fitting transactions, verify whether the query returns the correct result and gauge the time-consumption
Testing requirements:	○ The data used for benchmarking is the transaction flow data used in the previous scenario ○ The query logics is as follows: ● The transaction time of transactions along the path shows a time-series increasing relationship (trend), and the time difference between the two adjacent transactions is <24h ● The amount of each transaction in the path fluctuates within ±10% ● Multiple transactions are allowed to be split and combined in the subgraph (paths): for example, there are three transactions on the downstream link of a $1 million transaction that may be

Continued

TABLE 14 Complex path query requirements—cont'd

Testing items	Complex path (subnet) query

	split into \$300,000+\$300,000+\$400,000, of which \$300,000 may be further divided into \$100,000+\$150,000+\$50,000; the time difference between multiple splitting transactions is <10h; the transaction amount involved in the calculation of the combination also follows the principle of fluctuation within the range of ±10%.
	• The directional movement of cash flow is downstream (right)
	• The max query depth is 10 layers
	○ Specify data conditions for N transactions:
	  • (Optional) List of card numbers of starting vertices: XXXX
	  • (Optional) List of card numbers to terminal vertices: YYYY
	  • Trading time: 2022-XX-XX—2023-XX-XX
	  • Transaction amount range: >\$10,000

Benchmark standard	○ If the query is not supported, or the returned result is incorrect, no points will be awarded ○ Ranked by response time, explain ability of the queries (or algorithm), visualization of results and ease-of-use if encapsulated as a plug-in or stored procedure

TABLE 15 Loop queries (DFS).

Testing items	Loop query (DFS)
Testing purposes	Specify a certain type of starting or ending entities (such as the account numbers of the transaction initiator or receiver) and find out if there are circular paths formed involving each entity. Verify whether the returned result of the query is correct, and count the time consumption of identifying each single loop

Test requirements	○ Load the transaction flow data set
	○ The query logic are as follows:
	• The starting point of the path is also the ending point, such as $A \rightarrow B \rightarrow C \rightarrow D \rightarrow E \rightarrow F \rightarrow G \rightarrow A$
	• The time of each transaction in the path increases along the path
	  ■ Timestamps of transactions increase along the path when the query direction is downstream (outward)
	  ■ (Reverse query) When the query direction is upstream (inward), the timestamps decrease
	• The amount of two adjacent transactions in the path fluctuates between ±20%
	• Max traversal depth is ≤10 layers

Expected outcome	The query result is correct
Evaluation standard	○ If the query is not supported, or the returned result is incorrect, no points will be awarded ○ Ranked by response time, query intuitiveness (succinctness), or ease-of-use if encapsulated in a custom plug-in or algorithm

TABLE 16 DFS path queries.

Testing items	DFS path lookup
Testing purposes	Run the DFS path algorithm based on the loaded graph data, specify a certain type of transaction parties (ID list or card number set) that meet the conditions, query for directional transaction network (multiple paths) formed from each party, or paths in between any pair of parties

Test content	○ The data used for benchmarking is the transaction flow data as depicted in Table 10.
	○ Query logic:
	  • Starting from each eligible transaction account, perform upstream and downstream queries along the transaction flow (each path has a consistent transaction direction and filtering conditions as follows)
	  • The time (timestamp attribute) of each transaction along the path presents an increasing relationship (trend) along the path
	    ■ Forward query: if the direction is downstream (transfer out), the time sequence between path steps is increasing
	    ■ Reverse query: if the direction is upstream (transfer-in), the timing sequence between path steps is decreasing
	  • The fluctuation of the transaction amount of the previous and subsequent two transactions in the return result path should be within $\pm 10\%$ (or in a decreasing trend)
	  • The maximum query depth is ≤ 10 layers

Expected outcome	The query result is correct
Evaluation standard	○ If the query is not supported, or the returned result is incorrect, no points will be awarded ○ Ranked by response time, giving points where appropriate

development, and algorithm support. Table 17 lists the test requirements for typical and relatively simple customized model development capability.

Graph algorithms play a significant role in the evaluation of graph systems, and they are categorized into the following groups by global graph system manufacturers, third-party benchmark institutes, or enterprise-level customers:

- Self-assessment, academia, and third-party organizations:
  - BFS (Breadth-First Search).
  - CC (Connected Components).
  - PR (PageRank).
  - BC (Betweenness Centrality).
  - TC (Triangle Counting).
  - SSSP (Single Source Shortest Path).
  - LPA (Label Propagation Algorithm).

TABLE 17 Graph model development (secondary development).

Testing items	Custom model development capabilities
Testing purposes	Examine the ability to quickly respond to business needs and develop custom models and functions
Preconditions and test content	o Using the Business UBO Graph dataset o Query the information of the top 10 shareholders of the designated company, or all shareholders with a shareholding ratio of $\geq$1% o Query filtering conditions: • Query starting point 1: XXX Co., Ltd. • Query starting point 2: YYY Technology Group Co., Ltd. • Query starting point 3: ZZZ Financial Technology Co., Ltd. • Query starting point 4: ZZZ Investment Co., Ltd.
Expected outcome	Correctly execute the query and return with correct and complete results
Benchmark standard	o Rank by response time, and intuitiveness of the graph queries (ideally a one-liner or straightforward nested queries). If the model is implemented and packaged as a plug-in or stored procedure, inspect the explain ability and flexibility of the model

- Industry-specific algorithms (differing from the above):
  - Full graph K-hop.
  - Similarity algorithms.
  - Enhanced LPA (multilabeling).
  - Louvain (Community Detection Algorithm).
  - Random walk and graph embedding algorithms.
  - Parametric Computing (related to specific business use cases).

In Chapter 4 on graph algorithms, we have provided a detailed explanation of the logic and application scenarios for each type of graph algorithm. Now, let's focus on the benchmarking of two specific algorithms: Enhanced LPA (Table 18) and Node2vec (Table 19), based on the GAP benchmark standard. In the GAP benchmark, six algorithms (BFS, CC, PR, BC, TC, SSSP) and five datasets (Web, Twitter, Road, Kron, Urand) are used. Each participating graph system executes against each algorithm through multiple runs. It's important to note that the GAP benchmark has a strong academic background and may overlook some important indicators in the industries. For instance, it may ignore data loading time, dynamic data processing (GAP currently is limited to static data), and the need for persistent storage (GAP frameworks store all data in memory only). However, industrial graph vendors can learn from GAP benchmarks and incorporate features like:

TABLE 18 Graph algorithm capability (enhanced LPA).

Testing items	Graph algorithm capabilities
Testing purposes	Examine the support of LPA and its advanced variations by the current graph system
Preconditions and test content	o A synthetic dataset of 1.6 billion nodes and edges (600 million nodes and 1 billion edges, each carrying multiple attributes, all attribute fields combined are about 30 billion) o The graph algorithm logic is as follows: • Label Initialization Process: Randomly select 10,000 nodes whose degrees are [1,20] in the graph, and set the value of the attribute field named labelX to "1"; randomly select another set of 10,000 nodes with degrees standing at [21,40], and set the value of labelX to "2"; randomly select 10,000 nodes who degree falls between [41,60], set the value of labelX to "3"; randomly select 10,000 nodes with degree [61,80], set the value of labelX to "4"; randomly select 10,000 nodes whose degree value is [81,100], and set the value of labelX is "5"; set the labelX value of the randomly selected 10,000 nodes whose degree is in the range of [101,∞] to be "6" • Use the 60,000 nodes as the starting point to carry out five rounds of label propagation, reserve the Top-3 label values (propagation probability) for each vertex, write the results back to the database, and save the LPA results to the file system as disk file • The algorithm runs three times, recording each time, average time, maximum time, and minimum time Note: Different from the original (and primitive) LPA algorithm, where each vertex can only get 1 label value in the end, this advanced LPA variation allows multiple labels to be attached to each vertex at the end of the label propagation process
Expected outcome	Achieve the above-required functions, and the query results are correct
Evaluation standard	o Rank by the response time, label-propagation time, file-system/disk write-back time, and total time o Monitor the system resource consumption, if there is any crash, OOM, or downtime o Examining the time and effort taken for the customer-algorithm to be developed and supported (i.e., some vendors may support such feature out-of-box, while others may take weeks or months to implement the algorithm)

TABLE 19 Graph embedding algorithm capability (Node2vec).

Testing items	Graph embedding algorithm capabilities
Testing purposes	Examine the support of the basic graph embedding algorithm by the graph system
Preconditions and test content	○ Using the same dataset as in Table 18 ○ The graph algorithm logic is as follows: • Run the *node2vec* algorithm to calculate the embedding of the nodes in the whole graph. The number of random walks is 10, the depth of the walk is 80, the number of iterations is 10, and the dimensions number is 128. Other parameters are detailed below, and the first 1000 results are returned ▪ $q=0.8$ ▪ $p=0.2$ ▪ learning rate$=0.01$ ▪ min_learning_rate$=0.0001$ ▪ resolution$=10$ ▪ sub_sample_alpha$=0.75$ ▪ neg_num$=5$ ▪ min_frequency$=1$ • The algorithm runs three times, recording average time, maximum time, and minimum time Note: Full graph random walk and training must be implemented instead of only sampling 1000 paths—more specifically, after completing the full-graph random walks, return with 1000 embedded paths
Expected outcome	Achieve the above required functions, and the query results are correct (validation can be done on much smaller graphs with much simpler parameters for manual calculation)
Evaluation standard	○ Rank by response time, overall execution time, disk/file-system write-back time, and feature completeness (such as support of input parameter adjustment) ○ During the operation of the algorithm, record the consumption of system resources

- High Concurrency Architecture Design.
- Memory Computing Optimization.
- Low-latency graph traversal algorithm logic, etc.

Finally, we use a large matrix to list the benchmark indicators related to visualization, operation, maintenance, management and monitoring (M&M), and system security, as shown in Table 20.

TABLE 20 Visualization, M&M, and security.

Component name	Support?	Remarks
Visualized Graph Query	Y/N	Are 2D/3D viewing modes supported
Visualized Graph Data modification	Y/N	Support of modifying meta data interactively
Visual Data Addition	Y/N	Ditto
Visual Data Deletion	Y/N	Ditto
Visual Incremental Query	Y/N	Ability to continue to expand or run a new query based on a previous query result
Query Profiling and Optimization	Y/N	Query syntax prompting while typing (IDE), query profiling, explain, and optimization
Algorithm Management	Y/N	Supports operations such as direct algorithm execution, status check, and stop algorithm Is a hot-pluggable algorithm supported (Not requiring a restart of the system when a new algorithm is installed)?
Algorithm Extension	Y/N	Support of graph algorithm development with extensible API/SDKs
Algorithm Result Visualization	Y/N	Support of algorithm result visualization on selective algorithms (i.e., Louvain)
Graph Data Statistics	Y/N	Support front-end display of nodes, edges, schemas, and system resource consumption
User Permissions (RBAC)	Y/N	Provide user access control and permission management. Includes system-level and graph-set level permission management Provide policy and role definition
Quick Deployment	Y/N	Whether to support one-click deployment based on container packaging
DBMS Extension	Y/N	Plug-in, short-cut, or other types of extension supported
Persistent Storage	Y/N	Whether to support graph-data persistent storage
Data Import	Y/N	Online data import (production) Offline data import (trial/lab/offline scenarios)
LBS	Y/N	Support of system-wide load balancing

Continued

TABLE 20 Visualization, M&M, and security—cont'd

Component name	Support?	Remarks
HTAP	Y/N	Support of TP and AP operations within the same cluster, particularly instance-bound role assignment (leader, learner, algorithm-instance, etc.)
ACID & Data Consistency	Y/N	Support of levels of ACID (serialization, read-n-write data consistency)
Graph Query Language (GQL)	Y/N	Compliant with the upcoming GQL international standard? If using a bespoke/vendor-specific GQL, is the language intuitive, easy-to-use, and capable (feature-complete)? Evaluate the feature-completeness of online documentation
High-Availability	Y/N	Support of HA, Distributed Consensus, HTAP, or other forms of distributed cluster setup
Enterprise Support	Y/N	Enterprise-level support, whether it supports on-site, remote, ULA, SLA, and other advanced tech-support terms
Core R&D Staff Support	Y/N	Whether the original graph vendor support is available
Java SDK	Must support	With online documentation and samples
Python SDK	Must support	Ditto
Node SDK	Y/N	Ditto
Other language SDK	Y/N	Ditto
Restful API	Y/N	Ditto
Integration with Domain applications?	Y/N	Whether to support the upper-layer/user-facing application development
Training & Certification	Y/N	Are training supported? Online vs Offline Richness of training materials Are there certification programs?
Partnership Program	Y/N	Are there partnership programs? Existing industry partners, etc.

7.2.3 Graph query/algorithm results validation

The correctness of graph database queries and algorithms is of paramount importance in ensuring the accuracy and reliability of the results. In this section, we will discuss how to verify the correctness of graph database queries and algorithms.

There are mainly two types of data operations in any graph database:

○ Meta-data operation: Also known as an operation against vertex, edge, or their attributes. There are four main types of meta operations, specifically CURD (Create, Update, Read, or Delete).

○ High-dimensional data operation: By saying "high-dimensional," we are referring to high-dimensional data structures in the resulting datasets, for instance, a group of vertices or edges, a network of paths, a subgraph, or some mixed types of sophisticated results out of graph traversal—essentially, heterogeneous types of data can be mixed and assembled in one batch of result, therefore, examining such result both exciting and perplexing.

High-dimensional data manipulation poses as a key difference between graph databases and other types of DBMS, We'll focus on examining this unique feature of graph DB across the entire section. There are three main types of high-dimensional data operations, which are frequently encountered in most benchmark testing reports:

○ K-hop: K-hop query against a certain vertex will yield all vertices that are exactly K-hop away from the source vertex in terms of shortest-path length. K-hop queries have many variations, such as filtering by certain edge direction, vertex, or edge attribute. There is a special kind of K-hop query which is to run against all vertices of the entire graph, it is also considered a type of graph algorithm.

○ Path: Path queries have many variations, shortest-path queries are frequently used, with the addition of template-based path queries, circle-finding path queries, automatic-networking queries and auto-spread queries, etc.

○ Graph algorithm: Graph algorithms essentially are the combinations of meta-data, K-hop, path queries. There are rudimentary algorithms like degrees, but there are highly sophisticated algorithms like Louvain, and there are extremely complex, in terms of computational complexity, algorithms such as Betweeness-Centrality or full-graph K-hop, particularly when K is much larger than 1.

Most benchmark reports are reluctant to disclose how high-dimensional data operations are implemented, and this vagueness has created lots of trouble for readers and users to better understand graph technology. We hereby clarify, there are ONLY three types of implementations in terms of how high-dimensional graph data is traversed:

o BFS (Breadth First Search): Shortest Path, K-hop queries are typically implemented using the BFS method. It's worth clarifying that you may use DFS to implement, say, shortest path finding, but in most real-world applications, particularly with large volumes of data, BFS is guaranteed to be more efficient and logically more straightforward than DFS, period.
o DFS (Depth First Search): Queries like circle finding, auto-networking, template-based path queries, and random walks desire to be implemented using the DFS method. If you find it hard to understand the difference between BFS and DFS, find yourself a college textbook on Graph Theory.
o Combination of Both (BFS + DFS): There are scenarios where both BFS and DFS are applied, such as template-based K-hop queries, customized graph algorithms, etc.

Under BFS mode, if qualified first hop neighbors are not all visited first, the access to any second hop neighbor will not start, and traversal will continue in this way until all neighbors are visited hop after hop. Based on such a description, it's not hard to tell that if a certain K-hop or shortest-path query only returns a predefined limited number of neighbors or paths (say, 1 or 10), it's guaranteed that the query implementation is wrong! Because you do NOT know the total number of qualified neighbors or paths beforehand.

Note: Metadata-oriented graph database operations are similar to relational databases, and the correctness verification method will not be repeated here. This section focuses on the correctness verification of high-dimensional data queries unique to graph databases.

We take the Twitter-2010 dataset commonly used in graph database benchmark performance evaluation as an example to illustrate how to verify the correctness of queries on graphs. The download link[a] of the Twitter dataset (the number of vertices is 42 million, the number of edges is 1.47 billion, and the original data is 24.6 GB).

Before we start, let's get ourselves familiar with a few concepts in graph data modeling by using the Twitter dataset as an example:

o Directed graph: Every edge has a unique direction, In Twitter, an edge is composed of a starting vertex and an ending vertex, which map to the 2 IDs separated by TAB on each line of the data file, and the significance of the edge is that it indicates the starting vertex (person) follows the ending vertex (person). When modeling the edge in a graph database, the edge will be constructed twice, the first time as StartingVertex → EndingVertex, and the second time as EndingVertex → StartingVertex (a.k.a., the inverted edge), this is to allow traversing the edge from either direction. If the

---

a. Twitter Dataset: http://an.kaist.ac.kr/traces/WWW2010.html.

inverted edge is not constructed, queries' results will be inevitably wrong (We'll show you how later).

○ Simple-graph vs multigraph: If there are more than two edges of the same kind between a pair of vertices in either direction, it's a multigraph, otherwise, it's a simple-graph. Twitter and all social network datasets are considered simple graphs because it models a simple (singular) following relationship between the users. In financial scenarios, assuming user accounts are vertices and transactions are edges, there could be many transactions in between two accounts, therefore many edges. Some graph databases are designed to be simple-graph instead of multigraph, this will create lots of problems in terms of data modeling efficiency and query results correctness.

○ Node-edge attributes: Twitter data does NOT carry any node or edge attribute other than the designated direction of the edge. This is different from transactional graph in financial scenarios, where both node and edge may have many attributes so that filtering, sorting, aggregation, and attribution analysis can be done with the help of these attributes. There are so-called graph database systems that do NOT support the filtering by node or edge attributes, which are considered impractical and lack of commercial values.

We take K-hop queries as an example to verify the correctness of graph database query results. First, let's be clear about the definition of K-hop, there have been two types of K-hop:

○ Kth-Hop Neighbors, which are exactly K hops away from the source vertex.
○ All Neighbors from the First Hop all the way to the Kth Hop.

No matter which type of K-hop you are looking at, two essential points affect the correctness of the query:

○ K-hop should be implemented using BFS, instead of DFS.
○ Results deduplication: The results should NOT contain any duplicated vertices on the same hop or across different hops. (Some graph systems have this de-dup problem.)

Some graph systems use DFS to find the shortest paths, this approach has two major problems:

○ Outright Wrong: There is a high probability that the results are wrong as DFS can't guarantee a vertex belongs to the right hop (depth) which falls on the shortest-path.
○ Inefficient: On large and densely populated datasets, it's impractical to traverse all possible paths in DFS fashion, for instance, Twitter-2010 has many hotspot nodes having millions of neighbors, and any 2-hop or deeper queries mean astronomical computational complexity!

Let's validate the 1-Hop result of vertex ID = 27960125 in Twitter-2010, first, we start from the source file which shows eight edges (rows of connecting neighbors, as in Fig. 11), but, what exactly is its 1-hop?

```
[root@twitter-dataset-test twitter]# cat twitter.edges.csv | grep "27960125"
23396761,27960125
27498091,27960125
27960125,8488582
27960125,27498091
27960125,46799670
27960125,48962183
27960125,52502348
27960125,52697437
[root@twitter-dataset-test twitter]#
```

FIG. 11   Edges pertaining to an ID in the Twitter source file.

The correct answer is 7! Because the node 27960125 has a neighbor ID = 27498091 which appears twice (following each other), there are two edges in between these two vertices. If we deduplicate, we have 7. This can be validated by running an undirected K-hop against the node as in Fig. 12.

To verify the correctness of the results more accurately, the K-Hop query can also be filtered according to the direction of the edge, for example, query the outgoing edge, incoming edge, or bidirectional edge of the vertex "2796015" (Note: the default is to query bidirectionally). Fig. 13 shows how

```
>>> khop().src({_id == 27960125}).depth(1).boost() as nodes return nodes
2022/03/19 06:40:09 khop().src({_id == 27960125}).depth(1).boost() as nodes return nodes
+----+----------+---------+
| ID |   UUID   | Schema  |
+----+----------+---------+
|    | 36046807 | default |
|    | 30051355 | default |
|    | 27381488 | default |
|    | 25821004 | default |
|    | 21597395 | default |
|    | 11413328 | default |
|    | 10029388 | default |
+----+----------+---------+
>>>
```

FIG. 12   Verifying the neighbor result set of a vertex in CLI.

```
>>> use twitter
2022/03/18 17:55:37 Changed Current GraphSet to [twitter]
>>> khop().src({_id == 27960125}).depth(1).boost() as nodes return count(nodes)
2022/03/18 17:55:42 khop().src({_id == 27960125}).depth(1).boost() as nodes return count(nodes)
+--------------+
| count(nodes) |
+--------------+
| 7            |
+--------------+
>>> khop().src({_id == 27960125}).depth(1).direction(right).boost() as nodes return count(nodes)
2022/03/18 17:55:59 khop().src({_id == 27960125}).depth(1).direction(right).boost() as nodes return count(nodes)
+--------------+
| count(nodes) |
+--------------+
| 6            |
+--------------+
>>> khop().src({_id == 27960125}).depth(1).direction(left).boost() as nodes return count(nodes)
2022/03/18 17:56:10 khop().src({_id == 27960125}).depth(1).direction(left).boost() as nodes return count(nodes)
+--------------+
| count(nodes) |
+--------------+
| 2            |
+--------------+
>>>
```

FIG. 13   Launching K-hop from CLI.

FIG. 14 Data from a graph system's benchmark results (Github.com).

to use the graph query language to complete the corresponding work. Note that this vertex has six neighbors corresponding to outgoing edges and two neighbors corresponding to incoming edges, and the neighbor "27498091" is overlapping.

If you refer to the test result data file released by a graph system vendor on the Github website, there are obvious and extensive errors in the results of K-hop queries. For example, the 1-Hop result of vertex 27960125 in Fig. 14 only returns six neighbors!

There are three possible causes for the such K-hop problem:

○ **Data modeling mistake**: Each edge is stored only once (unidirectionally), so that the inverted edge traversal is impossible.

○ **Query method mistake**: The K-hop query is only conducted unidirectionally, instead of bidirectionally.

○ **Code implementation bug**: Such as results not being deduplicated, we'll continue our examination in multiple hop queries next.

Data modeling mistake is fatal; it means business logic will NOT be correctly reflected in the underpinning data models. Taking antifraud, AML, or BI scenarios as an example, account B receives a wire-transfer from account A, however, the system only stores an edge pointing from A to B (A → B) but not the inverted edge from B to A (B ← A), this would make it impossible to trace the transaction starting from account B. This clearly is unacceptable.

Similarly, query method and coding-logic bugs are also serious, query should go bidirectionally instead of unidirectionally, the former is exponentially more complex when traversing multiple hops. The same holds true for storage space usage and data ingestion time.

If we continue to track vertex ID = 27960125's 2-hop result, the mistakes may become more dramatic yet more difficult to validate. Note in Fig. 15, that 2-hop of the vertex returns 1128 neighbors, but there are duplicated neighbors, and the result is only based on outbound edges. The correct number of 2-hop neighbors should be 533108 (Fig. 16), the difference here is 473 times, 47,300%! Such query results carry three mistakes simultaneously: data-modeling mistake, query method mistake, and deduplication mistake.

FIG. 15   2-Hop results for traversing only outbound edges (Github).

```
>> use twitter
2022/03/20 10:11:34 Changed Current GraphSet to [twitter]
>>> khop().src({_id == 27960125}).depth(1:2).boost() as nodes return count(nodes)
2022/03/20 10:11:46 khop().src({_id == 27960125}).depth(1:2).boost() as nodes return count(nodes)
----------------
| count(nodes) |          1-Hop + 2-Hop Deduplicated
----------------
| 533115       |
----------------
>>> khop().src({_id == 27960125}).depth(2) as nodes return count(nodes)
2022/03/20 10:11:52 khop().src({_id == 27960125}).depth(2) as nodes return count(nodes)
----------------
| count(nodes) |          2-Hop Deduplicated
----------------
| 533108       |
----------------
>>> khop().src({_id == 27960125}).depth(1:2).direction(right).boost() as nodes return count(nodes)
2022/03/20 10:11:58 khop().src({_id == 27960125}).depth(1:2).direction(right).boost() as nodes return count(nodes)
----------------
| count(nodes) |          Outbound 1-Hop + 2-Hop Deduplicated
----------------
| 1133         |
----------------
>>> khop().src({_id == 27960125}).depth(2).direction(right).boost() as nodes return count(nodes)
2022/03/20 10:12:09 khop().src({_id == 27960125}).depth(2).direction(right).boost() as nodes return count(nodes)
----------------
| count(nodes) |          Outbound 2-Hop Deduplicated
----------------
| 1127         |
----------------
>>
```

FIG. 16   Four traversal query methods for K-hop query (Ultipa CLI).

Unfortunately, similar query result errors are not an isolated case in today's graph database market. We are seeing similar problems with other graph systems as well. For instance, some systems by default do NOT deduplicate K-hop results, if you force it to deduplicate, it will run exponentially slower. Another open-source graph system has a form of shortest-path query that returns only one path, which is fast but ridiculously wrong when there are many paths in the results pool.

The visualization tools integrated with the graph database can help us validate the correctness of the results more intuitively and conveniently. In Chapter 4 of Graph Algorithms, we introduce the relevant content in detail. Interested readers can refer to it for more information.

Above we have shown the correctness verification method of the basic K-hop query on the graph, as well as possible error situations. Many other graph operations also involve the problem of wrong results, but some basic methods can be used to verify the correctness of the results. Here are two more representative examples:

- Shortest path
- Graph algorithm

The shortest path can be regarded as a natural extension of the K-hop query, and it has two characteristics:

- High-dimensional results which are comprised of paths that are further composed of nodes and edges that are assembled in a certain sequence.
- All paths are to be returned: Only calculating and returning one path is NOT enough. In financial fraud detection, AML, and BI attribution analysis, there may exist many paths and returning just one is totally unacceptable.

Figs. 17 and 18 from above show that there are multiple shortest paths between ID 12 and 13 in, however, if you search by a certain direction, you may get different answers as search filtering logic kicks in. For simple cases like this, you may grep the source file to find out that there are two bidirectional edges (one inbound and one outbound) in between vertex ID 12 and 13, and if you do it recursively, you can validate more sophisticated queries. Note the keyword "recursively," it's exactly why GQL/Graph-DBMS will be mainstream and SQL/RDBMS will fade away, because SQL's weakest link is its inability to handle recursive queries, and GQL/Graph is automated on this.

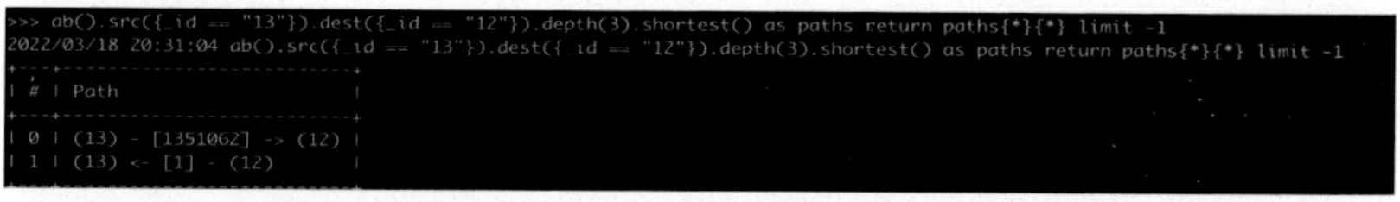

FIG. 17 Running shortest-path in CLI.

FIG. 18 Three ways of running shortest-path.

Next, we use Jaccard Similarity algorithm as an example to illustrate how to validate algorithm results. Taking the below diagram (Fig. 19) as an example, to calculate Jaccard Similarity of the green vertex and red vertex, you must find their common neighbors (2) and total neighbors (5), so that the similarity $= 2/5 = 0.4 = 40\%$.

If the Jaccard Similarity algorithm is integrated, launching it directly will churn out the correct answer of 0.4 (Fig. 20). If you desire to write several lines of GQL to implement it in an organic and white-box way, the code logic is as follows:

In Twitter-2010, computing Jaccard Similarity of any pair of vertices is to find each vertex's 1-hop neighbors, taking ID $= 12$ and ID $= 13$ as an example, they both are hotspot supernodes having over 1 million neighbors, it would be impractical to validate the algorithm results manually. However, as long as you understand the logics under the hood of the algorithm, there are ways to divide and conquer it. The steps below show two ways to validate the algorithm:

1. Launch the Jaccard algorithm (Fig. 21).

2. Validation method 1: Using multiple GQLs to implement the Jaccard algorithm (Fig. 22).

3. Validation method 2: Query the 1-hop neighbors of vertex 12 and 13 (Fig. 23).

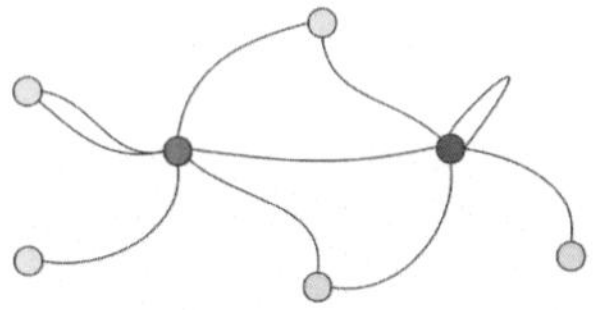

$$S_J(A, B) = \frac{|A \cap B|}{|A \cup B|} = \frac{|A \cap B|}{|A| + |B| - |A \cap B|}$$

FIG. 19 Jaccard Similarity algorithm dataset and formula.

```
khop().src({color == "blue"}).depth(1).boost() as node1
khop().src({color == "red"}).depth(1).boost() as node2
with collect(node1) as nodes1, collect(node2) as nodes2
with size(nodes1) as s1, size(nodes2) as s2, size(intersection(nodes1, nodes2)) as inters
return inters / (s1 + s2 - inters)
```

FIG. 20 Multisentence GQL/UQL for Jaccard Similarity.

```
>>> find().nodes({_id in [12]}) as node1 find().nodes({_id in [13]}) as node2 algo(jaccard).params({ids: [node1],
ids2: [node2]}).stream() as sim return sim
2022/03/19 10:03:23 find().nodes({_id in [12]}) as node1 find().nodes({_id in [13]}) as node2 algo(jaccard).params
({ids: [node1], ids2: [node2]}).stream() as sim return sim
+----------+----------+------------------+
|  node1   |  node2   |   similarity     |
+----------+----------+------------------+
| 20108436 | 17031860 | 0.15362655682654502 |
+----------+----------+------------------+
>>>
```

FIG. 21 Invoking Jaccard Similarity algorithm from CLI.

```
>>> khop().src({_id == 12}).depth(1).boost() as nodes return count(nodes)
2022/03/18 22:16:27 khop().src({_id == 12}).depth(1).boost() as nodes return count(nodes)
+---------------+
| count(nodes) |
+---------------+
| 1001159       |
+---------------+
>>> khop().src({_id == 13}).depth(1).boost() as nodes return count(nodes)
2022/03/18 22:16:35 khop().src({_id == 13}).depth(1).boost() as nodes return count(nodes)
+---------------+
| count(nodes) |
+---------------+
| 1031901       |
+---------------+
>>> khop().src({_id == 12}).depth(1).boost() as node1 khop().src({_id == 13}).depth(1).boost() as node2 with collect(node1) as nodes
1, collect(node2) as nodes2 with size(nodes1) as s1, size(nodes2) as s2, size(intersection(nodes1, nodes2)) as commons return common
s/(s1+s2-commons)
2022/03/18 22:16:42 khop().src({_id == 12}).depth(1).boost() as node1 khop().src({_id == 13}).depth(1).boost() as node2 with collect
(node1) as nodes1, collect(node2) as nodes2 with size(nodes1) as s1, size(nodes2) as s2, size(intersection(nodes1, nodes2)) as commo
ns return commons/(s1+s2-commons)
+----------------------------+
| commons/(s1+s2-commons) |
+----------------------------+
| 0.153626382480831          |
+----------------------------+
```

FIG. 22   Jaccard Similarity algorithm—validation method 1.

```
>>> khop().src({_id == 12}).depth(1).boost() as nodes return count(nodes)
2022/03/18 22:16:27 khop().src({_id == 12}).depth(1).boost() as nodes return count(nodes)
+---------------+
| count(nodes) |
+---------------+
| 1001159       |          1-hop of 12
+---------------+
>>> khop().src({_id == 13}).depth(1).boost() as nodes return count(nodes)
2022/03/18 22:16:35 khop().src({_id == 13}).depth(1).boost() as nodes return count(nodes)
+---------------+
| count(nodes) |
+---------------+
| 1031901       |          1-hop of 13
+---------------+
```

FIG. 23   Jaccard Similarity algorithm—validation method 2 part 1.

```
2022/03/18 22:54:51 n({_id == 12}).e().n({} as mid).e().n({_id == 13}) return count(distinct(mid))
+---------------------+
| count(distinct(mid)) |
+---------------------+
| 270739              |
+---------------------+
```

FIG. 24   Jaccard Similarity algorithm—validation method 2 part 2.

4. Validation method 2: Finding all distinct paths that are exactly 2-hop deep between ID 12 and 13, the number of paths is equal to the number of common neighbors of the two IDs (Fig. 24).

5. Verification method 2: The similarity = Results-from-Step-4/(12's 1-hop + 13's 1-hop—Results from step 4) = 0.15362638.

6. As you can see, the result from step 2 and step 5 are exact. We know the answer is right.

In this section, we introduce in detail how to verify the correctness of queries on graph databases. We hope it can broaden the minds of smart readers and achieve the effect of drawing inferences from one instance and eliminating the false while preserving the truth.

7.2.4 Graph system optimization

In the previous section, we summarized many graph system benchmark requirements based on real business scenarios. In this section, we will analyze the results of the benchmarks, and share with readers some noteworthy and nonobvious test contents, and give optimization suggestions.

Taking the node2vec embedding algorithm as an example, we used two graph systems' the implementation to compare how they handle the algorithm differently. The query statements of the two systems are as follows:

```
/* System-N*/
CALL gds.alpha.node2vec.stream(
  'my_native_graph',
{
    embeddingDimension: 128,
    iterations: 10,
    walkLength: 80,
    walksPerNode: 10,
    windowSize: 10,
    inOutFactor: 0.8,
    returnFactor: 0.2
})
YIELD
  nodeId
WITH
  gds.util.asNode(nodeId) AS node
RETURN node.v AS v
LIMIT 1000;
```

```
algo(node2vec).
params({
    buffer_size: 2000,
    walk_length: 80,
    walk_num: 10,
    p: 0.2,
    q: 0.8,
    window_size: 10,
    dimension: 128,
    learning_rate: 0.01,
    min_learning_rate: 0.0001,
    min_frequency: 1,
    sub_sample_alpha: 0.75,
    resolution: 10,
    neg_num: 5,
    loop_num: 2,
    limit: 1000
}) as results
return results
```

```
/* System-U */
```

From a purely grammatical point of view, we cannot tell the underlying logic of the two systems processing the algorithm, but we know that System-N defaults to sampling 1000 paths (walkBufferSize = 1000), and returns directly after such sampling is complete. The requirement for the benchmark is to return 1000 paths after the full graph sampling is completed. On a billion-scale graph, the difference in algorithm complexity between the two is 1,000,000 times! Therefore, the partial sampling of the System-N is of typical implementation or invocation error. If it takes 1 s to sample 1000 paths, it would take 12 days to finish sampling 1 billion paths. Now the question should be: how to optimize a graph system to have it run exponentially faster given the existing hardware?

As we have introduced in the previous chapters, two major optimization points to contemplate with:

- High Concurrency Architecture.
- Low Latency Data Structures.

The above two points implicitly express the following optimization pathways:

- Minimize disk I/O.
- Take advantage of in-memory computing.
- Adopt a multilevel storage and caching strategy.
- Adopt a low-latency, high-concurrency-friendly programming language and computing architecture.

Obviously, if you use Python to implement node2vec, the probability of completing the above node2vec benchmark is 0%. The main reason is that its execution efficiency is low and serial, which is completely unsuitable as a database-level system programming language. However, is it possible to switch to Java? The author believes that Java has built a huge and mature framework system at the application layer, but as the underlying language of the database system, Java has obvious disadvantages compared with C/C++/Go/Rust, especially when it requires high computing power and low delays. In a low-level environment, neither the JVM nor the GC is conducive to completing the benchmarks of complex algorithms such as node2vec.

Similarly, in some algorithm benchmarking, such as PageRank, more complex (and closer to real application scenarios) benchmark requirements require global iterative calculation of rank values, and then sorting the full results and return Top-100. From the perspective of algorithm correctness, the correct implementation of the PageRank algorithm requires the completion of global iterations to obtain the rank value of each vertex. In the worst case, the global iteration must also be completed within the connected component where each vertex is currently located. If a connected component has far greater than 50 vertices, it is an obvious algorithm logic error to calculate the rank value of only 50 vertices. For algorithms that are meant to iterate through the entire graph for multiple times, if the algorithm returns in a very short time and the runtime system resources (CPU, memory, etc.) consumption is very low, it's a sign that there is a systemic error in the algorithm, be it a case of implementational bug or intentionally-cheating. (Note: We have seen cases that in some institutional benchmarks, vendors use preprocessing, caching or prestored static data to return illegitimate query (or algorithm) results instantly instead of genuinely computing for authentic results on the fly. This may warrant a dedicated discussion separated from the optimization topic we focus on here.)

In the benchmarking of the LPA algorithm, there are two modes, one is native LPA, and the other is enhanced LPA. The operation logic of the two algorithms is as follows, and the results are shown in Table 21.

- Native LPA: All nodes are initiated with a certain label, and propagated for classification, After five rounds of iteration, each node will have a final and propagated label, the result is written back.
- Advanced LPA: Nodes are initiated with different labels, propagate for classification, After five iterations, each vertex retains three labels with the highest probabilities (%), and writes the results back.

TABLE 21 Comparison of benchmark results of LPA algorithms.

Benchmark item	Performance on billion-scale graph dataset
Native LPA time-consumption (full graph computing)	Computing time: 61 s Write-back: 65 s
Enhanced LPA time-consumption (full graph computing, with two labels)	Computing time: 116 s Write-back: 84 s

```
/* Regular LPA */
algo(lpa).params
({
  loop_num:5,
  node_label_property:"label",
  k:1
}).write()
```

```
/* Enhanced LPA */
algo(lpa).params
({
  loop_num:5,
  node_label_property:"labels",
  k:3
}).write()
```

The calling methods of the two algorithms are very similar with the difference being in the number of labels propagatable to each node. In Fig. 25, the timeliness of the two LPA algorithms are recorded, because the complexity and amount of write-back data of the enhanced LPA are higher than those of ordinary LPA, so the computing time and write-back time are proportionally higher (2.5×: 2 labels vs 5 labels).

Figs. 26–29 list the benchmark results of two graph databases (System-N and System-U) conducted by a Fortune-100 company's AI Labs on graph algorithms and queries. The benchmark was conducted on the Alimama

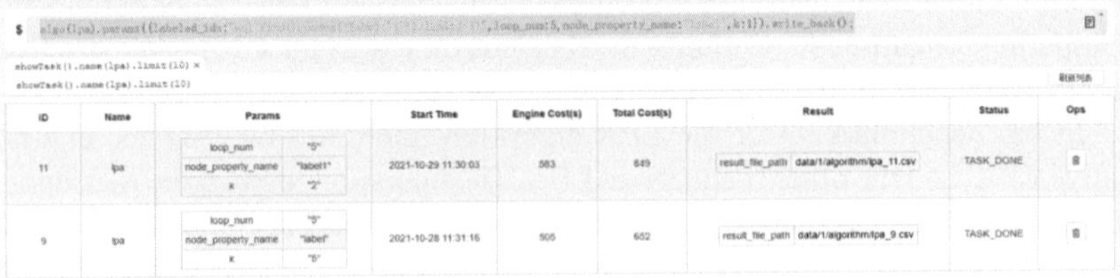

FIG. 25 Enhanced LPA vs native LPA.

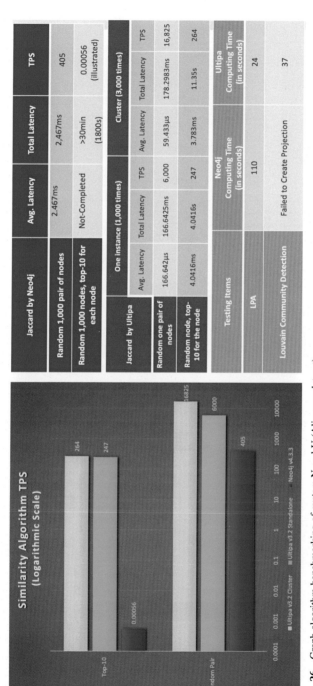

Jaccard by Neo4j	Avg. Latency	Total Latency	TPS
Random 1,000 pair of nodes	2.467ms	2,467ms	405
Random 1,000 nodes, top-10 for each node	Not-Completed	>30min (1800s)	0.00056 (illustrated)

Jaccard by Ultipa	One Instance (1,000 times)			Cluster (3,000 times)		
	Avg. Latency	Total Latency	TPS	Avg. Latency	Total Latency	TPS
Random one pair of nodes	166.642μs	166.6425ms	6,000	59.433μs	178.2983ms	16,825
Random node, top-10 for the node	4.0416ms	4.0416s	247	3.783ms	11.35s	264

Testing Items	Neo4j Computing Time (in seconds)	Ultipa Computing Time (in seconds)
LPA	110	24
Louvain Community Detection	Failed to Create Projection	37

FIG. 26   Graph algorithm benchmarking of system N and U (Alimama dataset).

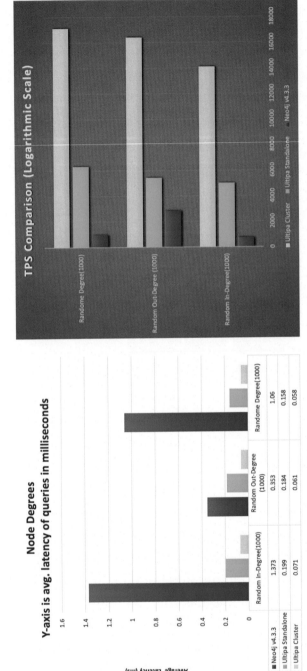

FIG. 27 Degree computing performance by System-U vs System-N.

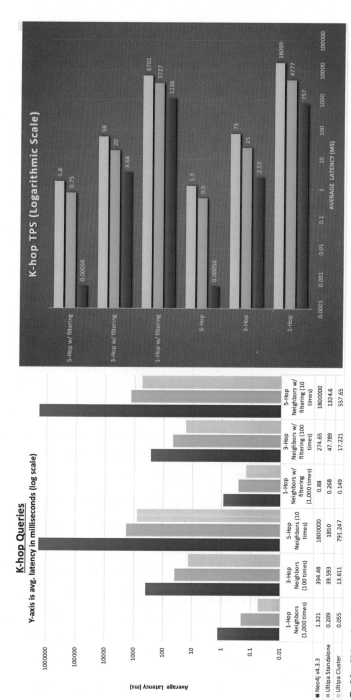

FIG. 28 K-hop benchmark System-U vs System-N.

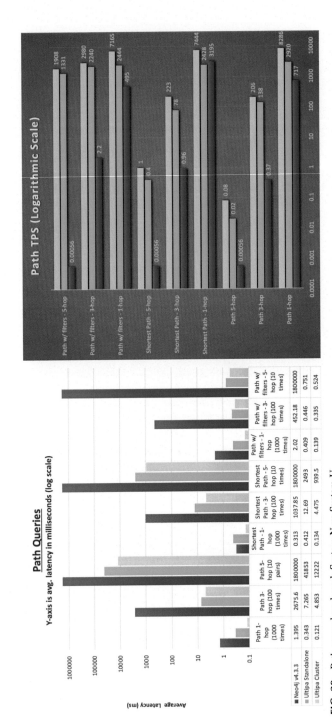

FIG. 29 Path query benchmark System-N vs System-U.

e-commerce dataset with 105 million nodes and edges (on average each node is connected to 40 neighbors, implying the graph is densely connected).

From the above results, we can observe the following:

○ System-N faces significant challenges when dealing with complex algorithms (like Louvain) and large data volumes (hundred-million or above), often struggling to complete the projection (mapping data from the disk persistence layer to the memory data structure) required for iterative algorithm computations.

○ System-U's 3-node cluster mode achieves a TPS (transactions per second) approximately three times higher than that of a single node, indicating the system's ability to scale linearly with resource expansion. This scalability is particularly evident in load-balanced deep query and metadata query scenarios.

In the benchmarking of the full-graph degree algorithm (Fig. 27), we performed 1000 random calculations in each of the following three aspects: in-degree, out-degree, and degree. We can see that the single-instance benchmarking performance difference between System-U vs System-N is more than 10 times, and the cluster difference is more than 30 times, and this is only the simplest and shallow metadata-level calculations. When conducting deep-traversal algorithms, the performance gap would be enlarged.

In K-hop and path benchmarks (as shown in Figs. 28 and 29), from shallow to deep queries, the performance gap between System-N and System-U is 5–10 times on the 1-hop, 30 times on 3-hop, and 1000+ times on 5-hop or beyond System-N couldn't return within 1800s on 5-hop or deeper queries.

In the K-hop benchmark, there are six scenarios to test the average delay of query operations on layers 1, 3, and 5 with and without filtering. The Alimama dataset used in the test exhibits a high degree of connectivity $(2^*|E|/|V| \geq 40$, indicating a high ratio of vertices to edges). When performing a 5-hop query starting from each vertex, the computational complexity of traversing the entire graph of 100 million plus nodes and edges is significant. As a result, System-N experiences a sudden increase in delay from an average of 400 ms for a 3-hop query (without filtering) to no returning anything for 5-hop queries. However, the System-U system shows an increase in delay from 14–17 ms to 558–791 ms. Theoretically, the computational complexity increases approximately 1600 times (complexity increase: $O((2^*|E|/|V|)^2)$) from 3-hop to 5-hop queries. In practice, due to the dynamic pruning of the vertex filter and variations in vertex connectivity, the increase in time consumption is around 40–50 times, the System-U numbers from Fig. 28 reflect such a modest increase.

In the path benchmark (Fig. 29), nine scenarios were used to compare the time consumption of both systems in queries with depths of 1, 3, and 5 hops with and without filtering conditions, as well as the shortest path queries. It can be observed that in the 1-layer path query, both System-N and System-U exhibit slightly longer time consumption with filtering conditions compared to queries without filtering conditions. However, in queries with depths of ≥ 2 layers, filtering conditions result in a significant reduction in time consumption compared to queries without filtering conditions (System-U: 15x; System-N: 5x).

The characteristics mentioned above can be explained as follows:

- In 1-hop path queries, the filter condition increases the amount of calculation, leading to increased time consumption (System-U: 20%; System-N: 40%).
- In 2-hop and deeper path queries, filtering dynamically prunes the graph data set to be traversed, resulting in more efficient returns and significantly reduced time consumption (System-U: 15x; System-N: 5x).
- In deep path queries with depths ≥ 5 hops, due to dynamic pruning and the connectivity characteristics of the graph (there are 11 connected components, but 99.999% of nodes belong to the largest CC, which has a diameter of 8, but most nodes are connected within five to six hops), System-U with filtering maintains submillisecond level time consumption (0.524 ms), while queries without filtering conditions take 12 s to return the result due to the escalating computational complexity (there are sometimes tens of millions of paths in between any pair of nodes). The System-N is unable to return results from such depth.

In scenarios with large data volumes and complex queries or algorithms, System-N often struggles to return results successfully. This indicates that the core of System-N faces significant challenges when dealing with high computing power requirements. However, a closer examination of the underlying architecture of System-N reveals that its system, based on the Java/JVM architecture, can achieve approximately 10%–15% of the performance of the System-U system (written in C/C++) in metadata and shallow computing. This already represents the limit of Java systems, as it is generally believed that the "computing power" of a Java program is about 1/7 of that of a C++ program on the same hardware basis. When comparing System-N to other Java-based systems on the market, other systems' performance would likely be even lower than that of System-N.

In the GAP performance evaluation, five systems (GraphBLAS, Galois, GraphIt, GKC, NWGraph) were benchmarked against the GAP reference system. Analyzing the benchmarking results and open-source code, we can gather the following information (Fig. 30):

- GAP outperforms the other systems in three algorithms: Breadth First (BFS), Single Source Shortest Path (SSSP), and Connected Component (CC), demonstrating clear performance advantages.
- The other graph systems may have better performance than the GAP system in the remaining three algorithms (PageRank (PR), Betweenness Centrality (BC), or Triangle Counting (TC)).

| | | Baseline (speedup over GAP reference) | | | | | Optimized (speedup over GAP reference) | | | | |
| | | Real Graphs | | | Synthetic Graphs | | Real Graphs | | | Synthetic Graphs | |
		Web	Twitter	Road	Kron	Urand	Web	Twitter	Road	Kron	Urand
SuiteSparse GraphBLAS	BFS	39.98%	60.50%	13.74%	58.14%	51.09%	36.38%	54.04%	8.02%	53.71%	46.48%
	SSSP	8.50%	32.23%	0.35%	32.10%	40.51%	5.84%	31.18%	0.43%	23.95%	32.56%
	CC	12.66%	18.87%	7.40%	20.13%	43.45%	11.08%	15.65%	6.30%	15.96%	33.05%
	PR	92.86%	87.92%	137.50%	91.04%	91.45%	85.02%	91.21%	173.42%	96.53%	97.81%
	BC	54.00%	70.93%	3.96%	80.38%	92.40%	42.69%	69.64%	3.46%	85.74%	84.95%
	TC	48.76%	31.92%	12.86%	34.01%	61.51%	55.53%	34.49%	12.47%	37.46%	61.04%
Galois	BFS	54.18%	44.77%	351.04%	57.14%	8.93%	58.55%	41.88%	220.92%	62.16%	77.85%
	SSSP	46.13%	55.94%	54.40%	41.76%	49.47%	26.62%	45.11%	67.37%	58.06%	53.53%
	CC	64.43%	114.02%	84.11%	85.22%	66.06%	113.94%	75.16%	90.16%	85.53%	49.16%
	PR	157.54%	84.36%	331.66%	106.15%	117.35%	154.67%	108.96%	456.72%	110.63%	125.71%
	BC	102.90%	68.88%	54.66%	71.36%	30.88%	105.52%	73.18%	43.83%	72.87%	75.12%
	TC	113.14%	108.29%	111.57%	98.02%	81.26%	235.19%	140.02%	130.04%	106.39%	90.62%
GraphIt	BFS	64.24%	86.40%	37.14%	84.29%	88.59%	54.11%	83.92%	74.34%	88.59%	95.14%
	SSSP	106.50%	110.96%	94.74%	112.40%	107.56%	86.17%	104.35%	93.88%	96.13%	106.48%
	CC	19.60%	8.86%	0.17%	7.06%	16.92%	16.10%	19.55%	0.45%	16.45%	27.85%
	PR	194.40%	109.23%	307.38%	102.72%	101.64%	149.14%	196.47%	350.03%	211.61%	186.20%
	BC	73.23%	100.23%	45.98%	224.15%	272.49%	75.85%	189.21%	34.67%	223.41%	251.01%
	TC	99.30%	108.45%	67.67%	113.89%	101.73%	98.72%	107.06%	98.41%	106.97%	104.38%
Graph Kernel Collection (GKC)	BFS	68.68%	67.33%	157.85%	61.20%	67.47%	74.44%	60.29%	83.29%	56.75%	64.35%
	SSSP	113.22%	89.68%	18.38%	86.72%	119.25%	115.98%	98.23%	18.53%	77.29%	118.17%
	CC	31.87%	26.53%	14.29%	32.95%	295.12%	27.69%	19.76%	10.82%	23.46%	214.27%
	PR	191.32%	105.56%	358.54%	136.28%	142.03%	125.03%	104.14%	324.19%	137.15%	150.24%
	BC	106.98%	100.30%	101.55%	101.60%	102.33%	106.23%	97.49%	77.15%	101.34%	102.76%
	TC	107.36%	157.92%	149.43%	197.51%	123.19%	106.98%	160.46%	176.41%	187.20%	113.98%
NWGraph	BFS	23.78%	65.85%	53.02%	65.34%	42.54%	26.59%	66.57%	33.97%	67.28%	48.74%
	SSSP	47.62%	85.35%	4.61%	114.69%	54.25%	46.33%	109.46%	6.58%	102.53%	55.39%
	CC	59.89%	69.09%	62.36%	61.50%	99.63%	49.60%	64.33%	60.34%	57.21%	87.41%
	PR	230.67%	110.38%	373.94%	108.16%	120.65%	175.33%	119.14%	499.59%	112.20%	124.68%
	BC	139.07%	135.88%	41.49%	163.21%	92.44%	117.33%	139.02%	38.15%	151.84%	90.77%
	TC	249.06%	132.30%	60.61%	108.27%	124.01%	228.14%	129.97%	51.35%	109.45%	112.77%

FIG. 30 The GAP matrix of six algos and five datasets.

- All systems heavily rely on in-memory computing and optimize the utilization of contiguous memory storage to achieve optimal algorithm implementation. However, this approach also means that the data structures used are static, read-only, and cannot be modified after loading.
- Customization and optimization of data structures are necessary for each algorithm, requiring separate loading of each dataset into the graph. This means that 30 separate loadings are needed for the 6 algorithms on the 5 datasets (5 * 6).

These findings highlight the varying performance strengths of different graph systems in specific algorithms, the importance of memory computing, and the need for tailored data structures for optimal algorithm implementation.

The last point above is particularly hard to accept. In actual business scenario applications, it is impossible to reload the data set for each algorithm, otherwise, the efficiency would be too low. Therefore, the entire test of GAP is filled with a strong academic atmosphere, but most graph systems originating from academia have similar problems, and GAP is not a special case.

In enterprise setup, one data set can be loaded (into persistent storage) only once, and from there all graph algorithms shall be able to execute against the data set—we may call this "Load Once and Run Many Algos." In Fig. 31, we benchmarked the GAP reference system against the System-U. It takes 1–2 h for each of the larger datasets of GAP to load, but it only takes 5–27 min for Ultipa, an acceleration of 7–19 times is achieved, and saving of 77%–95% of data loading time.

Table 22 summarizes the optimizations that GAP and other benchmarking systems have put in.

Indeed, the performance optimization items mentioned in the GAP reference system rely on certain preconditions that may not always hold true in real business scenarios. These preconditions include static read-only data, negligible preprocessing delay, continuous storage data structures in memory, and the assumption of a single connected component in the graph.

In real-world scenarios, these assumptions may not be realistic or feasible. For example, data in business applications is often dynamic and subject to frequent updates, requiring support for read-write operations and concurrent access. Preprocessing data may introduce additional delays and may not be practical in time-sensitive applications. Continuous storage and sequential access may not be possible due to data distribution and storage constraints. Additionally, real-world graphs often consist of multiple connected components, assuming a single connected component is unrealistic.

While the GAP reference system achieves high performance in specific scenarios, it is important to consider the trade-offs and limitations when applying these optimization techniques to real-world graph systems. Adapting these techniques to handle dynamic data, concurrent access, and graph structures with multiple connected components poses additional challenges that need to be addressed in practical implementations.

Readers may be curious about the differences between a general-purpose graph database system and the GAP reference system after some analysis of the GAP system. We have made some simple modifications to the System-U (mainly adding the implementation of the SSSP algorithm, turning off the write-back function of the algorithm, and adjusting the maximum concurrent scale of the system in the ULA license to make full use of the multicore of the underlying hardware concurrency capability), the benchmarking results are shown in Fig. 32. The following points are worth mentioning:

- BFS shortest path and BC algorithms: The GAP reference system maintains its advantage in these algorithms, with performance ranging from 30% to eight times better than the modified System-U.
- TC algorithm: The System-U demonstrates a significant advantage, with performance ranging from 2 to 76 times better than the GAP reference system.
- PR, SSSP, and CC algorithms: In these algorithms, both systems have their own strengths. The System-U performs better on small and medium data sets in the PR algorithm, while the GAP system has advantages in other scenarios. The performance difference between the two systems may vary depending on the specific data set size and characteristics.

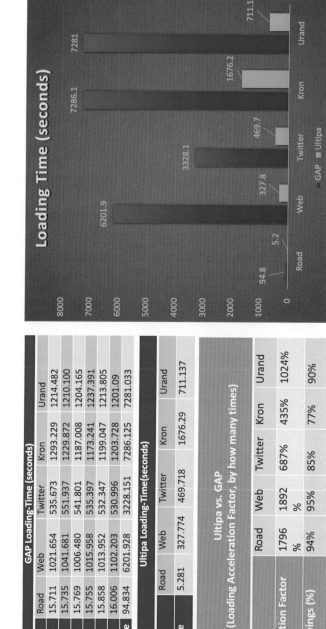

GAP Loading-Time (seconds)

	Road	Web	Twitter	Kron	Urand
BFS	15.711	1021.654	535.673	1293.229	1214.482
SSSP	15.735	1041.681	551.937	1229.872	1210.100
CC	15.769	1006.480	541.801	1187.008	1204.165
PR	15.755	1015.958	535.397	1173.241	1237.391
BC	15.858	1013.952	532.347	1199.047	1213.805
TC	16.006	1102.203	530.996	1203.728	1201.09
Total Time	94.834	6201.928	3228.151	7286.125	7281.033

Ultipa Loading-Time(seconds)

	Road	Web	Twitter	Kron	Urand
Total Time	5.281	327.774	469.718	1676.29	711.137

Ultipa vs. GAP (Loading Acceleration Factor, by how many times)

	Road	Web	Twitter	Kron	Urand
Acceleration Factor (%)	1796%	1892%	687%	435%	1024%
Time Savings (%)	94%	95%	85%	77%	90%

FIG. 31 GAP data load test—GAP vs System-U.

TABLE 22 GAP reference system optimizations.

GAP reference system optimization items	Pros	Cons
Read-only data structure	High query efficiency	Cannot be changed dynamically
Index-free Adjacency/Storage	Graph traversal and query efficiency are high	Very different from common linked list programming logic
Customize data structures for each algorithm	High query efficiency	Poor code reusability, changing algorithms requires reloading data
Parallel computing (OpenMP)	Algorithms run efficiently	Code implementation logic is complex
Data preprocessing (edge deduplication, neighbor sorting)	High computing and query efficiency	Unable to meet the needs of the industry
Preallocated memory space	Elevated efficiency	Unable to adapt to real scenarios in the industry
Shallow graph computing acceleration and optimization	1-Hop (nearest neighbor) computation greatly optimized (preallocation, preloading, etc.)	2-Hop and above depth and complex computing performance will drop significantly

GAP faster   [System-U faster]

	Road		Web		Twitter		Kron		Urand	
BFS	0.352	0.482	0.349	1.065	0.132	0.985	0.228	1.767	0.421	1.421
SSSP	0.227	2.940	0.685	1.278	2.007	▮	2.982	▮	4.129	▮
CC	0.042	0.044	0.185	0.194	0.128	▮	0.321	▮	0.906	▮
PR (16 loop)	0.269	▮	2.76	▮	3.841	4.288	7.504	10.350	9.597	18.222
BC	2.708	12.095	2.711	9.841	4.733	15.033	16.52	27.512	23.292	38.264
TC	0.030	▮	12.60	▮	39.368	▮	247.486	▮	12.303	▮

FIG. 32 GAP vs System-U in GAP benchmarking.

In summary, the following points can be highlighted regarding the planning, evaluation, and optimization of graph systems:

1. Accuracy and approximation: While some graph algorithms may use approximate calculations, ensuring accuracy is crucial for all query and calculation operations in graph systems.
2. Centralized and distributed systems: Distributed systems are more suitable for shallow and simple queries, while centralized systems excel in deep and complex queries. The integration of both approaches is a growing trend in graph system development.
3. Alignment with real business scenarios: Architecture and services should align with real business needs and scenarios. Solutions that are disconnected from actual business requirements are not beneficial for long-term system development.
4. Incremental planning and scenario adoption: It is advisable to start utilizing the graph system in incremental and innovative scenarios first, and then plan for the replacement of existing scenarios.
5. Expansion, user experience, and graph thinking: The success and acceptance of a graph system depend on two key factors. First, the system and its tools should be user-friendly, intuitive, and efficient. Second, promoting a graph thinking mindset is crucial to shift users from traditional two-dimensional relational thinking to the multidimensional graph thinking world.

These considerations contribute to the effective planning, evaluation, and optimization of graph systems, enabling them to meet the needs of real-world scenarios and users while ensuring accuracy and usability.

The evolving interaction between graph (database) systems and SQL databases suggests a process of mutual displacement. Over the past 40 years, SQL/RDBMS has gained widespread adoption across various industries, with low barriers to entry. However, most implementations have remained at a surface level. As application scenarios become more complex, the development and programming of SQL codes (and applications) have become exponentially more challenging, leading to a decline in operational efficiency. This is evident in the prevalence of $T+N$ ($N \geq 1$) batch processing scenarios.

In contrast, GQL and Graph DBMS offer a unique characteristic: as the application scenario complexity increases, the development and programming complexity decreases compared to SQL. This implies a convergence point between the two approaches, as illustrated in Fig. 33. Beyond this convergence point, GQL/Graph is expected to gain significant traction, eventually establishing GQL/Graph as the new generation mainstream database. The future awaits to witness this transformation.

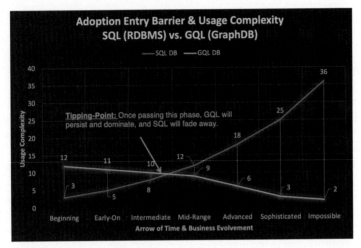

FIG. 33   SQL vs a turning point for GQL.

It is worth noting that graph's ability to handle complex application scenarios with relative ease compared to SQL is a driving force for its potential widespread adoption. As businesses face increasingly intricate data challenges, graph's strengths in managing graph-oriented data and facilitating intuitive programming will likely contribute to its rise as a prominent database solution.

Index

Note: Page numbers followed by *f* indicate figures and *t* indicate tables.

A

Active-standby system, 247
Adjacency matrix, 43–45, 43*t*
Aladdin, 18
Amdahl's Law, 227–228
Anti entropy, 238–239
Antimoney laundering, 24, 287–290
Artificial general intelligence (AGI), 1, 12
Artificial intelligence (AI), 1, 26–27, 312–313, 340
Asset and liability management (ALM), 284–308, 298–299*f*, 303–308*f*
Attribution analysis, 138
Automatic networking (subgraph formation), 90, 91*f*
autoNet(), 92
Auxiliary data, 71

B

Benchmarking, graph systems, 330–380
content, 338–356
fixed mode modeling, 340, 341*t*
flexible modeling, 340–356, 342–356*t*
environment, 332–338, 332*t*, 333*f*, 334*t*, 335*f*, 336*t*
graph database queries and algorithms, 357–365, 360–365*f*
Betweenness centrality (BC) algorithm, 31, 170–172, 171–173*f*, 376
BFS. *See* Breadth-first search (BFS)
Big data and database technologies, evolution of, 19
from data to big data to deep data, 19–22, 20*f*
relational database *vs.* graph database, 22–23, 23*f*
Big-data framework-based graph systems, 290
Black-box, 301
Blocking protocol, 240
Bloom filter, 64, 65*f*
Boratvka algorithm. *See* Minimum spanning tree (MST) algorithm

Breadth-first search (BFS), 35, 37–38, 38*f*, 89, 346–347, 358
Breadth-first search shortest path algorithm, 376
Brute-force partitioning, 127
B-tree based database storage engines, 60–63, 62*f*
B-tree-type data structure, 113, 114*f*
Business continuity/disaster recovery (BC/DR) plan, 134
Business relationship graph, 71
Butterfly effect, 309
Byzantine general problem, 249

C

Cache, 106, 107–109*f*
Capacity planning, 324–330, 328*f*
Cartesian Product, 11–12
Cassandra Query Language (CQL), 82–83
Centrality algorithm, 166–172
betweenness centrality, 170–172, 171–173*f*
closeness centrality, 168–169, 169*t*
disadvantages, 172
graph centrality, 167–168, 167*f*
ChatGPT, 24–25, 26*f*
Clock synchronization, 254
Closeness centrality, 168–169, 169*t*
Cloud-based high-performance computing environment, 47
Cloud computing, 224
Collaborative filtering (CF), 292–293, 293*f*
Commodity classification data, 292
Community compression, 199, 200*f*, 202
Community computing, 194–206
Louvain community recognition, 197–206, 197–201*f*, 203–206*f*, 204*t*
triangle counting, 194–197, 195–196*f*
Community detection, 202–203
Community recognition based recommendation (CRBR), 293–295
Community weight, 197–198, 198*f*
Complete random walk algorithm, 207–210, 207*f*, 209*f*, 209*t*

Complex path query, 349–350t
Complex theorem-proving software, 14–15
Compute engine query, 80
Concurrent read-only, 240
Concurrent writing, 240
Connected components algorithm, 179–181,
    179–180f, 376
Connectivity, 178–183, 178f
    connected component, 179–181, 179–180f
    minimum spanning tree (MST), 181–183,
        182t, 183f
Consistency, availability, and partition
    tolerance (CAP) theorem, 236
Copy-on-write mode, 112
Cosine similarity, 176–178, 176–177f
Counterparty credit risk, 309
COW technology, 113
CQL. *See* Cassandra Query Language
    (CQL)
Credit loan application, 272–275, 273–274f,
    275t, 276f

D
Damping factor, 184–185
Data access, 342–343, 343t
Database programmers, 87
Database query language
    concepts, 81–87, 82–84f, 86–87f
    evolution, 85, 86f
Database storage engine, 60
Database technology evolution, 84, 84f
Data connectivity, 278
Data connector, 321
Data exploration, 279
Data import tool, 321
Data modeling, 123, 321–322
Data mutability, 112, 113–114f
Data presentation, 278
Datasets, 88, 88f, 335–338, 336t
Data sorting, 109, 110–112f
Data structures
    classification, 39f
    and computational efficiency, 38
    vector-based, 46
    of vertices and edges, 65–66
Data types, 121, 121t
Data workflow, 116, 117f
Decoupling, 235
Deduplication algorithm, 142–143, 268–269
Deep data processing capability, 22
Deep graph query, 346–347
Deep random walk process, 295

DeepWalk, 218–220, 219f
Degree algorithm, 162–166, 163t,
    163–166f
Denominator, 46
Depth-first search (DFS) algorithm, 35, 37–38,
    38f, 89, 346–347, 358–359
Depth-first search path queries, 351t
Dijkstra's algorithm, 17–18, 33, 49–51, 56–57
Direct Mail, 238
Disaster recovery distributed high availability
    (DR-DHA), 130
Distributed algorithm, 235–236
Distributed consensus system, 248–255, 249f,
    251–253f, 255f
Distributed graph system, 269
Distributed high-availability cluster
    (DHA-HTAP), 130
Draft International Standard (DIS) mode, 18
Dual-center cluster backup scheme, 132, 133f
Dynamic graph database, 322, 322t
Dynamic programming (DP) problems, 56–57
Dynamic scheme, 119–120

E
eBay, 18
Edge data query, 144–145, 145f
Edge query, 265
External memory, 105–106

F
Facebook's technical framework, 17
Fast leader election (FLE), 252
Fixed mode modeling, 340, 341t
Flexible modeling, 340–356, 342–356t
Fraud detection, 282–284, 282–284f
Fraud, types of, 282
Full data synchronization, 239
Full-text search, integration of, 97
Functional testing, 331–332

G
GAP benchmark, 346–347, 352–354
GAP performance evaluation, 373–375
Gartner five-layer model, 29, 30f
General database storage engine, 60
Goldman Sachs, 18
Google File System (GFS), 19–20
GQL. *See* Graph query language (GQL)
Graph algorithms, 357
    capability, 353t
    centrality, 161, 166–172

betweenness centrality, 170–172,
  171–173*f*
closeness centrality, 168–169, 169*t*
graph centrality, 167–168, 167*f*
community computing, 161, 194–206
  Louvain community recognition, 197–206,
    197–201*f*, 203–206*f*, 204*t*
  triangle counting, 194–197, 195–196*f*
connectivity, 161, 178–183, 178*f*
  connected component, 179–181, 179–180*f*
  minimum spanning tree (MST), 181–183,
    182*t*, 183*f*
degree, 161–166, 163*t*, 163–166*f*
graph embedding computing, 161, 206–213
  complete random walk algorithm,
    207–210, 207*f*, 209*f*, 209*t*
  Struc2Vec algorithm, 210–213, 210–211*f*,
    212*t*, 213*f*
and interpretability, 213–221, 214–219*f*
propagation computing, 161, 189–194
  hot attenuation and node preference
    (HANP) algorithm, 192–194, 193*t*, 194*f*
  label propagation algorithm (LPA),
    189–192, 191*f*, 191*t*
ranking, 161, 184–189
  PageRank, 184–186, 185*t*, 186*f*
  SybilRank, 187–189, 187–188*f*, 188*t*
similarity, 161, 172–178
  cosine similarity, 176–178, 176–177*f*
  Jaccard similarity, 173–176, 174–175*f*
visualization, 159
Graph-based liquidity risk management system,
  305, 306*f*
Graph centrality, 167–168, 167*f*
Graph computing, 1, 33
  applicable scenarios of, 56–60, 57*f*, 59*f*
  concepts of, 34–55, 36–39*f*, 41*f*, 43*f*, 43–44*t*,
    47–48*f*, 50–51*f*, 52–53*t*, 53–54*f*
  development of, 56
  in internet of things (IoT) era, 23–24
    graph computing *vs.* graph database,
      29–32, 32*t*
    unprecedented capabilities, 24–29, 25–28*f*,
      30*f*
  origin of, 13
Graph database, 1, 3–4, 12–13, 16–17, 23–24,
  29
  for connected/networked data analysis,
    24–25, 25*f*
  core functions of, 29
  development of, 19
  vs. graph computing, 29–32, 32*t*

products, 19
purpose of, 22–23
in real industrial scenarios, 45
real-time, 25–28
vs. relational database, 22–23
scenarios, 59*f*
storage engine of, 61
technology, 1
  foundation, 13
Graph data modeling, for card transaction, 76,
  78*f*
Graph data structures, 42
Graph, definition of, 1
Graph embedding based recommendation
  (GEBR), 293, 296*f*
Graph embedding computing, 206–213
  capability, 354*t*
  complete random walk algorithm, 207–210,
    207*f*, 209*t*, 209*f*
  Struc2Vec algorithm, 210–213, 210–211*f*,
    212*t*, 213*f*
Graph labeling, 57, 57*f*
Graph modeling mechanisms, 76–77, 78*f*
Graph powered asset and liability
  management, 299, 300*t*
Graph query and analysis framework design,
  134–159
  K neighbor query, 142–144, 142–144*f*
  language design ideas, 135–159, 135*f*,
    137–138*f*
  metadata query, 144–147, 145–147*f*
  path query, 139–142, 139–141*f*
Graph query language (GQL), 18, 87–100, 88*f*,
  90–93*f*, 95–96*f*, 98–99*f*, 135
  code, 89
  compiler, 147–153, 148–150*f*
    graph visualization, 150–153, 152–153*f*
  evolution of, 81–100
Graph storage, 60–81
  concepts of, 60–68, 61–66*f*, 67*t*, 68*f*, 69*t*
  data structure and modeling, 68–81, 70*f*,
    72–73*f*, 74*t*, 76–77*t*, 78–80*f*
  inefficiency, 45
Graph storage engine, 60, 61*f*, 66
Graph systems
  awareness of, 318
  capacity planning, 329–330
  magic quadrant, 318, 318*f*
  multiple genres, 319
  software environments, 334, 334*t*
  tools, 320–321, 321*f*
Graph technology development, 13–19, 14–16*f*

Graph theory, 13, 33
  Euler's invention of, 13–14
  Königsberg and, 13–14, 14f
  traditionally in, 92–93
Graph thinking, 1–13, 2f, 5–9f
GraphX, 125–126, 126f
Graph-XAI ALM system, 305–308, 308f
Grid, 238

H

Hadoop project, 19–20
HANP algorithm. *See* Hot attenuation and node
    preference (HANP) algorithm
Heterogeneous graph, 323–324, 325f
Hierarchical data storage, 323f, 327, 328f
High-density parallel graph computing, 52
High-dimensional data operation, 357
High-performance graph computing, 290
  architecture, 114–134
  data structure, 124–125, 125f
  failure and recovery, 132–134, 133f
  graph database schema, 118–122, 119f,
    121t
  high availability, 129
  partitioning, 125–129, 126f, 129f
  real-time graph computing, 115–118,
    116–118f
  scalability, 131
High-performance graph storage architecture,
    101–114, 102f
  access efficiency, 102–103
  characteristics, 102–104, 103f
  complexity, 103–104
  core pillars of, 101
  cost-effectiveness, 103
  database management system architecture,
    101, 102f
  design ideas, 105–114
    cache, 105–106, 107–109f
    data mutability, 105, 112, 113–114f
    data sorting, 105, 109, 110–112f
    storage and memory occupation ratio, 105
  hierarchical storage, logic of, 104
  layers of, 103–104, 103f
  storage efficiency, 102
  update efficiency, 102–103
Horizontally distributed system, 256–269,
    257–261f, 263f, 264t, 266–267t
Horizontal partitioning, 259, 259f
Horizontal scalability, 131, 223, 233–244, 233t,
    234f, 237–239f, 242f

Hot attenuation and node preference (HANP)
    algorithm, 192–194, 193t, 194f
Human brain, network model of, 1–2, 2f
Hybrid transactional and analytical processing
    (HTAP), 29

I

Incidence matrix, 44, 44t
Index-free adjacency (IFA), 12–13, 110, 122
Industry-bound graph benchmarking, 331
Instance failure, graph databases, 132
Intelligent recommendation, 290–297
Interconnected financial risk management,
    309–312, 311–312f
Internal memory, 105–106
Internet of things (IoT), 23–24
  graph computing *vs.* graph database, 29–32,
    32t
  unprecedented capabilities, 24–29, 25–28f,
    30f
Interpretability, 213–221, 214–219f
Inverse ratio, 109, 109f
Iteration, 202

J

Jaccard similarity algorithm, 173–176,
    174–175f, 324, 364–365, 364–365f

K

K-hop
  benchmark, 373
  and count of results, 90, 90f
  Neo4j *vs.* Ultipa, 53t
  operations, types, 36, 37f
  problem, 361
  query parallelization, 52, 53f
  System-N *vs.* System-U, 54f
  using DFS, 37–38
K-hop queries, 142–144, 142–144f, 165, 166f,
    347, 347t, 357, 359, 360f
Knowledge graph, 1, 4, 16–18, 24–26, 284–297,
    286f
Königsberg and graph theory, 13–14, 14f

L

Label data, 292
Label propagation algorithm (LPA), 189–192,
    191t, 191f, 367–368, 368f, 368t
Lexical analyzer, 148
Limit/stress testing, 345

Liquidity coverage ratio (LCR), 299–302, 304, 306*f*
Liquidity risk management (LRM), 297–308, 304*f*, 308*f*
Lock-free record updating, 113, 113*f*
Logical vertex storage, 263, 264*t*
Log-structured merge-tree (LSMT), 63, 63–64*f*, 110–111
Long-chain tasks, 257, 257*f*
Loop query, 140, 141*f*, 350*t*
Louvain community recognition, 51, 58, 98–99, 197–206, 197–201*f*, 203–206*f*, 204*t*, 275
   community compression, 199, 200*f*
   community detection, 202–203, 203*f*, 274–275, 276*f*, 294*f*
   example, 204–205, 204–205*f*
   input parameters, 204, 204*t*
   modularity, 199–203, 200*f*
   modularity gain, 200–201, 201*f*
   visualization, 205–206, 206*f*
   weighted degree, 197–199, 197–199*f*
Low-cost horizontal scalability, 234, 234*f*
LPA. *See* Label propagation algorithm (LPA)

M

Management workflow, 116
Map coloring problem, 15, 15*f*
MapReduce, 19–20
Mass data loading, 344–345, 345*t*
Master-slave HA, 244–248, 246–248*f*
Master-slave server (MSS) cluster, 129
Master-slave-slave system (MSS), 246, 247*f*
Max flow minimum cut theorem, 178
Mechanical hard disk, 105–109, 107–108*f*
MemC3/Cuckoo hash implementation, 47, 47–48*f*
Memory database, 105
Metadata, 71
   benchmarking, 346, 346*t*
   display, 154, 155*f*
   operation, 357
   query, 144–147, 145–147*f*
   storage of, 71
Minimum spanning tree (MST) algorithm, 181–183, 182*t*, 183*f*
Modularity, 199–203, 200*f*
Modularity gain, 200–201, 201*f*
Multigraph, 77
   data modeling, 79
   edge cutting, 263, 263*f*
   vs. simple graph, 79–80, 79*f*

Multilabel propagation, 190, 191*f*
Multiversion concurrency, 242
Multiversion concurrency control (MVCC/MCC), 241–242, 242*f*

N

Native graph, 12, 49, 52
   computing engine, 71
   deep query and visualization with, 71, 72*f*
   heterogeneous (paths) data results on, 71, 73*f*
   vs. nonnative graph, 66, 68, 69*t*
   query complexity of, 69*t*
   query steps on, 68
Native graph query, 68, 68*f*
Neo4j's Cypher solution, 137–138, 137*f*
Network analysis, 22–25, 29
Network communication modes, 237–238, 239*f*
Networked data analysis, 24–25, 25*f*
Network search, 285–287
Network visualization, 154, 156–158*f*
Node2Vec, 219–220, 219*f*, 295
Node2vec, 367
Node weight, 197, 197*f*
Nonnative graph
   vs. native graph, 66, 68, 69*t*
   storage query mode, 66–67, 66*f*
NoSQL databases, 16–17, 29, 84–85

O

Offloading, 235
One-hot vector method, 215–216
Online analytical processing (OLAP) system, 21–22
Online fraud detection, 276–278, 277*t*
Online transactional processing (OLTP), 29, 31
Optimistic lock, 241
Optimization, graph system, 366–380, 366*f*, 368–372*f*, 368*t*, 375*f*, 377–378*f*, 378*t*, 380*f*
Oracle LRM, 302–304, 304*f*

P

PageRank, 17, 19–20, 58, 184–186, 185*t*, 186*f*, 367, 376
Parse tree, 148, 149*f*
Partitioning, 125–129, 126*f*, 129*f*, 236, 258–259, 259*f*, 329
Path benchmarks, 373–374
Path query, 139–142, 139–141*f*, 154–158, 156–158*f*, 347, 348–350*t*, 357
Paxos consensus algorithm, 236–238, 237–238*f*
Pessimistic concurrency control, 240

PG. *See* Property graph (PG)
Pipeline, 235
Planning, graph system, 317–330, 318*f*
    capacity planning, 324–330, 328*f*
    data and modeling, 319–324, 320–321*f*, 322*t*,
        323–326*f*
POS machines, 35, 36*f*
PreCommit phase, 240
Primitive data structure, 40
Propagation computing, 189–194
    hot attenuation and node preference (HANP)
        algorithm, 192–194, 193*t*, 194*f*
    label propagation algorithm (LPA), 189–192,
        191*f*, 191*t*
Property graph (PG), 2–3, 18–19

Q

Query computational workflow, 116, 117*f*
Queue, 235
Quick termination, 249–250

R

RAFT algorithm, 129, 253–255
Range queries, 110
Ranking, 184–189
    PageRank, 184–186, 185*t*, 186*f*
    SybilRank, 187–189, 187–188*f*, 188*t*
RDBMS. *See* Relational database management
    system (RDBMS)
RDF. *See* Resource Description Framework
    (RDF)
RDMA. *See* Remote direct memory access
    (RDMA)
Read synchronization, 238
Read-write mixed mode, 240
Real-time decision system, 272–278
    credit loan, 272–275, 273–274*f*, 275*t*, 276*f*
    features, 277, 277*t*
    online fraud detection, 276–278, 277*t*
Real-time graph computing, 115–118, 116–118*f*
Real-time liquidity coverage ratio
    back-tracing, 305, 307*f*
Real-world graph-powered application,
    271–272
    antimoney laundering, 287–290
    asset and liability management (ALM),
        284–308, 298–299*f*, 303–308*f*
    BFSI industries, 271
    business decisionmaking and intelligence,
        272–278
    credit loan, 272–275, 273–274*f*, 275*t*, 276*f*

in financial field, 312–315
    fraud detection, 282–284, 282–284*f*
    in interconnected financial risk
        management, 309–312, 311–312*f*
    internet, 271
    internet of things (IoT), 271
    knowledge graph, 284–297, 286*f*
    liquidity management, 297–308
    online fraud detection, 276–278, 277*t*
    real-time decision system, 272–278
    supply chain finance, 271
    telecom operators, 271
    ultimate beneficiary owners (UBOs),
        278–281, 279–280*f*, 281*t*
Relational database
    design ideas of, 85
    vs. graph database, 22–23
Relational database management system
    (RDBMS), 16–17, 29, 115, 298*f*, 299*t*
Remote direct memory access (RDMA),
    230–231, 230*f*
Remote disaster recovery model, 247, 248*f*
Resource Description Framework (RDF), 16, 18
Resource description framework (RDF), 16, 18
Rumor Mongering, 238

S

Scalability, 131
Scalable graph database, 223–244
    architectural design schemes, 244–269,
        245*f*
        distributed consensus system, 248–255,
            249*f*, 251–253*f*, 255*f*
        horizontally distributed system, 256–269,
            257–261*f*, 263*f*, 264*t*, 266–267*t*
        master-slave HA, 244–248, 246–248*f*
    horizontal scalability, 233–244, 233*t*, 234*f*,
        237–239*f*, 242*f*
    vertical scalability, 224–232, 224*f*, 228*f*,
        229*t*, 230–232*f*
Schema, 118–122, 119*f*, 121*t*
Schema-free scheme, 87, 119–120
Schema management, 154, 155*f*
Self-assessment, 330–331
Self-benchmarking, 330
Serialization, 122
Shards, 258–259, 267*t*, 269, 329
Short-chain transactions, 256–257, 257*f*
Shortest path queries, 349*t*, 363, 363*f*
Similarity algorithm, 172–178
    cosine similarity, 176–178, 176–177*f*
    Jaccard similarity, 173–176, 174–175*f*

Simple graph (single-edge graph), 31, 77, 79–80, 79*f*
Single-cluster spanning, 133, 133*f*
Single label propagation, 190
Single-source shortest paths (SSSP), 31
Skip-gram algorithm, 215
Smart partitioning, 128
Social network graph, 9, 10*f*
SoftMax training method, 215
Solid-state drive (SSD), 105–109
    access delay decomposition, 107, 108*f*
    characteristics of, 111
    storage units, 108*f*
Spark SQL, 83, 83*f*
SPARQL, 18, 136
Spread() operation, 89
SQL standard formation, 135–136, 135*f*
SSSP algorithm, 376
Static graph computing, 322, 322*t*
Static graph computing database, 322, 322*t*
Storage engine, 33, 60, 80–81
    database, 60
    design for graphs, 80
    general database, 60
    graph, 60, 61*f*
        architecture design of, 66
    for graph data, 66–67
    of graph databases, 61
    traditional, 71
    of traditional database, 64, 65*f*, 66
Strategic Data Department (SDS), 83–84
Strongly connected component (SCC), 179, 179–180*f*
Struc2Vec algorithm, 210–213, 210–211*f*, 212*t*, 213*f*
Subgraph query, 139–140
Supercomputing systems, 333–334, 333*f*, 335*f*
Structured query language (SQL), 16–17, 85, 115
    databases, 68–70, 242–244
Switch-on-delete, 110, 110*f*
SybilRank, 187–189, 187–188*f*, 188*t*

T
Template path queries, 347, 348*t*
Three-phase commit (3PC), 240
Tigergraph's GSQL solution, 137–138, 137*f*
Time-series-based partitioning, 128
Tinkerpop Gremlin, 136
Traditional asset and liability management, 299, 300*t*

Traditional storage engines, 71
Transactional synchronization extensions (TSX), 48–49
Transaction failure, graph databases, 132
Transaction network, 323–324, 323–324*f*
Traversal optimizations, 124
Triangle counting (TC) algorithm, 194–197, 195–196*f*, 376
Two-general communication problem, 245–246, 246*f*
Two-instance cluster system (HA), 129

U
Ultimate beneficiary owners (UBOs), 278–281, 279–280*f*, 281*t*
Ultipa graph, 90, 91*f*
Ultipa liquidity risk management, 302–304, 304*f*
Unanimity, 249
User-specified partitioning, 128

V
Validity, 249
Vector-based data structures, 46
Vertex attributes, 88, 88*f*
Vertex-cutting method, 125–126, 126*f*
Vertex data query, 144–145, 145*f*
Vertex storage structure, 71, 74*t*, 76
Vertical partitioning, 259, 259*f*
Vertical scalability, 131, 223–232, 224*f*, 228*f*, 229*t*, 230–232*f*
Vertices, 285
Virtualization, 234–235
Visualization, 302
Visualized full-text search, 305, 307*f*

W
Weakly connected component (WCC), 179, 179*f*
Web page ranking algorithms, 58
Weighted degree, 197–199, 197–199*f*
White-box process, 218
WiscKey, 110–111, 111–112*f*, 225–226
Word2Vec method, 215, 216*f*, 217–218
Write synchronization, 239

Z
Zookeeper atomic broadcast (ZAB), 250–253

Printed in the United States
by Baker & Taylor Publisher Services